Progress in Molecular and Subcellular Biology

Series Editors: W.E.G. Müller (Managing Editor),
Ph. Jeanteur, Y. Kuchino, A. Macieira-Coelho, R. E. Rhoads

38

Progress in Molecular and Subcellular Biology

Volumes Published in the Series

Volume 13
Molecular and Cellular Enzymology
Ph. Jeanteur (Ed.)

Volume 14
Biological Response Modifiers: Interferons, Double-Stranded RNA and 2´,5´-Oligoadenylates
W.E.G. Müller and H.C. Schröder (Eds.)

Volume 15
Invertebrate Immunology
B. Rinkevich and W.E.G. Müller (Eds.)

Volume 16
Apoptosis
Y. Kuchino and W.E.G. Müller (Eds.)

Volume 17
Signaling Mechanisms in Protozoa and Invertebrates
G. Csaba and W.E.G. Müller (Eds.)

Volume 18
Cytoplasmic Fate of Messenger RNA
Ph. Jeanteur (Ed.)

Volume 19
**Molecular Evolution:
Evidence for Monophyly of Metazoa**
W.E.G. Müller (Ed.)

Volume 20
Inhibitors of Cell Growth
A. Macieira-Coelho (Ed.)

Volume 21
Molecular Evolution: Towards the Origin of Metazoa
W.E.G. Müller (Ed.)

Volume 22
Cytoskeleton and Small G Proteins
Ph. Jeanteur (Ed.)

Volume 23
**Inorganic Polyphosphates:
Biochemistry, Biology, Biotechnology**
H.C. Schröder and W.E.G. Müller (Eds.)

Volume 24
Cell Immortalization
A. Macieira-Coelho (Ed.)

Volume 25
Signaling Through the Cell Matrix
A. Macieira-Coelho (Ed.)

Volume 26
**Signaling Pathways for Translation:
Insulin and Nutrients**
R.E. Rhoads (Ed.)

Volume 27
**Signaling Pathways for Translation:
Stress, Calcium, and Rapamycin**
R.E. Rhoads (Ed.)

Volume 28
Small Stress Proteins
A.-P. Arrigo and W.E.G. Müller (Eds.)

Volume 29
Protein Degradation in Health and Disease
M. Reboud-Ravaux (Ed.)

Volume 30
Biology of Aging
A. Macieira-Coelho

Volume 31
Regulation of Alternative Splicing
Ph. Jeanteur (Ed.)

Volume 32
Guidance Cues in the Developing Brain
I. Kostovic (Ed.)

Volume 33
Silicon Biomineralization
W.E.G. Müller (Ed.)

Volume 34
Invertebrate Cytokines and the Phylogeny of Immunity
A. Beschin and W.E.G. Müller (Eds.)

Volume 35
RNA Trafficking and Nuclear Structure Dynamics
Ph. Jeanteur (Ed.)

Volume 36
Viruses and Apoptosis
C. Alonso (Ed.)

Volume 37
**Marine Molecular Biotechnology:
Sponges (Porifera)**
W.E.G. Müller (Ed.)

Volume 38
Epigenetics and Chromatin
Ph. Jeanteur (Ed.)

Philippe Jeanteur (Ed.)

Epigenetics and Chromatin

With 29 Figures, 7 in Color, and 3 Tables

 Springer

Professor Dr. PHILIPPE JEANTEUR
Institute of Molecular Genetics
of Montpellier
1919 Route de Mende
34293 Montpellier Cedex 05
France

ISSN 0079-6484
ISBN 3-540-23372-5 Springer-Verlag Berlin Heidelberg New York

Library of Congress Control Number: 2004112648

Springer-Verlag is a part of Springer Science+Business Media
springeronline.com

© Springer Berlin Heidelberg 2005
Printed in Germany

The use of general descriptive names, registered names, trademarks, etc. in this publication does not imply, even in the absence of a specific statement, that such names are exempt from the relevant protective laws and regulations and therefore free for general use.

Product liability: The publishers cannot guarantee the accuracy of any information about dosage and application contained in this book. In every individual case the user must check such information by consulting the relevant literature.

Production and typesetting: Friedmut Kröner, Heidelberg, Germany
Cover design: *design & production* GmbH, Heidelberg, Germany

Printed on acid free paper   39/3150 YK 5 4 3 2 1 0

Preface

Although their scopes definitely overlap, the terms epigenesis and epigenetics are commonly used in different contexts. Epigenesis is etymologically derived from genesis and as such includes everything that touches upon development. Its scope is extremely wide and covers not only somatic, but also mental development. It is not surprising that its oldest and most rudimentary formulation dates back to early embryologists. The term epigenetics is a more recent and apparently more focused concept since it explicitly refers to genes and chromosomes and via which mechanisms this genetic information can be heritably repressed, or activated, in specific lineages or transmitted from one generation to the next.

In parallel with our increasing understanding of gene function, it became overwhelmingly obvious that not everything, albeit hereditary, is defined by the DNA sequences of our genes as we can now read them in complete genomic databanks. In the limited context of this book, we will examine how modifications of DNA and associated proteins can heritably impinge on its packaging into chromatin and subsequently on fundamental DNA transactions – replication and transcription – which marks the first step towards gene expression.

The first chapter by P. Varga-Weisz gives a comprehensive survey of the already highly documented field of chromatin assembly and remodeling factors which accompany DNA replication and the deposition of new nucleosomes. It also emphasizes the key role of PCNA in the general problem of epigenetic inheritance and reviews the role of some of the above factors in the hereditary transmission of chromatin states. This same issue is tackled by Déjardin and Cavalli who question how Polycomb- and trithorax-induced chromatin states can be maintained across a specific differentiation program, allowing cells to remember their identity throughout.

The next chapter by Caron and coworkers addresses the specific problems posed by sperm chromatin, histone modification and histone variants, the role of transition proteins in histone removal and subsequent protamine deposition, and the reorganization of chromatin after meiosis.

Chromatin modifications are central to the problem of dosage compensation raised by the presence of two copies of the X chromosome in females. There are three known epigenetic mechanisms which can compensate for this allelic imbalance and thus achieve comparable levels of X chromosome genes in males and females. In mammalian organisms, a mechanism has evolved to inactivate one copy of the X chromosome and this very peculiar mechanism is reviewed by Cohen and coworkers. The reciprocal possibility is to upregulate the single X copy in males. This is the mechanism chosen by *Drosophila* where hypertranscription is achieved by an RNA protein-containing compensation complex which is described in detail in the chapter by Taipale and Akhtar. The intermediate possibility, i.e., downregulating to 50 % expression both X copies in females, as is the case in the nematode *Caenorhabditis elegans*, has not been considered in this book.

In many instances, be they physiological or pathological, DNA methylation is the causal event in gene silencing. Razin and Kantor give a general survey of the various facets of this mechanism and its involvement in genomic imprinting and in disease, of which cancer provides many well-documented examples. The chapter by Ballestar and Esteller follows along this latter line and focuses on epigenetic mechanisms leading to cancer and raises the possibility of therapeutic reactivation of silenced genes.

Developmental regulation of the β-globin gene cluster has been a long-standing, highly relevant model to study the influence of chromatin structure on gene expression. However, the recognition of epigenetic control of developmental β-globin gene expression has only recently emerged and this new aspect is reviewed by Chakalova and coworkers.

The last two chapters cover genomic imprinting, i.e., the fact that some (few) genes are expressed from only one parental allele, and this is the last, but not least, facet of epigenetic regulation. Weber and coworkers give an in-depth analysis of this mechanism in mammals which they adequately place in an evolutionary perspective. Köhler and Grossniklaus conclude with a comprehensive review of this phenomenon in plants.

Philippe Jeanteur

Contents

Chromatin Remodeling Factors and DNA Replication 1
P. Varga-Weisz

1   Introduction 1
2   Chromatin and Chromatin Remodeling Factors 2
2.1   Chromatin 2
2.2   Histone Modification Enzymes 3
2.3   ATP-Dependent Chromatin Remodeling Factors 4
3   Chromatin Structure and DNA Replication 5
4   Chromatin Assembly 7
4.1   Histone Chaperones 7
4.2   Chromatin Assembly Factor 1,
      a Replication-Coupled Histone Chaperone 9
4.3   CAF-1 Functions in the Inheritance of Chromatin States . . 9
4.4   Histone Chaperone ASF1 10
4.5   Histone Chaperones and Heterochromatin Replication . . . 11
4.6   Histone Variants 11
4.7   DNA Replication-Independent Chromatin Assembly 12
5   Histone Modifications and Chromatin Replication 13
5.1   Histone Deacetylation During Chromatin Replication 13
5.2   Histone Acetylation at the Replication Site 13
6   ATP-Dependent Chromatin Remodeling Factors
      in Chromatin Replication 14
6.1   ISWI Complexes in Chromatin Assembly in Vitro 14
6.2   ISWI Complexes and Their Role in Chromatin Replication
      in Vivo 17
7   The Assembly of Higher Order Chromatin Structures 19
8   PCNA, a Central Coordinator of Epigenetic Inheritance . . . 20
9   Mechanisms of Epigenetic Inheritance
      Through Chromatin: Conclusion 21
      References 21

**Epigenetic Inheritance of Chromatin States Mediated by Polycomb
and Trithorax Group Proteins in *Drosophila***   31
J. Déjardin and G. Cavalli

1   Introduction   31
2   Proteins of the Polycomb Group of Genes   33
2.1   PcG Complexes   33
2.1.1   The ESC-E(Z) Complex   33
2.1.2   The Polycomb Repressive Complex 1   35
2.1.3   Other Identified PcG Proteins and Partners   36
2.2   Targeting of PcG-Mediated Repression   37
2.2.1   PcG Response Elements   37
2.2.2   Chromatin Determinants Associated with Targeting   39
2.3   Mechanisms of Repression   43
2.3.1   Spreading or Looping?   43
2.3.2   Proposed Silencing Mechanisms   44
3   Proteins of the Trithorax Group   46
3.1   trxG Complexes   48
3.1.1   The Trithorax Acetylation Complex (TAC1)   48
3.1.2   The Brahma Complex   48
3.1.3   The GAF–FACT Complex   49
3.1.4   Other trxG Complexes and Partners   49
3.2   Targeting of trxG Complexes at TREs   49
3.3   Mechanisms of Action   51
4   Modes of Inheritance   54
5   Concluding Remarks   55
           References   56

How to Pack the Genome for a Safe Trip   65
C. Caron, J. Govin, S. Rousseaux and S. Khochbin

1   Introduction   65
2   Synthesis of Histone Variants   67
2.1   Non-Testis-Specific Core Histone Variants   68
2.2   Testis-Specific Histone Variants   69
2.2.1   Linker Histones   69
2.2.2   Core Histones   70
3   Histone Modifications   71
3.1   Acetylation   72
3.2   Ubiquitination   73
3.3   Phosphorylation   73
3.4   Methylation   74
4   Transition Proteins   74
5   Final Components of the Sperm Chromatin   75

5.1   Protamines   76
5.2   Histones   77
6   Mechanisms Controlling Post-Meiotic
         Chromatin Reorganization: A General Discussion   77
6.1   Active Transcription Followed by Repression
         in Round Spermatids   78
6.2   Functional Link Between Histone Acetylation
         and Chromatin Condensation and Histone Replacement . .   79
6.3   Does Histone Ubiquitination Play a Role
         in Spermatid-Specific Chromatin Remodeling?   80
6.4   Is There a Spermiogenesis-Specific Histone Code?   81
6.5   Do Histone Variants Play a Role
         in Spermatid-Specific Chromatin Remodeling?   82
7   Concluding Remarks   82
         References   84

Chromatin Modifications on the Inactive X Chromosome   91
H.R. Cohen, M.E. Royce-Tolland, K.A. Worringer and B. Panning

1   Introduction   91
2   Features of Xi Chromatin   92
2.1   Histone H3 Lysine 9 Methylation   92
2.2   Histone H3 Lysine 27 Methylation   94
2.3   Methylation at Other Histone H3 Residues   95
2.4   Histone Acetylation   96
2.5   Histone MacroH2A   97
2.6   Other Variant Histones   98
2.7   Nucleosome Position   98
2.8   Shape of the Xi   98
2.9   DNA Methylation   99
2.10   Late Replication Timing   100
2.11   *Xist* RNA   101
2.12   Redundant Mechanisms Maintain Silencing   101
3   Chromatin at the *Xic*   101
3.1   Histone Modifications   102
3.2   DNA Methylation   102
3.3   Replication Timing   103
4   Genes that Escape X-Inactivation   103
4.1   Histone Modifications   103
4.2   DNA Methylation   104
4.3   Replication Timing   104
4.4   Chromosome Organization   104
5   Developmental Regulation of X-Inactivation   105

5.1   Three Stages of X-Inactivation   105
5.2   Embryonic Stem Cells   107
5.3   Extraembryonic Cells   108
5.4   Reactivation of the Xi   109
6   Chromatin Features of the X Chromosomes Prior
        to X-Inactivation   111
6.1   Imprinted X-Inactivation   111
6.2   Random X-Inactivation   112
7   Conclusion   114
        References   115

Chromatin Mechanisms in *Drosophila* Dosage Compensation   123
M. Taipale and A. Akhtar

1   Introduction   123
2   The MSL Complex   124
2.1   MSL-1   126
2.2   MSL-2   127
2.3   MSL-3   127
2.4   MOF   128
2.5   MLE   129
2.6   JIL-1   131
2.7   *roX1* and *roX2*   131
2.7.1   *roX* Genes as Non-Coding RNAs   132
2.7.2   *roX* Loci as Chromatin Entry Sites   132
3   Targeting, Assembly and Spreading of the MSL Complex . .   133
3.1   Targeting and Assembly   133
3.2   Spreading   135
4   Cracking the Code X   136
4.1   Establishing the Code   136
4.2   Reading the Code   137
5   Molecular Mechanism of Dosage Compensation   138
5.1   Initiation Versus Elongation   138
5.2   The Inverse Effect Hypothesis   139
6   The Origin and Evolution of the MSL Complex   141
        References   143

DNA Methylation in Epigenetic Control of Gene Expression   151
A. Razin and B. Kantor

1   Introduction   151
2   Changes in Gene-Specific Methylation Patterns
        During Early Embryo Development   153

3   Effect of Methylation on Gene Expression   154
3.1   Direct Transcription Inhibition   154
3.2   Indirect Transcription Inhibition   156
4   DNA Methylation and Genomic Imprinting   159
5   DNA Methylation and Disease   162
6   Concluding Remarks   163
       References   163

The Epigenetic Breakdown of Cancer Cells:
From DNA Methylation to Histone Modifications   169
E. Ballestar and M. Esteller

1   Introduction   169
2   What Is Responsible for DNA Methylation
       and for How Deregulation Occurs?   171
3   Is Methylation Specific to the Tumor Type?   173
4   Connecting DNA Methylation Changes with Transcription:
       Chromatin Mechanisms   174
5   Can We Reactivate Epigenetically Silenced Genes?
       Towards Epigenetic Therapy   178
       References   178

Developmental Regulation of the β-Globin Gene Locus   183
L. Chakalova, D. Carter, E. Debrand, B. Goyenechea, A. Horton,
J. Miles, C. Osborne and P. Fraser

1   Introduction   183
2   The β-Globin Clusters and Their Ontogeny   184
3   Models for Studying the β-Globin Locus   185
4   The LCR Is Required for High-Level Expression   186
5   The Role of Individual HS   188
6   Gene Competition and the LCR Holocomplex   189
7   The β-Globin Locus Resides in a Region
       of Tissue-Specific Open Chromatin   190
8   The Role of Insulators   191
9   Intergenic Transcription   192
10   Intergenic Promoters   193
11   Histone Modification
       and Developmental Globin Gene Expression   194
12   The Role of Intergenic Transcription   195
13   The Cell Cycle Connection   196

14   The Corfu Deletion   197
15   Higher Order Folding and Long-Range Regulation   198
16   Nuclear Organization   199
17   Summary Model   200
          References   201

Epigenetic Regulation of Mammalian Imprinted Genes:
From Primary to Functional Imprints   207
M. Weber, H. Hagège, N. Aptel, C. Brunel, G. Cathala and T. Forné

1   Introduction   207
2   Imprinting Evolution   208
2.1   Conservation of Parental Genomic Imprinting
          in Therian Mammals   208
2.2   Theories on the Evolution of Parental Genomic Imprinting .   209
2.2.1   The Parental Conflict Theory   210
2.2.2   Alternative Theories   211
3   Characteristics of Mammalian Imprinted Genes   211
4   Epigenetic Control of Imprinted Genes   212
4.1   DNA Methylation   212
4.2   Histone Modifications   214
4.3   Asynchronous DNA Replication Timing   215
4.4   Chromatin Architecture   216
5   The Parental Genomic Imprinting Cycle   216
5.1   Erasure   216
5.2   Establishment   218
5.2.1   Primary Imprinting Marks   218
5.2.2   Imprinting Centres   219
5.3   Maintenance   221
5.4   Monoallelic Expression of Imprinted Genes   222
5.4.1   Formatting for Gene Expression   223
5.4.2   Acquisition of Functional Imprints   223
6   Conclusion   225
          References   226

Seed Development and Genomic Imprinting in Plants   237
C. Köhler and U. Grossniklaus

1   Introduction   237
2   Seed Development in Angiosperms   238
3   Development and Function of the Endosperm   238

4   A Role for Genomic Imprinting in Seed Development?   . . .   241
5   The Discovery of Genomic Imprinting in Maize   242
6   Studies on Other Potentially Imprinted Genes in Maize   . . .   243
7   Maternal Control of Early Seed Development in *Arabidopsis*   244
8   Intragenomic Parental Conflict
         and the Evolution of Genomic Imprinting   246
9   Imprinting of the *MEDEA* Locus in *Arabidopsis*   247
10   Function of *MEDEA* During Gametophyte
         and Seed Development   249
11   Imprinting of the *FWA* Locus in the Female Gametophyte   .   251
12   The Role of Imprinting During Gametophyte
         and Seed Development   252
13   Imprinting and Apomixis   253
14   Possible Epigenetic Marks
         Distinguishing Maternal and Paternal Alleles   254
14.1   Chromatin Structure   254
14.2   DNA Methylation During Gametogenesis   255
14.3   DNA Methylation During Seed Development   256
15   Conclusions   257
References   257

Subject Index   263

Chromatin Remodeling Factors and DNA Replication

Patrick Varga-Weisz

Abstract   Chromatin structures have to be precisely duplicated during DNA replication to maintain tissue-specific gene expression patterns and specialized domains, such as the centromeres. Chromatin remodeling factors are key components involved in this process and include histone chaperones, histone modifying enzymes and ATP-dependent chromatin remodeling complexes. Several of these factors interact directly with components of the replication machinery. Histone variants are also important to mark specific chromatin domains. Because chromatin remodeling factors render chromatin dynamic, they may also be involved in facilitating the DNA replication process through condensed chromatin domains.

1
Introduction

Inheritable traits are not only encoded in the sequence of DNA, but also determined by factors 'on top' of the DNA, the epigenetic information ('epi' is classical Greek for 'on top'). Epigenetic phenomena play an important role in the maintenance of gene expression patterns through cell generations, for example in tissue-specific gene expression. A striking example of epigenetic regulation is found in X-chromosome inactivation in mammalian cells, where one of the two X chromosomes is maintained in an inactive, highly condensed state throughout development. In many organisms epigenetic regulation is mediated by DNA methylation, but epigenetic phenomena are also found in organisms where DNA methylation does not take place. The eukaryotic genome is packaged and organized by a plethora of proteins forming the superstructure chromatin that is a major facet of epigenetics. It is very important, therefore, for chromatin structures to be faithfully duplicated during DNA replication to maintain epigenetic information. There is accumulating

P. Varga-Weisz
Marie Curie Research Institute, The Chart, Oxted, Surrey RH8 0TL, UK,
e-mail: p.varga-weisz@mcri.ac.uk

Progress in Molecular and Subcellular Biology
P. Jeanteur (Ed.)
Epigenetics and Chromatin
© Springer-Verlag Berlin Heidelberg 2005

evidence that chromatin remodeling factors play a key role in facilitating and regulating DNA replication through chromatin and the propagation of epigenetic information during DNA replication. This chapter summarizes our current knowledge about chromatin remodeling factors that have been linked directly to the DNA and chromatin replication process.

2
Chromatin and Chromatin Remodeling Factors

2.1
Chromatin

The most abundant of chromatin proteins are the histones that together with DNA assemble the basic building block of chromatin, the nucleosome. The nucleosome is basically a spool, a histone octamer around which 147 base pairs of DNA are wrapped in almost two superhelical turns (reviewed in Luger 2003). The octamer is composed of two H2A-H2B dimers interacting with a core of an H3-H4 tetramer. In the human genome, nucleosomes occur on every 180 base pairs of DNA on average, forming periodic arrays ('beads-on-a-string' fiber). Histone H1 interacts with the linker DNA at the entry–exit point of the nucleosome and stabilizes higher levels of folding of nucleosome arrays. Little is known about the molecular mechanisms of further levels of compaction of the chromatin fiber, but proteins that regulate fiber–fiber interactions, such as heterochromatin protein 1 (HP1), cohesins, condensins and topoisomerases, play an important role in chromatin organization (reviewed in Gasser 1995).

The existence of different levels of chromatin folding or organization is evident at the microscope level in interphase nuclei, where one can differentiate between the highly condensed structures called heterochromatin and the more 'loose' euchromatin (Hennig 1999). Heterochromatin and related structures have been clearly linked to gene silencing (Wallrath 1998). One refers to a gene as silenced when it is shut-off through subsequent cell generations.

Nucleosomes occlude much of the surface of the DNA wrapped around them and limit the access of many factors. In this way, chromatin is involved in the regulation of many processes, including transcription activation. Our understanding of chromatin received a major boost with the discovery of enzymes that render chromatin highly dynamic and facilitate access of cellular factors to the DNA. These enzymes, chromatin remodeling factors, are involved in all processes of DNA metabolism and are integral parts of the transcription regulatory machinery (reviewed in Narlikar et al. 2002). Two major classes of chromatin remodeling factors can be distinguished: histone modification enzymes and ATP-dependent chromatin remodeling factors.

2.2
Histone Modification Enzymes

Histones are evolutionarily highly conserved proteins and yet there is clear evidence that nucleosomes act as mediators of epigenetic information (reviewed in Jenuwein and Allis 2001). Epigenetic information is stored via chemical modifications of the histones, especially in their N-terminal tails which span about 25 amino acids. The histone tails protrude from each histone out of the nucleosome core body and can be recipients of many alterations, including acetylation, phosphorylation, methylation, ADP-ribosylation and ubiquitination. These various modifications occur at multiple sites within the tails and result in a great nucleosome heterogeneity. Figure 1 illustrates some of these modifications in histones H3 and H4. An idea has been developed that histone modifications form a 'bar code' for each nucleosome, which defines and regulates its interactions with other chromatin components and carries information about the transcriptional status of the gene that it is part of (Strahl and Allis 2000).

Enzymes that mediate histone modifications have been identified only relatively recently, and many of them are important transcriptional regulators. Histone acetyltransferases (HATs) have usually been linked to transcriptional

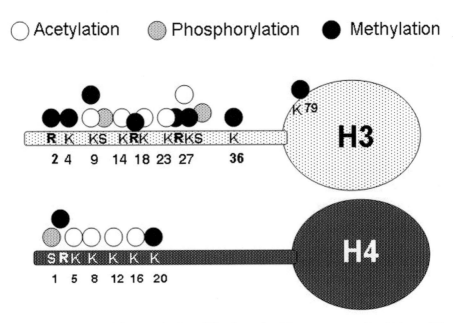

Fig. 1. A summary of characterized modifications of residues on mammalian histones *H3* and *H4*. The extended rod symbolizes the N-terminal tail-domain. Lysine residues can be mono-, di- and tri-methylated. Lysine 9 methylation of histone H3 is associated with transcriptional repression and is not compatible with acetylation of the same residue (which is linked to transcriptional activation)

activation, whereas histone deacetylases (HDAcs) have a major role in transcriptional repression. However, this is an oversimplification, and there are examples where a HAT is involved in gene silencing (see below) or an HDAc in activation (de Rubertis et al. 1996). Often, these enzymes are found in complexes with other proteins that facilitate histone acetylation or deacetylation within the nucleosome (Grant et al. 1997; Tong et al. 1998).

The complexity of the histone code is illustrated by the different functional associations of histone methylation. Methylation of lysine 9 (K9) of histone H3 by the histone methyltransferase (HMT) SU(VAR) 3–9 and its homologues is linked to heterochromatin formation and gene silencing, whereas methylation of lysine 4 (K4) is linked to transcriptional activity (see, for example, Noma et al. 2001, reviewed in Grewal and Elgin 2002). In addition, in budding yeast, dimethylation of K4 is linked to potential transcriptional activity, whereas trimethylation of the same residue occurs when the gene is actually actively transcribed (Santos-Rosa et al. 2002).

2.3
ATP-Dependent Chromatin Remodeling Factors

The nucleosome is a relatively stable entity. A class of enzymes use the energy gained by ATP-hydrolysis to move or disrupt nucleosomes efficiently. These enzymes are usually complexes of diverse proteins, but they have in common ATPases that resemble a specific class of DNA helicases. Helicase activity has not been demonstrated for any of these ATPases, but there is evidence that they function by distorting DNA structure to some degree (Havas et al. 2000). Figure 2 illustrates the major classes of well-characterized nucleosome remodeling ATPases and some of their complexes. SWI2/SNF2-type ATPases are highly conserved and are involved in transcriptional regulation, the *Drosophila* homologue is called Brahma, and mammalian cells contain two closely related homologues, Brg1 and Brm. The nucleosome remodeling ATPase Mi2 and its related proteins have been linked to transcriptional repression; they are found in complexes containing histone deacetylases. ISWI (Imitation Switch) was originally identified in *Drosophila* where it is the core of the NURF, CHRAC and ACF chromatin remodeling complexes. In mammalian cells there are two isoforms of ISWI called SNF2H and SNF2L. Several recent reviews cover the biology and biochemistry of ATP-dependent nucleosome remodeling factors (Becker and Hörz 2002; Narlikar et al. 2002).

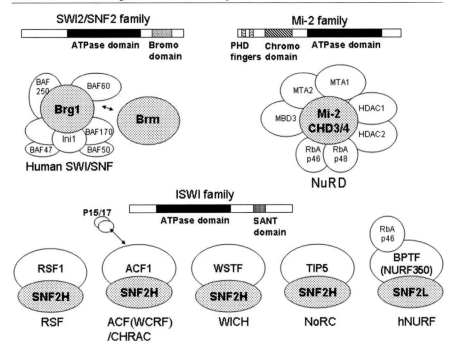

Fig. 2. Summary cartoon of the best characterized human nucleosome remodeling ATPases with their specific domain architecture and their respective complexes. *Brg1* and *Brm* are highly related to the yeast SWI2/SNF2 and STH1 ATPases. The complexes formed by Brg1 and Brm are very similar. However, Brg1 is also found in a complex containing a protein called Polybromo (BAF180) instead of BAF250 and this complex may be the mammalian counterpart of the yeast RSC complex (reviewed in Muchardt and Yaniv 2001). ISWI interacts with a number of different large proteins, forming distinct complexes with diverse biological functions. This scheme does not include recently identified ISWI complexes with cohesin subunits (Hakimi et al. 2002) and with TBP-related factor (Hochheimer et al. 2002). The CHRAC complex differs from ACF by the presence of a pair of histone-fold proteins that enhance the nucleosome sliding and chromatin assembly activity of the ACF1/ISWI core complex. (Kukimoto et al. 2004)

3
Chromatin Structure and DNA Replication

Chromatin limits the accessibility to DNA, and this raises the questions: How does the replication machinery interact with chromatin? Does replication occur through nucleosomes? Does it disrupt chromatin structure? These questions have been studied extensively in the SV40 system. SV40 is a double-stranded DNA animal virus whose DNA is packaged by nucleosomes. Firing of DNA replication in this system is dependent on the binding of T-antigen to the origin of replication from where replication initiates. If the origin is occupied by a nucleosome, DNA replication firing is prohibited and necessitates nucleosome remodeling to facilitate DNA replication (Alexiadis et al. 1998). Crosslinking DNA by psoralen allows the identification of nucleosomal DNA structures on single molecules by electron microscopy. This approach visualized the disruption and reformation of the nucleosomal array structure at the replication fork of SV40 (Gasser et al. 1996): The passage of the replication machinery destabilizes the nucleosomal organization over a distance of 650–1,100 base pairs. In front of the fork, an average of two nucleosomes are disrupted. On daughter strands, the first nucleosome is detected at about 260 base pairs from the elongation point. Thus the disruption of the nucleosomal array structure by the replication machinery is confined to relatively small segments of chromatin. However, newly replicated chromatin is characterized by an altered structure compared to the parental chromatin or chromatin of non-replicating cells. This altered structure shows greater nuclease sensitivity, indicating a more open chromatin structure (Cusick et al. 1983; Pulm and Knippers 1984). Newly replicated chromatin then matures into a more compact structure by a process that is so far poorly characterized (Schlaeger et al. 1983). It is likely that changes in histone modifications are part of this process (see below). The limitations of the chosen model systems leave many questions unanswered that are relevant in somatic cells of higher eukaryotes: How does the replication machinery interact with higher orders of chromatin structures, such as heterochromatin? How does the replication machinery bind to chromatin?

Several studies indicate a close link between the DNA replication machinery and the establishment of specific chromatin structures such as heterochromatin (see, for example, Ehrenhofer-Murray et al. 1999; Zhang et al. 2000; Ahmed et al. 2001; Nakayama et al. 2001). The impact of chromatin structure on DNA replication is most evident through its role in regulating the replication programme, the coordinated, ordered firing of replication units (replicons) throughout the genome (reviewed in Bailis and Forsburg 2003; McNairn and Gilbert 2003). There is a correlation between the timing of replication with chromatin structure and gene activity: usually, active genes are replicated early in S phase, whereas heterochromatin and silenced genes are replicated late in S phase. The role of chromatin components in establishing replication timing has been studied in budding yeast, where deletion or mistargeting of various regulators of chromatin structure cause an alteration of the replication timing of specific loci (Stevenson and Gottschling 1999; Cosgrove et al. 2002; Vogelauer et al. 2002; Zappulla et al. 2002). The timing of replication itself is in turn a major determinant of gene activity and chromatin structure (Zhang et al. 2002).

4
Chromatin Assembly

DNA replication-coupled chromatin assembly is essential for epigenetic inheritance through the propagation of chromatin states. The duplication of chromatin requires the assembly of new nucleosomes onto the nascent chromatin. Chromatin assembly occurs immediately after DNA replication. Parental histones are segregated onto the two nascent DNA duplexes randomly (Sogo et al. 1986; Gruss et al. 1993; reviewed in Gruss and Sogo 1992; Krude 1999), and these 'parental' histones may transmit information about the chromatin structure of the template to the daughter strands. However, at least half of the complement of histones has to be supplied from newly synthesized ones.

Nucleosomes can be assembled in vitro by mixing histones with DNA in high levels of salt (~ 2 M NaCl) and then slowly lowering the salt concentration by dialysis. This procedure prevents the non-specific aggregation of histones with DNA, which would otherwise occur through strong charge interaction. Alternatively, acidic polymers such as polyglutamic acid or RNA can mediate nucleosome assembly in vitro. In the cell, histone chaperones play an important role in this process. A critical part of their function is to neutralize the charges on the histones, thus preventing non-specific aggregation with DNA (Akey and Luger 2003).

4.1
Histone Chaperones

Several proteins with histone chaperone activity have been identified that mediate nucleosome assembly under physiological conditions, and all of these proteins are highly charged (reviewed in Akey and Luger 2003). Many histone chaperones show a preference in the interaction with either the histone H3-H4 tetramer, such as chromatin assembly factor-1 (CAF-1; Kaufman et al. 1995), antisilencing function 1 protein (ASF1; Tyler et al. 1999), or histone regulatory A (HIRA) (Ray-Gallet et al. 2002), while others preferentially interact with the H2A-H2B dimer, such as nucleosome assembly protein-1 (NAP-1; Ishimi et al. 1987). Therefore, several histone chaperones may be involved in the sequential assembly of the nucleosome particle, for example histones H3/H4 are deposited first by CAF-1, and the nucleosome is then completed by the deposition of two H2A/H2B dimers carried by NAP-1. Histone chaperones may also have roles in the nuclear import of histones, histone storage (e.g. nucleoplasmin in the frog oocyte), sperm decondensation and transcriptional activation (reviewed in Akey and Luger 2003). In fact, an important principle one should keep in mind is that factors involved in nucleosome assembly may just as well catalyze nucleosome disassembly, and in this way may be involved in transcriptional activation. This principle has been

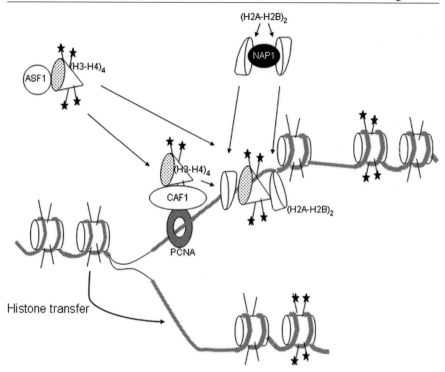

Fig. 3. Pathways of nucleosome assembly at the replication fork. Half of the histone complement is provided by histone transfer from parental DNA to both new DNA strands. These histones may carry modifications that transmit epigenetic information to the newly synthesized DNA. The rest of the histones are derived from a newly synthesized pool that carries specific acetylation patterns on histones H3 and H4 (*stars*) mediated by a cytoplasmic acetyltransferase. Deposition of these histones is facilitated by chaperones such as CAF1 or ASF1 for H3/H4 or NAP1 for H2A/H2B. CAF-1 is tethered directly to the replication site by its interaction with the sliding clamp of PCNA

illustrated in vitro with the histone chaperone nucleoplasmin that facilitates transcription factor binding on nucleosomal templates (Chen et al. 1994; Owen-Hughes and Workman 1996). For some histone chaperones, including CAF-1 and ASF1, a role in epigenetic phenomena has been demonstrated (see below). Figure 3 illustrates chromatin assembly at the replication fork by the two major processes involved, histone transfer from the parental DNA strand and deposition of newly synthesized histones by histone chaperones CAF-1 and ASF1.

4.2
Chromatin Assembly Factor 1, a Replication-Coupled Histone Chaperone

One of the most intensely studied histone chaperones is CAF-1. It contains three subunits (p150, p60 and p48 in human) and is required for deposition of newly synthesized histone H3/H4 tetramers specifically onto replicating DNA (Smith and Stillman 1989, 1991), either during S phase or during nucleotide excision repair (Gaillard et al. 1996). This deposition is followed by H2A/H2B association with the H3/H4 tetramer, which is mediated by other histone chaperones such as NAP-1 (Ito et al. 1996). Co-localization of CAF-1 with DNA replication and repair foci supports its role during chromatin replication and DNA repair (Krude 1995; Marheineke and Krude 1998; Martini et al. 1998). CAF-1 is evolutionarily conserved and homologues have been described in yeast, plants, insects and vertebrates. In budding yeast, CAF-1 was purified in a complex (known as CAC) with histones H3 and H4, where H4 has an acetylation pattern similar to that of newly synthesized histones (Kaufman et al. 1995; Verreault et al. 1996). CAF-1 interacts with proliferating cell nuclear antigen (PCNA, a central molecule in the replication process; see below) and is targeted to replication sites through this interaction (Shibahara 1999).

4.3
CAF-1 Functions in the Inheritance of Chromatin States

Even though there is evidence for an essential role of CAF-1 in cell proliferation and genome stability in animal cells (Quivy et al. 2001; Hoek and Stillman 2003; Ye et al. 2003), deletion of CAF-1 in budding yeast is not lethal. In this organism, CAF-1 deletion uncovers its role in epigenetic inheritance: without CAF-1 gene silencing at telomeres, mating type loci and ribosomal DNA is impaired (Enomoto et al. 1997; Kaufman et al. 1997; Monson et al. 1997; Enomoto and Berman 1998; Smith et al. 1999). Telomeres, mating type loci and ribosomal DNA, are normally organized into heterochromatin-like structures, which may point to a role of CAF-1 in heterochromatin assembly. In budding yeast, the CAC1 gene coding for the p150 homologue is required for the normal distribution of the telomere-binding protein Rap1 within the nucleus (Enomoto et al. 1997). Rap1 protein binds to telomere DNA repeats and controls telomere length and telomeric silencing. The perturbation of Rap1p localization reflects defects in the organization of telomeric chromatin. In cells lacking CAF-I, the silent mating type loci are derepressed partially and CAF-I contributes to the maintenance but not the re-establishment (after a transient derepression) of silencing of these loci (Enomoto and Berman 1998).

Together, these data suggest that CAF-I plays a role in generating specific chromatin structures in vivo (Enomoto et al. 1997). In higher eukaryotes,

CAF-1 may also play a role in heterochromatin assembly and gene silencing. This is seen, for example, in a mammalian cell culture model designed to monitor DNA methylation-dependent gene silencing and its reversal to gene activation upon DNA demethylation. Expression of a truncated version of the human p150 subunit of CAF-1, but not the full-length subunit, dramatically increased the frequency at which transcriptional silencing of a reporter gene was reversed in this system (Tchenio et al. 2001). This suggests that the truncated p150 acts in a dominant negative manner and that mammalian CAF-1 has a role in the maintenance of silencing. CAF-1 is associated with heterochromatin component HP1 in mammalian cells and it has been shown to mediate the chromatin deposition of HP1 during chromatin assembly in vitro (Murzina et al. 1999). In plants, CAF-1 subunits are also involved in epigenetic regulation of gene expression. This is observed in the apical meristems, groups of cells that give rise to post-embryonic tissues (such as shoots and roots). CAF-1 subunit homologues of p150 and p60 are the gene products of FASCIATA1 and 2 (*FAS1* and *FAS2*) that are essential for the cellular and functional organization of both shoot and root apical meristems (Kaya et al. 2001). Mutants in the FAS genes show a non-uniform misexpression of key genes involved in the control of the stem cells (Kaya et al. 2001). This non-uniform, 'variegated' phenotype, whereby genes are misregulated in a clonally inheritable manner, is often observed when components of chromatin are mutated.

4.4
Histone Chaperone ASF1

ASF1 was initially identified as a gene product that affects transcriptional silencing (Le et al. 1997). The characterization of activities involved in replication-dependent chromatin assembly lead to the identification of fly and human homologues of ASF1 as histone H3/4 chaperone that deposits newly synthesized histones onto replicated DNA (Tyler et al. 1999; Munakata et al. 2000). ASF1, in complex with newly synthesized histones (RCAF complex), collaborates synergistically with CAF-1 in chromatin assembly in vitro (Tyler et al. 1999; Mello et al. 2002). Whereas deletion of ASF1 in budding yeast results in no significant impairment of silencing, the disruption of both ASF1 and CAF-1 function enhances the silencing defects of CAF-1 deletion, and this points to a cooperative role of both factors in the assembly of silent chromatin (Tyler et al. 1999; Krawitz et al. 2002). In *Drosophila* the p105 subunit of CAF-1, the fly homologue to human p60, interacts directly with ASF1 in vitro and co-localizes in polytene chromosomes consistent with a functional synergy between these factors (Tyler et al. 2001). Furthermore, mutations of ASF1 also counteract gene silencing of a reporter gene that has been placed close to a heterochromatic environment in the fly (Moshkin et al. 2002).

4.5
Histone Chaperones and Heterochromatin Replication

CAF-1 is apparently involved in chromatin assembly throughout the genome, both in euchromatin and heterochromatin (Taddei et al. 1999), and this may be true for ASF1, too. How could histone chaperones that seem to be involved in chromatin assembly in general affect gene silencing at particular loci? One suggestion that has been put forward is that such factors ensure rapid and efficient chromatin assembly and in this way reinforce heterochromatin maintenance (Enomoto and Berman 1998; Kaya et al. 2001). If chromatin assembly is delayed or impaired during the replication of heterochromatin, factors that counteract heterochromatin formation, such as transcription factors, may have an opportunity to bind and win the upper hand, e.g. by recruiting histone acetyltransferases. Because chromatin states are inheritable, this disrupted state would then be perpetuated in subsequent rounds of chromatin replication. It has been also argued that the lack of transcription within heterochromatin domains makes these structures more vulnerable to defects during chromatin assembly compared to transcriptional active chromatin, where transcription-coupled chromatin assembly or remodeling may allow the 'repair' of chromatin assembly defects during DNA replication (Mello and Almouzni 2001). Another possibility might be that specific histone chaperones mediate the preferential deposition of histone variants that in turn define specific chromatin structures.

4.6
Histone Variants

Histone variants differ from the canonical histones to various extents. This may involve a few amino acids (e.g. H3.3 versus the 'canonical' H3.1) or substantial domains outside of the histone fold structure (e.g. CENP-A, macro H2A). Histone variants mark specific chromatin domains, such as the centromere (CENP-A), the inactive X chromosome (macro H2A; Costanzi and Pehrson 1998) and transcriptionally active regions (Histone H3.3), and are of great functional importance, deletion of some having grave consequences on the viability of the organism (for example, see van Daal and Elgin 1992, and below).

The importance of histone variants has been demonstrated by their role in centromeres, which are the chromosome domains at which the kinetochore assembles to separate sister chromatids at mitosis. Centromeres are marked by a unique histone H3-like protein (reviewed in Sullivan 2001), named CENP-A in human cells (Earnshaw et al. 1986; Palmer et al. 1991), Cse4p in budding yeast (Stoler et al. 1995) and Cid in *Drosophila* (Malik and Henikoff 2001). CENP-A and its homologues play essential roles in centromere function, but the precise role is unclear (reviewed in Mellone and Allshire 2003).

CENP-A interacts with histones H2A, H2B and H4 to form a nucleosome-like structure (Yoda et al. 2000). How is the association of CENP-A limited to the centromere? Analysis of the timing of centromere replication and expression of CENP-A protein in some organisms led to the proposal that CENP-A deposition is regulated by its distinct, timed expression, away from the bulk histone H3 expression and deposition (Shelby et al. 1997; Csink and Henikoff 1998). However, there is evidence that targeting of CENP-A to centromeres does not require centromere replication (Shelby et al. 2000; Ahmad and Henikoff 2001; Sullivan and Karpen 2001; Ouspenski et al. 2003). It has been suggested that heterochromatin provides a special, spatially segregated domain for CENP-A deposition (Henikoff et al. 2000). HIRA is a histone chaperone related to the budding yeast proteins Hir1p and Hir2p (Lorain et al. 1998) and acts in replication-independent chromatin assembly (Ray-Gallet et al. 2002). Interestingly, in budding yeast the Hir proteins and CAF-1 are required for proper deposition of the CENP-A homologue, Cse4 (Sharp et al. 2002). However, a specific chaperone for CENP-A has not been identified yet.

4.7
DNA Replication-Independent Chromatin Assembly

Chromatin assembly occurs independently of DNA replication during DNA repair, recombination and transcription. Replication-independent chromatin assembly involves histone variants such as H2A.Z and H3.3 that are synthesized throughout the cell cycle rather than just in S phase. Specific histone chaperones (Ray-Gallet et al. 2002) and chromatin remodeling factors (Mizuguchi et al. 2004) mediate the deposition of these 'replacement histones'. Indeed, deposition of the major histone H3 (H3.1) is coupled to DNA replication and possibly DNA repair, whereas histone variant H3.3 serves as the replacement variant for the DNA-synthesis-independent deposition pathway (Ahmad and Henikoff 2002a,b). H3.1 and H3.3 assembly complexes have recently been purified and shown to contain distinct histone chaperones, CAF-1 together with H3.1 for DNA-synthesis-dependent nucleosome assembly and HIRA with H3.3 for DNA synthesis-independent nucleosome assembly. Strikingly, these complexes possess one molecule each of H3.1 or H3.3 in association with H4, suggesting that histones H3 and H4 exist as dimeric intermediates in nucleosome formation (Tagami et al. 2004). This finding may provide new insights into how epigenetic information could spread from the parental to the daughter DNA strands during DNA replication (Tagami et al. 2004): if the histone transfer from the parental strands to the daughter strands also occurs with dimeric units of H3/H4, then this would possibly facilitate the transmission of epigenetic information by ensuring the presence of 'old' histones in every daughter nucleosome. The 'old' H3/H4 dimers would then enforce their modification pattern onto the new units by recruiting the relevant enzymes. For example, H3 K9 methylation could be enforced by the recruitment of the histone methyltransferase SU(VAR)3–9 that forms a complex with HP1 that in turn binds to the methylation site of an 'old' H3 tail.

5
Histone Modifications and Chromatin Replication

5.1
Histone Deacetylation During Chromatin Replication

Histone modifications are important determinants of heritable chromatin states such as heterochromatin whose domains are marked by the overall underacetylation of histone H4 (Jeppesen and Turner 1993; O'Neill and Turner 1995; Belyaev et al. 1996; Braunstein et al. 1996). The incorporation of newly deposited histones into the nascent chromatin, therefore, necessitates the resetting of the modification status of these histones to match the parental situation. Newly synthesized histone H4 is acetylated at lysines 5 and 12 (Sobel et al. 1995) and deposited in this form throughout the genome (reviewed in Annunziato and Hansen 2000). The role of this acetylation is unclear and is apparently not required for chromatin assembly per se (Shibahara et al. 2000). These histones have to be deacetylated after deposition to maintain the heterochromatin-specific signature. In fact, the transient deposition of histone H4 acetylated at K5 and K12 within heterochromatin and its subsequent deacetylation has been well documented in mammalian cells (Taddei et al. 1999). Inhibition of deacetylation leads to a disruption of heterochromatin structure with subsequent consequences on genome stability (Ekwall et al. 1997; Taddei et al. 2001). The responsible deacetylase in mammalian cells has not been identified, but the specific accumulation of HDAC2 at replication sites during the late stages of S phase, when heterochromatin is replicated, proposes a likely candidate (Rountree et al. 2000). Histone deacetylases have been shown by genetic means to be essential for centromeric gene silencing in fission yeast, supporting a role in heterochromatin assembly (Grewal et al. 1998; Bjerling et al. 2002). One key reason why histone deacetylation may be important for heterochromatin assembly around the centromeres is that it allows the subsequent methylation of key lysine residues of histone H3, thus creating a specific binding site for HP1 and mediating heterochromatin assembly (see below).

5.2
Histone Acetylation at the Replication Site

The re-establishment of the acetylation state of nucleosomes after DNA replication requires both deacetylation and acetylation of specific lysine residues on newly incorporated histones. Histone modifying enzymes interact with replication proteins such as the Origin Recognition Complex (ORC) that marks replication origins in eukaryotes and the Minichromosome Maintenance Complex (MCM), a multisubunit helicase composed of proteins MCM2–7 involved in DNA replication. In budding yeast, a histone acetyltransferase called Sas2 (something about silencing 2) acetylates lysine 16 of histone H4 and is important to counteract gene silencing at the HMR locus and the rDNA but has opposite effects on silencing at the HML, and telomeric loci (Meijsing and Ehrenhofer-Murray 2001; Osada et al. 2001). Indeed, a mutation in histone H4 replacing K16 with R has a phenotype very similar to that of the sas2 deletion (Meijsing and Ehrenhofer-Murray 2001; Osada et al. 2001). This HAT interacts with the large subunit of CAF-1 (Cac1p) and ASF1 (Meijsing and Ehrenhofer-Murray 2001; Osada et al. 2001). These findings suggest that the role of Cac1p in epigenetic inheritance may be, at least in part, a result of its interaction with Sas2. In human cells, a highly related acetyltransferase named HBO1 (Histone acetyltransferase Bound to ORC 1) interacts with the DNA replication proteins ORC1 (Iizuka and Stillman 1999) and MCM2 (Burke et al. 2001). In *Drosophila* a protein called Chameau is the fly counterpart of HBO1 according to sequence analysis, and this putative histone acetyltransferase operates in epigenetic silencing mediated by pericentromeric heterochromatin and the Polycomb group transcriptional repressors (Grienenberger et al. 2002). The transcriptional co-activator p300 is yet another factor with histone acetyltransferase activity that interacts with a central component of the DNA replication and repair machinery, PCNA, and this interaction has been specifically linked to nucleotide excision repair-coupled DNA synthesis (Hasan et al. 2001). Together, these findings illustrate a close link between the replication machinery and histone acetyltransferases involved in gene regulation. Caveats to this interpretation may be that histone modifying enzymes may also modify other components of chromatin including the replication machinery itself, and that replication proteins such as the ORC complex may have functions outside of DNA replication (see, for example, Prasanth et al. 2002; Bailis and Forsburg 2003).

6
ATP-Dependent Chromatin Remodeling Factors in Chromatin Replication

6.1
ISWI Complexes in Chromatin Assembly in Vitro

Crude extracts of frog oocytes and early *Drosophila* embryos support the efficient assembly of chromatin in vitro. The analysis of these chromatin assembly systems identified an ATP requirement for the creation of regular, periodic nucleosomal arrays (Glikin et al. 1984; Almouzni and Méchali 1988; Becker and Wu 1992) which led to the identification of a role of ATP-dependent nucleosome remodeling factors in chromatin assembly. The requirement is demonstrated by a partial digestion of the reconstituted chromatin with micrococcal nuclease that preferentially cleaves between nucleosomes. Such a digest will result in a ladder of DNA fragments representing mono-nucleosomes, di-nucleosomes, etc. in native chromatin. In the absence of ATP during chromatin assembly, only a 'smear' of DNA of non-discrete size distribution will be observed following digestion (see, for example, Varga-Weisz et al. 1997).

Fractionation of *Drosophila* extracts lead to the purification of ISWI (Imitation Switch)-containing chromatin remodeling complexes as mediators of the ATP-dependent nucleosome array formation (Ito et al. 1997; Varga-Weisz et al. 1997). ISWI is a member of a family of ATPases that are involved in various aspects of chromatin remodeling. ISWI was subsequently found to be the core 'engine' of various complexes supporting nucleosome assembly in organisms ranging from budding yeast, frog to human (see above and Fig. 2; reviewed in Längst and Becker 2001).

ISWI itself is a potent nucleosome remodeling factor in vitro, and the major activity of this ATPase seems to be the sliding of nucleosomes along DNA without a major disruption of the nucleosome structure (reviewed in Längst and Becker 2001; see also Fazzio and Tsukiyama 2003). ISWI is found in complexes with subunits that enhance and regulate its nucleosome remodeling activity, including the ACF1 subunit in ACF (ATP-dependent Chromatin remodeling and assembly Factor) and Chrac (Chromatin Accessibility; Ito et al. 1999; Eberharter et al. 2001), NURF301 in NURF (Nucleosome Remodeling Factor; Xiao et al. 2001), TIP5 in NorC (Nucleolar remodeling Complex; Strohner et al. 2001) and RSF1 in RSF (Remodeling and Spacing Factor; Loyola et al. 2001). WSTF (Williams Syndrome Transcription Factor) is strongly related to ACF1 and it interacts with ISWI to form nucleosome 'spacing' factor WICH (WSTF-ISWI chromatin remodeling factor; Bozhenok et al. 2002; MacCallum et al. 2002; see Figure 2 for a summary of human ISWI complexes). While most ISWI complexes mediate chromatin assembly in vitro, they also render chromatin accessible to incoming proteins (Varga-Weisz et al. 1997), and this allows gene activation by transcription factors (Ito et al. 1997; LeRoy et al. 1998) or DNA replication firing by T-antigen on chromatin templates (Alexiadis et al. 1998). RSF, composed of the RSF1 subunit and human ISWI isoform SNF2H, was originally identified as a factor that facilitates transcription from chromatin templates. It mediates chromatin assembly and nucleosome spacing without the assistance of histone chaperones and, indeed, RSF1 interacts with histones directly (Loyola et al. 2001, 2003). In contrast to RSF, ACF, composed of ACF1 and ISWI, requires additional histone chaperones such as NAP1 or CAF-1 for chromatin assembly (Ito et al. 1997, 1999). Once ACF initiates chromatin assembly on a DNA molecule, it will be 'committed' to this template and assemble nucleosomes in localized arrays (Fyodorov and Kadonaga 2002). This observation supports a 'tracking'-type mechanism by which this complex moves along the DNA during chromatin assembly. How could an ATP-dependent chromatin remodeling factor such as ACF or RSF be involved in the formation of periodic nucleosomal arrays? Figure 4 illustrates two possible models based on ideas developed in two reviews on this subject (Varga-Weisz 1998; Haushalter and Kadonaga 2003).

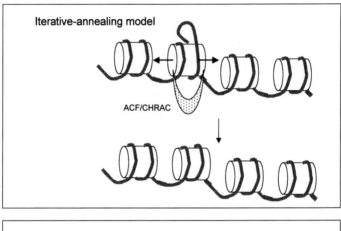

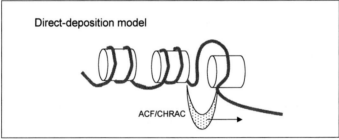

Fig. 4. Schematic representation of two possible, mutually non-exclusive, mechanisms by which ISWI complexes may be involved in ATP-dependent chromatin assembly. These models are based on ideas developed in two reviews on this subject (Varga-Weisz 1998; Haushalter and Kadonaga 2003). In the iterative-annealing model, defects that arise during the early steps in nucleosome assembly, such as unbound DNA within the nucleosome, chaperones associated with histones or gaps between nucleosome, are 'repaired' by the transient, repetitive disruption of DNA–histone interactions. This allows the removal of kinetic traps that may arise in the early steps of nucleosome assembly. The same transient disruption of histone–DNA contacts may drive nucleosome movement along the DNA, which would be important in the creation of regular nucleosomal arrays. Regular arrays, in turn, may be required for formation of the higher order fiber structure. In the direct deposition model, deposition of nucleosomes is directly connected to translocation of the chromatin remodeling factor along the naked DNA. This could be mediated by formation of a specific DNA structure, such as a DNA loop, during the tracking motion, and would facilitate the processive formation of nucleosome arrays

6.2
ISWI Complexes and Their Role in Chromatin Replication in Vivo

ISWI complexes have a role in chromatin assembly, but they also render chromatin accessible. What in vivo function does their 'dual role' reflect? ISWI is essential in *Drosophila* and genetic deletion of ISWI leads to global alterations of chromosome structure, whereby the male X chromosome that is subject to dosage compensation is especially affected (Deuring et al. 2000). The role of ACF1/ISWI and WSTF/ISWI complexes has been studied in chromatin assembly extracts from frog eggs (Demeret et al. 2002; MacCallum et al. 2002). Here, depletion of ISWI from the extracts affected nucleosome spacing, but not DNA replication or chromosome condensation. Genetic deletion of ACF1 confirms the role of ACF1-ISWI complexes in chromatin assembly in early fly embryos (Fyodorov et al. 2004): extracts prepared from ACF1 -/- fly embryos are somewhat impaired in chromatin assembly and the chromatin of these embryos shows some, albeit minor, global changes in nucleosome spacing or density. Homozygous deletion of ACF1 is lethal to 75% of the mutant fly embryos, but 25% surprisingly escape with only minor defects and are fertile and essentially normal. Mutant fly embryos that survive show alterations in polycomb and pericentromeric heterochromatin-mediated gene silencing that reflect changes in chromatin structure. Some of these changes are in line with a defect in chromatin assembly, while others point to a role in counteracting chromatin condensation. One reason why ACF1/ISWI complexes are important in the formation of regular nucleosomal arrays in the very rapidly dividing early fly and frog nuclei may be that these organisms lack a canonical linker Histone H1 (Dimitrov et al. 1993; Ner and Travers 1994) that may mediate nucleosome spacing in somatic cells.

In mammalian cells, ACF1 interacts with the SNF2H isoform of ISWI (LeRoy et al. 1998, 2000; Bochar et al. 2000; Poot et al. 2000) and appears to be targeted specifically to replication foci in pericentromeric heterochromatin in mouse and human cells (Collins et al. 2002). SNF2H is also targeted to these foci (Collins et al. 2002). Depletion of ACF1 by RNA interference impairs progression through the late stages of S phase, when the pericentromeric heterochromatin is replicated (Collins et al. 2002). This impairment can be alleviated when the cells are treated with the DNA methylation inhibitor aza-deoxycytidine at doses that cause a specific decondensation of heterochromatin (Haaf 1995; Collins et al. 2002). These results imply a role of ACF1 in facilitation of DNA replication through condensed chromatin. Depletion of SNF2H causes a general impairment in S phase progression that again can be alleviated by aza-deoxycytidine treatment (Collins et al. 2002). It is not clear how the ACF1/SNF2H complex facilitates DNA replication through heterochromatin. One could imagine that this factor is involved in setting up functional origins of replication within condensed chromatin or that it facilitates the rapid access of the replication machinery to the DNA within heterochromatin. Interestingly, the SWI/SNF complex is important for the function of specific origins of replication in budding yeast (Flanagan and Peterson 1999). One could also imagine that the ACF1/ISWI complex assembles a chromatin structure that on one side is condensed but on the other side is readily 'unzipped' to allow DNA replication.

The suggested role of ACF1 and ISWI in heterochromatin replication in mammalian cells is not necessarily in contradiction to those findings obtained from the experiments with frog egg extracts (Demeret et al. 2002; MacCallum et al. 2002) where it was shown that these complexes do not have a role in DNA replication through chromatin per se. This is because chromatin assembled in these extracts lacks the heterochromatin structures found in somatic nuclei. However, findings of the effect of ACF1 on DNA replication in *Drosophila* are in marked contrast to those obtained from mammalian cell culture. Here homozygous deletion of the ACF1 gene results in an apparent accelerated S phase progression that has been linked to a more loose chromatin structure, similar to the accelerated S phase obtained after deletion of a histone gene cluster (Fyodorov et al. 2004). These findings may point to substantial differences between the embryonal *Drosophila* cells that have been studied and the mammalian cells in culture. More studies are needed in both systems to obtain a comprehensive picture of ACF1 function and mechanism of action.

WSTF is also targeted to replication foci in pericentromeric heterochromatin (Bozhenok et al. 2002). An extraction protocol using a detergent and high salt (350 mM NaCl)-containing buffer allowed visualization of a general retention of WSTF and SNF2H in replication foci throughout S phase. Furthermore, there is strong evidence that WSTF is targeted to these sites by interacting with PCNA and, then, in turn recruits SNF2H (Poot et al., unpubl.). WSTF depletion leads to a more condensed structure of newly replicated chromatin and a general increase in markers of heterochromatin, such as HP1. This is accompanied by a delay in S phase progression.

ISWI (SNF2H) is also found in association with subunits of cohesin, a conserved four-subunit complex that mediates sister chromatid interaction during mitosis. The role of ISWI here is to facilitate the binding of cohesin to heterochromatin (Hakimi et al. 2002). These findings are consistent with the dual function of ISWI-chromatin remodeling factors observed in vitro in both chromatin assembly and chromatin accessibility. A recent study on the function of the yeast CHRAC complex supports this notion: whereas earlier studies pointed to a role of subunits of this complex in the assembly of repressive promoter structures of specific genes (Goldmark et al. 2000; Kent et al. 2001), it can also counteract heterochromatin at telomeres by maintaining a more open configuration (Iida and Araki 2004).

ATP-dependent nucleosome remodeling factors may alter chromatin structure not only through their direct action on nucleosomes, but also by mediating changes in histone and DNA modifications. Indeed, recent findings support an important role of ISWI complexes in specifying histone modifications for transcription in budding yeast (Morillon et al. 2003) and mammalian cells

(Santoro et al. 2002; Zhou et al. 2002). Furthermore, a factor related to ISWI is required for global DNA methylation and histone H3 K9 methylation in the pericentromeric heterochromatin in mammalian cells (Dennis et al. 2001; Yan et al. 2003).

In conclusion, the studies of ISWI and ACF1 function highlight the possible close link between processes that are involved in chromatin assembly and those that may actually counteract chromatin structure. What unifies both processes is that they impart a dynamic nature to chromatin.

7
The Assembly of Higher Order Chromatin Structures

The assembly of ordered, periodic nucleosome arrays mediated by ISWI complexes could be necessary for the formation of higher order chromatin structures. This may be because such ordered nucleosome arrays facilitate the correct folding of the fiber and/or the interaction of chromatin proteins that are involved in higher order structures.

Histone H1 has a major role in the stabilization of higher order folding structures of the chromatin fiber (Thoma and Koller 1977). Further levels of chromatin compaction may depend on the interaction of chromatin fiber segments with distal segments. An important protein that may mediate these interactions is HP1 and its homologues, conserved from fission yeast (where it is called Swi6) to human. HP1 proteins are important for heterochromatin formation around the centromeres (pericentromeric heterochromatin) and at telomeres and affect position-effect variegation, a form of transcriptional silencing that depends on the spread of heterochromatin (reviewed in Grewal and Elgin 2002). The interaction of HP1 with heterochromatin is regulated by the methylation of histone H3 at lysine 9, which is mediated by histone methyltransferase SU(VAR)3–9 and its homologues (SUV39H1 and SUV39H2 in mammals, Clr4 in fission yeast; Rea et al. 2000; Bannister et al. 2001; Lachner et al. 2001; Peters et al. 2001). HP1 interacts with the ORC complex in *Drosophila* and this interaction is required to load HP1 to heterochromatin (Pak et al. 1997). Swi6p, the fission yeast homologue of HP1, interacts directly with DNA polymerase α and there is evidence that this interaction is required to mediate Swi6-dependent silencing (Ahmed et al. 2001; Nakayama et al. 2001). Swi6 is important for chromosome cohesion between replicated sister chromatids by mediating the interaction of cohesin with centromeres (Bernard et al. 2001; Nonaka et al. 2002). It is not known whether SU(VAR)3–9 and its homologues are cell cycle-regulated. The functional interaction of Swi6 with centromeres and its role in cohesion is coupled to S phase through the action of the hsk1 kinase, the Cdc7-related kinase in fission yeast that regulates replication initiation (Bailis et al. 2003). This may be through direct phosphorylation of Swi6 by the kinase. This finding illustrates how a cell cycle regulator may be involved in the coordination of the assembly of higher order structures of the chromatin fiber with progression through S phase.

The association of HP1 with heterochromatin depends on an unidentified RNA component (Maison et al. 2002), and, indeed, HP1 associates with RNA directly (Muchardt et al. 2002). This RNA connection may be linked to the role of the RNA interference pathway in gene silencing mediated by Swi6, Clr4 and its homologues (Volpe et al. 2003; Pal-Bhadra et al. 2004; Verdel et al. 2004). However, specific non-coding RNA molecules may also have a structural role in establishing higher order chromatin, as, for example, the Xist RNA in the inactive X chromosome (reviewed in Brockdorff 2002), and RNA may be part of chromatin remodeling factors themselves, for example, the dosage compensation complex in *Drosophila* (Akhtar 2003). It will be insightful to find out how the expression and localization of these non-coding RNAs are linked to the cell cycle and chromatin assembly.

8
PCNA, a Central Coordinator of Epigenetic Inheritance

How is chromatin replication coupled to the DNA replication process? One important facet in this process is clearly PCNA. PCNA is an essential processivity factor for replicative DNA polymerases by assembling a closed ring structure (a sliding clamp) around the DNA duplex, but it interacts with many factors that have a role in DNA replication, including Okazaki fragment processing, DNA repair, translesion DNA synthesis, DNA methylation, chromatin remodeling, chromatin assembly and cell cycle regulation (reviewed in Maga and Hubscher 2003). PCNA is, therefore, a central coordinator of the DNA and chromatin replication process. PCNA is involved in epigenetic inheritance in yeast by marking replicated DNA and recruiting CAF-1 (Zhang et al. 2000). Mutations in PCNA disrupt silencing of an integrated gene near the telomeres. PCNA mutations in *Drosophila* also affect position-dependent silencing (PEV; Henderson et al. 1994). DNA methylation, important for the inheritance of repressed chromatin structures in higher eukaryotes (Kass and Wolffe 1998), is maintained through replication by the interaction of the DNA methyltransferase DNMT1 with PCNA (Leonhardt et al. 1992). We find that a WSTF/ISWI complex is also recruited by PCNA, but, unlike CAF-1 and DNMT1, it seems to be involved in the maintenance of open chromatin structures (Poot et al., unpublished). This may suggest that WSTF facilitates or promotes epigenetic inheritance of transcriptionally active chromatin. How all the PCNA-interacting proteins access PCNA in a coordinated manner and how PCNA organizes their function in DNA and chromatin replication are important unresolved questions in molecular biology.

9
Mechanisms of Epigenetic Inheritance Through Chromatin: Conclusion

In this chapter, we have described key components of the mechanisms of propagation of chromatin states through DNA replication. These include histone modifications and the responsible enzymes, histone chaperones, chromatin remodeling factors, RNA components, DNA methylation and the DNA replication machinery itself. Some of the known interactions suggest pathways by which histone modifications (for example, histone methylation at K9 of H3) recruit factors [e.g. HP1-SU(VAR)3–9 complexes] that propagate the same modification at adjacent sites. The segregation of parental histones to the daughter DNA strands during DNA replication may, therefore, mediate maintenance of the histone code of specific domains. One requirement for this scenario is that the dispersal of the parental histones to the new DNA strands has to be kept localized. One could imagine a molecular conveyor belt, possibly consisting of the DNA replication machinery itself, histone chaperones and remodeling factors, which secures the localized transmission of histones.

What may become obvious from this chapter is that while we know several components that are involved in epigenetic inheritance, we are far from understanding the mechanisms of this important aspect of biology. This is likely because, unlike the duplication of DNA, there is no single mechanism that mediates the duplication of chromatin states, but many different mechanisms exist, depending on chromatin domain, organism and function. This field will remain a major intellectual challenge for the foreseeable future.

Acknowledgements. Work in my laboratory was funded by Marie Curie Cancer Care. I wish to thank Sofia Aligianni, Ludmila Bozhenok, Sarah Elderkin, Margaret Grimaldi, Raymond Poot, Ana Neves Costa and Joao Ferreira for comments that improved the manuscript. Thanks also to Jim Kadonaga and Paul Kaufman for preprints of manuscripts.

References

Ahmad K, Henikoff S (2001) Centromeres are specialized replication domains in heterochromatin. J Cell Biol 153:101–110

Ahmad K, Henikoff S (2002a) The histone variant H3.3 marks active chromatin by replication-independent nucleosome assembly. Mol Cell 9:1191–1200

Ahmad K, Henikoff S (2002b) Histone H3 variants specify modes of chromatin assembly. Proc Natl Acad Sci USA 99 (Suppl 4):16477–16484

Ahmed S, Saini S, Arora S, Singh J (2001) Chromodomain protein Swi6-mediated role of DNA polymerase alpha in establishment of silencing in fission yeast. J Biol Chem 276:47814–47821

Akey CW, Luger K (2003) Histone chaperones and nucleosome assembly. Curr Opin Struct Biol 13:6–14

Akhtar A (2003) Dosage compensation: an intertwined world of RNA and chromatin remodelling. Curr Opin Genet Dev 13:161–169

Alexiadis V, Varga-Weisz PD, Becker PB, Gruss C (1998) In vitro chromatin remodelling by chromatin accessibility complex (CHRAC) at the SV40 origin of DNA replication. EMBO J 17:3428–3438

Almouzni G, Méchali M (1988) Assembly of spaced chromatin. Involvement of ATP and DNA topoisomerase activity. EMBO J 7:4355–4365

Annunziato AT, Hansen JC (2000) Role of histone acetylation in the assembly and modulation of chromatin structures. Gene Expr 9:37–61

Bailis JM, Forsburg SL (2003) It's all in the timing: linking S phase to chromatin structure and chromosome dynamics. Cell Cycle 2:303–306

Bailis JM, Bernard P, Antonelli R, Allshire RC, Forsburg SL (2003) Hsk1-Dfp1 is required for heterochromatin-mediated cohesion at centromeres. Nat Cell Biol 5:1111–1116

Bannister AJ, Zegerman P, Partridge JF, Miska EA, Thomas JO, Allshire RC, Kouzarides T (2001) Selective recognition of methylated lysine 9 on histone H3 by the HP1 chromo domain. Nature 410:120–124

Becker P, Hörz W (2002) ATP-dependent nucleosome remodeling. Annual Reviews, Palo Alto

Becker PB, Wu C (1992) Cell-free system for assembly of transcriptionally repressed chromatin from *Drosophila* embryos. Mol Cell Biol 12:2241–2249

Belyaev ND, Keohane AM, Turner BM (1996) Histone H4 acetylation and replication timing in Chinese hamster chromosomes. Exp Cell Res 225:277–285

Bernard P, Maure JF, Partridge JF, Genier S, Javerzat JP, Allshire RC (2001) Requirement of heterochromatin for cohesion at centromeres. Science 294:2539–2542

Bjerling P, Silverstein RA, Thon G, Caudy A, Grewal S, Ekwall K (2002) Functional divergence between histone deacetylases in fission yeast by distinct cellular localization and in vivo specificity. Mol Cell Biol 22:2170–2181

Bochar DA, Savard J, Wang W, Lafleur DW, Moore P, Cote J, Shiekhattar R (2000) A family of chromatin remodeling factors related to Williams syndrome transcription factor. Proc Natl Acad Sci USA 97:1038–1043

Bozhenok L, Wade PA, Varga-Weisz P (2002) WSTF-ISWI chromatin remodeling complex targets heterochromatic replication foci. EMBO J 21:2231–2241

Braunstein M, Sobel RE, Allis CD, Turner BM, Broach JR (1996) Efficient transcriptional silencing in *Saccharomyces cerevisiae* requires a heterochromatin histone acetylation pattern. Mol Cell Biol 16:4349–4356

Brockdorff N (2002) X-chromosome inactivation: closing in on proteins that bind Xist RNA. Trends Genet 18:352–358

Burke TW, Cook JG, Asano M, Nevins JR (2001) Replication factors MCM2 and ORC1 interact with the histone acetyltransferase HBO1. J Biol Chem 276:15397–15408

Chen H, Li B, Workman JL (1994) A histone-binding protein, nucleoplasmin, stimulates transcription factor binding to nucleosomes and factor-induced nucleosome disassembly. EMBO J 13:380–390

Collins N, Poot RA, Kukimoto I, Garcia-Jimenez C, Dellaire G, Varga-Weisz PD (2002) An ACF1-ISWI chromatin-remodeling complex is required for DNA replication through heterochromatin. Nat Genet 32:627–632

Cosgrove AJ, Nieduszynski CA, Donaldson AD (2002) Ku complex controls the replication time of DNA in telomere regions. Genes Dev 16:2485–2490

Costanzi C, Pehrson JR (1998) Histone macroH2A1 is concentrated in the inactive X chromosome of female mammals. Nature 393:599–601

Csink AK, Henikoff S (1998) Something from nothing: the evolution and utility of satellite repeats. Trends Genet 14:200–204

Cusick ME, Lee KS, DePamphilis ML, Wassarman PM (1983) Structure of chromatin at deoxyribonucleic acid replication forks: nuclease hypersensitivity results from both prenucleosomal deoxyribonucleic acid and an immature chromatin structure. Biochemistry 22:3873–3884

Demeret C, Bocquet S, Lemaitre JM, Francon P, Mechali M (2002) Expression of ISWI and its binding to chromatin during the cell cycle and early development. J Struct Biol 140:57–66

Dennis K, Fan T, Geiman T, Yan Q, Muegge K (2001) Lsh, a member of the SNF2 family, is required for genome-wide methylation. Genes Dev 15:2940–2944

De Rubertis F, Kadosh D, Henchoz S, Pauli D, Reuter G, Struhl K, Spierer P (1996) The histone deacetylase RPD3 counteracts genomic silencing in *Drosophila* and yeast. Nature 384:589–591

Deuring R, Fanti L, Armstrong JA, Sarte M, Papoulas O, Prestel M, Daubresse G, Verardo M, Moseley SL, Berloco M, Tsukiyama T, Wu C, Pimpinelli S, Tamkun JW (2000) The ISWI chromatin-remodeling protein is required for gene expression and the maintenance of higher order chromatin structure in vivo. Mol Cell 5:355–365

Dimitrov S, Almouzni A, Dasso M, Wolffe AP (1993) Chromatin transitions during early *Xenopus* embryogenesis: changes in histone H4 acetylation and linker histone type. Dev Biol 160:214–227

Earnshaw W, Bordwell B, Marino C, Rothfield N (1986) Three human chromosomal autoantigens are recognized by sera from patients with anti-centromere antibodies. J Clin Invest 77:426–430

Eberharter A, Ferrari S, Langst G, Straub T, Imhof A, Varga-Weisz P, Wilm M, Becker PB (2001) Acf1, the largest subunit of CHRAC, regulates ISWI-induced nucleosome remodelling. EMBO J 20:3781–3788

Ehrenhofer-Murray AE, Kamakaka RT, Rine J (1999) A role for the replication proteins PCNA, RF-C, polymerase epsilon and Cdc45 in transcriptional silencing in *Saccharomyces cerevisiae*. Genetics 153:1171–1182

Ekwall K, Olsson T, Turner BM, Cranston G, Allshire RC (1997) Transient inhibition of histone deacetylation alters the structural and functional imprint at fission yeast centromeres. Cell 91:1021–1032

Enomoto S, Berman J (1998) Chromatin assembly factor I contributes to the maintenance, but not the re-establishment, of silencing at the yeast silent mating loci. Genes Dev 12:219–232

Enomoto S, McCune-Zierath PD, Gerami-Nejad M, Sanders MA, Berman J (1997) RLF2, a subunit of yeast chromatin assembly factor-I, is required for telomeric chromatin function in vivo. Genes Dev 11:358–370

Fazzio TG, Tsukiyama T (2003) Chromatin remodeling in vivo: evidence for a nucleosome sliding mechanism. Mol Cell 12:1333–1340

Flanagan JF, Peterson CL (1999) A role for the yeast SWI/SNF complex in DNA replication. Nucleic Acids Res 27:2022–2028

Fyodorov DV, Kadonaga JT (2002) Dynamics of ATP-dependent chromatin assembly by ACF. Nature 418:897–900

Fyodorov DV, Blower MD, Karpen GH, Kadonaga JT (2004) Acf1 confers unique activities to ACF/CHRAC and promotes the formation rather than the disruption of chromatin in vivo. Genes Dev 18(2):170–183

Gaillard PL, Martini EM, Kaufman PD, Stillman B, Moustacchi E, Almouzni G (1996) Chromatin assembly coupled to DNA repair: a new role for chromatin assembly factor-1. Cell 86:887–896

Gasser R, Koller T, Sogo JM (1996) The stability of nucleosomes at the replication fork. J Mol Biol 258:224–239

Gasser SM (1995) Chromosome structure. Coiling up chromosomes. Curr Biol 5:357–360

Glikin GC, Ruberti I, Worcel A (1984) Chromatin assembly in *Xenopus* oocytes: in vitro studies. Cell 37:33–41

Goldmark JP, Fazzio TG, Estep PW, Church GM, Tsukiyama T (2000) The Isw2 chromatin remodelling complex represses early meiotic genes upon recruitment by Ume6p. Cell 103:423–433

Grant PA, Duggan L, Cote J, Roberts SM, Brownell JE, Candau R, Ohba R, Owen-Hughes T, Allis CD, Winston F, Berger SL, Workman JL (1997) Yeast Gcn5 functions in two multi-subunit complexes to acetylate nucleosomal histones: characterization of an Ada complex and the SAGA (Spt/Ada) complex. Genes Dev 11:1640–1650

Grewal SI, Elgin SC (2002) Heterochromatin: new possibilities for the inheritance of structure. Curr Opin Genet Dev 12:178–187

Grewal SI, Bonaduce MJ, Klar AJ (1998) Histone deacetylase homologs regulate epigenetic inheritance of transcriptional silencing and chromosome segregation in fission yeast. Genetics 150:563–576

Grienenberger A, Miotto B, Sagnier T, Cavalli G, Schramke V, Geli V, Mariol MC, Berenger H, Graba Y, Pradel J (2002) The MYST domain acetyltransferase Chameau functions in epigenetic mechanisms of transcriptional repression. Curr Biol 12:762–766

Gruss C, Sogo JM (1992) Chromatin replication. Bioessays 14:1–8

Gruss C, Wu J, Koller T, Sogo JM (1993) Disruption of nucleosomes at the replication fork. EMBO J 12:4533–4545

Haaf T (1995) The effects of 5-azacytidine and 5-azadeoxycytidine on chromosome structure and function: implications for methylation-associated cellular processes. Pharmacol Ther 65:19–46

Hakimi MA, Bochar DA, Schmiesing JA, Dong Y, Barak OG, Speicher DW, Yokomori K, Shiekhattar R (2002) A chromatin remodelling complex that loads cohesin onto human chromosomes. Nature 418:994–998

Hasan S, Hassa PO, Imhof R, Hottiger MO (2001) Transcription coactivator p300 binds PCNA and may have a role in DNA repair synthesis. Nature 410:387–391

Haushalter KA, Kadonaga JT (2003) Chromatin assembly by DNA-translocating motors. Nat Rev Mol Cell Biol 4:613–620

Havas K, Flaus A, Phelan M, Kingston R, Wade PA, Lilley DM, Owen-Hughes T (2000) Generation of superhelical torsion by ATP-dependent chromatin remodeling activities. Cell 103:1133–1142

Henderson DS, Banga SS, Grigliatti TA, Boyd JB (1994) Mutagen sensitivity and suppression of position-effect variegation result from mutations in mus209, the *Drosophila* gene encoding PCNA. EMBO J 13:1450–1459

Henikoff S, Ahmad K, Platero JS, van Steensel B (2000) Heterochromatic deposition of centromeric histone H3-like proteins. Proc Natl Acad Sci USA 97:716–721

Hennig W (1999) Heterochromatin. Chromosoma 108:1–9

Hochheimer A, Zhou S, Zheng S, Holmes MC, Tjian R (2002) TRF2 associates with DREF and directs promoter-selective gene expression in *Drosophila*. Nature 420:439–445

Hoek M, Stillman B (2003) Chromatin assembly factor 1 is essential and couples chromatin assembly to DNA replication in vivo. Proc Natl Acad Sci USA 100:12183–12188

Iida T, Araki H (2004) Noncompetitive counteractions of DNA polymerase epsilon and ISW2/yCHRAC for epigenetic inheritance of telomere position effect in *Saccharomyces cerevisiae*. Mol Cell Biol 24:217–227

Iizuka M, Stillman B (1999) Histone acetyltransferase HBO1 interacts with the ORC1 subunit of the human initiator protein. J Biol Chem 274:23027–23034

Ishimi Y, Kojima M, Yamada M, Hanaoka F (1987) Binding mode of nucleosome-assembly protein (AP-I) and histones. Eur J Biochem 162:19–24

Ito T, Bulger M, Kobayashi R, Kadonaga JT (1996) *Drosophila* NAP-1 is a core histone chaperone that functions in ATP-facilitated assembly of regularly spaced nucleosomal arrays. Mol Cell Biol 16:3112–3124

Ito T, Bulger M, Pazin MJ, Kobayashi R, Kadonaga JT (1997) ACF, an ISWI-containing and ATP-utilizing chromatin assembly and remodeling factor. Cell 90:145–155

Ito T, Levenstein ME, Fyodorov DV, Kutach AK, Kobayashi R, Kadonaga JT (1999) ACF consists of two subunits, Acf1 and ISWI, that function cooperatively in the ATP-dependent catalysis of chromatin assembly. Genes Dev 13:1529–1539

Jenuwein T, Allis CD (2001) Translating the histone code. Science 293:1074–1080

Jeppesen P, Turner BM (1993) The inactive X chromosome in female mammals is distinguished by a lack of histone H4 acetylation, a cytogenetic marker for gene expression. Cell 74:281–289

Kass SU, Wolffe AP (1998) DNA methylation, nucleosomes and the inheritance of chromatin structure and function. Novartis Found Symp 214:22–35; discussion 36–50

Kaufman PD, Kobayashi R, Kessler N, Stillman B (1995) The p150 and p60 subunits of chromatin assembly factor I: a molecular link between newly synthesized histones and DNA replication. Cell 81:1105–1114

Kaufman PD, Kobayashi R, Stillman B (1997) Ultraviolet radiation sensitivity and reduction of telomeric silencing in *Saccharomyces cerevisiae* cells lacking chromatin assembly factor-I. Genes Dev 11:345–357

Kaya H, Shibahara KI, Taoka KI, Iwabuchi M, Stillman B, Araki T (2001) FASCIATA genes for chromatin assembly factor-1 in *Arabidopsis* maintain the cellular organization of apical meristems. Cell 104:131–142

Kent NA, Karabetsou N, Politis PK, Mellor J (2001) In vivo chromatin remodeling by yeast ISWI homologs Isw1p and Isw2p. Genes Dev 15:619–626

Krawitz DC, Kama T, Kaufman PD (2002) Chromatin assembly factor I mutants defective for PCNA binding require Asf1/Hir proteins for silencing. Mol Cell Biol 22:614–625

Krude T (1995) Chromatin assembly factor 1 (CAF-1) colocalizes with replication foci in HeLa cell nuclei. Exp Cell Res 220:304–311

Krude T (1999) Chromatin assembly during DNA replication in somatic cells. Eur J Biochem 263:1–5

Kukimoto I, Elderkin S, Grimaldi M, Oelgeschläger T, Varga-Weisz P (2004) The histone-fold protein complex CHRAC-15/17 enhances nucleosome sliding and assembly mediated by ACF. Mol Cell 13(2):265–277

Lachner M, O'Carroll D, Rea S, Mechtler K, Jenuwein T (2001) Methylation of histone H3 lysine 9 creates a binding site for HP1 proteins. Nature 410:116–120

Längst G, Becker P (2001) Nucleosome mobilization and positioning by ISWI-containing chromatin-remodeling factors. J Cell Sci 114:2561–2568

Le S, Davis C, Konopka JB, Sternglanz R (1997) Two new S-phase-specific genes from *Saccharomyces cerevisiae*. Yeast 13:1029–1042

Leonhardt H, Page AW, Weier HU, Bestor TH (1992) A targeting sequence directs DNA methyltransferase to sites of DNA replication in mammalian nuclei. Cell 71:865–873

LeRoy G, Orphanides G, Lane WS, Reinberg D (1998) Requirement of RSF and FACT for transcription of chromatin templates in vitro. Science 282:1900–1904

LeRoy G, Loyola A, Lane WS, Reinberg D (2000) Purification and characterization of a human factor that assembles and remodels chromatin. J Biol Chem 275:14787–14790

Lorain S, Quivy JP, Monier-Gavelle F, Scamps C, Lecluse Y, Almouzni G, Lipinski M (1998) Core histones and HIRIP3, a novel histone-binding protein, directly interact with WD repeat protein HIRA. Mol Cell Biol 18:5546–5556

Loyola A, LeRoy G, Wang YH, Reinberg D (2001) Reconstitution of recombinant chromatin establishes a requirement for histone-tail modifications during chromatin assembly and transcription. Genes Dev 15:2837–2851

Loyola A, Huang JY, LeRoy G, Hu S, Wang YH, Donnelly RJ, Lane WS, Lee SC, Reinberg D (2003) Functional analysis of the subunits of the chromatin assembly factor RSF. Mol Cell Biol 23:6759–6768

Luger K (2003) Structure and dynamic behavior of nucleosomes. Curr Opin Genet Dev 13:127–135

MacCallum DE, Losada A, Kobayashi R, Hirano T (2002) ISWI remodeling complexes in *Xenopus* egg extracts: identification as major chromosomal components that are regulated by INCENP-aurora B. Mol Biol Cell 13:25–39

Maga G, Hubscher U (2003) Proliferating cell nuclear antigen (PCNA): a dancer with many partners. J Cell Sci 116:3051–3060

Maison C, Bailly D, Peters AH, Quivy JP, Roche D, Taddei A, Lachner M, Jenuwein T, Almouzni G (2002) Higher-order structure in pericentric heterochromatin involves a distinct pattern of histone modification and an RNA component. Nat Genet 30:329–334

Malik HS, Henikoff S (2001) Adaptive evolution of Cid, a centromere-specific histone in *Drosophila*. Genetics 157:1293–1298

Marheineke K, Krude T (1998) Nucleosome assembly activity and intracellular localization of human CAF-1 changes during the cell division cycle. J Biol Chem 273:15279–15286

Martini E, Roche DM, Marheineke K, Verreault A, Almouzni G (1998) Recruitment of phosphorylated chromatin assembly factor 1 to chromatin after UV irradiation of human cells. J Cell Biol 143:563–575

McNairn AJ, Gilbert DM (2003) Epigenomic replication: linking epigenetics to DNA replication. Bioessays 25:647–656

Meijsing SH, Ehrenhofer-Murray AE (2001) The silencing complex SAS-I links histone acetylation to the assembly of repressed chromatin by CAF-I and Asf1 in *Saccharomyces cerevisiae*. Genes Dev 15:3169–3182

Mello JA, Almouzni G (2001) The ins and outs of nucleosome assembly. Curr Opin Genet Dev 11:136–141

Mello JA, Sillje HH, Roche DM, Kirschner DB, Nigg EA, Almouzni G (2002) Human Asf1 and CAF-1 interact and synergize in a repair-coupled nucleosome assembly pathway. EMBO Rep 3:329–334

Mellone BG, Allshire RC (2003) Stretching it: putting the CEN(P-A) in centromere. Curr Opin Genet Dev 13:191–198

Mizuguchi G, Shen X, Landry J, Wu WH, Sen S, Wu C (2004) ATP-driven exchange of histone H2AZ variant catalyzed by SWR1 chromatin remodeling complex. Science 303:343–348

Monson EK, de Bruin D, Zakian VA (1997) The yeast Cac1 protein is required for the stable inheritance of transcriptionally repressed chromatin at telomeres. Proc Natl Acad Sci USA 94:13081–13086

Morillon, A, Karabetsou N, O'Sullivan J, Kent N, Proudfoot N, Mellor J (2003) Isw1 chromatin remodeling ATPase coordinates transcription elongation and termination by RNA polymerase II. Cell 115:425–435

Moshkin YM, Armstrong JA, Maeda RK, Tamkun JW, Verrijzer P, Kennison JA, Karch F (2002) Histone chaperone ASF1 cooperates with the Brahma chromatin-remodelling machinery. Genes Dev 16:2621–2626

Muchardt C, Yaniv M (2001) When the SWI/SNF complex remodels...the cell cycle. Oncogene 20:3067–3075

Muchardt C, Guilleme M, Seeler JS, Trouche D, Dejean A, Yaniv M (2002) Coordinated methyl and RNA binding is required for heterochromatin localization of mammalian HP1alpha. EMBO Rep 3:975–981

Munakata T, Adachi N, Yokoyama N, Kuzuhara T, Horikoshi M (2000) A human homologue of yeast anti-silencing factor has histone chaperone activity. Genes Cells 5:221–233

Murzina N, Verreault A, Laue E, Stillman B (1999) Heterochromatin dynamics in mouse cells: interaction between chromatin assembly factor 1 and HP1 proteins. Mol Cell 4:529–540

Nakayama J, Allshire RC, Klar AJ, Grewal SI (2001) A role for DNA polymerase alpha in epigenetic control of transcriptional silencing in fission yeast. EMBO J 20:2857–2866

Narlikar GJ, Fan H-Y, Kingston RE (2002) Cooperation between complexes that regulate chromatin structure and transcription. Cell 108:475–487

Ner SS, Travers AA (1994) HMG-D, the *Drosophila melanogaster* homologue of HMG 1 protein, is associated with early embryonic chromatin in the absence of histone H1. EMBO J 13:1817–1822

Noma K, Allis CD, Grewal SI (2001) Transitions in distinct histone H3 methylation patterns at the heterochromatin domain boundaries. Science 293:1150–1155

Nonaka N, Kitajima T, Yokobayashi S, Xiao G, Yamamoto M, Grewal SI, Watanabe Y (2002) Recruitment of cohesin to heterochromatic regions by Swi6/HP1 in fission yeast. Nat Cell Biol 4:89–93

O'Neill LP, Turner BM (1995) Histone H4 acetylation distinguishes coding regions of the genome from heterochromatin in a differentiation-dependent but transcription-independent manner. EMBO J 14:3946–3957

Osada S, Sutton A, Muster N, Brown CE, Yates JR III, Sternglanz R, Workman JL (2001) The yeast SAS (something about silencing) protein complex contains a MYST-type putative acetyltransferase and functions with chromatin assembly factor ASF1. Genes Dev 15:3155–3168

Ouspenski II, van Hooser AA, Brinkley BR (2003) Relevance of histone acetylation and replication timing for deposition of centromeric histone CENP-A. Exp Cell Res 285:175–188

Owen-Hughes T, Workman JL (1996) Remodeling the chromatin structure of a nucleosome array by transcription factor-targeted trans-displacement of histones. EMBO J 15:4702–4712

Pak DT, Pflumm M, Chesnokov I, Huang DW, Kellum R, Marr J, Romanowski P, Botchan MR (1997) Association of the origin recognition complex with heterochromatin and HP1 in higher eukaryotes. Cell 91:311–323

Pal-Bhadra M, Leibovitch BA, Gandhi SG, Rao M, Bhadra U, Birchler JA, Elgin SC (2004) Heterochromatic silencing and HP1 localization in *Drosophila* are dependent on the RNAi machinery. Science 303:669–672

Palmer DK, O'Day K, Trong HL, Charbonneau H, Margolis RL (1991) Purification of the centromere-specific protein CENP-A and demonstration that it is a distinctive histone. Proc Natl Acad Sci USA 88:3734–3738

Peters AH, O'Carroll D, Scherthan H, Mechtler K, Sauer S, Schofer C, Weipoltshammer K, Pagani M, Lachner M, Kohlmaier A, Opravil S, Doyle M, Sibilia M, Jenuwein T (2001) Loss of the Suv39h histone methyltransferases impairs mammalian heterochromatin and genome stability. Cell 107:323–337

Poot RA, Dellaire G, Hulsmann BB, Grimaldi MA, Corona DF, Becker PB, Bickmore WA, Varga-Weisz PD (2000) HuCHRAC, a human ISWI chromatin remodelling complex contains hACF1 and two novel histone-fold proteins. EMBO J 19:3377–3387

Prasanth SG, Prasanth KV, Stillman B (2002) Orc6 involved in DNA replication, chromosome segregation, and cytokinesis. Science 297:1026–1031

Pulm W, Knippers R (1984) Chromatin structure and DNA replication. Adv Exp Med Biol 179:127–141

Quivy JP, Grandi P, Almouzni G (2001) Dimerization of the largest subunit of chromatin assembly factor 1: importance in vitro and during *Xenopus* early development. EMBO J 20:2015–2027

Ray-Gallet D, Quivy JP, Scamps C, Martini EM, Lipinski M, Almouzni G (2002) HIRA is critical for a nucleosome assembly pathway independent of DNA synthesis. Mol Cell 9:1091–1100

Rea S, Eisenhaber F, O'Carroll D, Strahl BD, Sun ZW, Schmid M, Opravil S, Mechtler K, Ponting CP, Allis CD, Jenuwein T (2000) Regulation of chromatin structure by site-specific histone H3 methyltransferases. Nature 406:593–599

Rountree MR, Bachman KE, Baylin SB (2000) DNMT1 binds HDAC2 and a new co-repressor, DMAP1, to form a complex at replication foci. Nat Genet 25:269–277

Santoro R, Li J, Grummt I (2002) The nucleolar remodeling complex NoRC mediates heterochromatin formation and silencing of ribosomal gene transcription. Nat Genet 32:393–396

Santos-Rosa H, Schneider R, Bannister AJ, Sherriff J, Bernstein BE, Emre NC, Schreiber SL, Mellor J, Kouzarides T (2002) Active genes are tri-methylated at K4 of histone H3. Nature 419:407–411

Schlaeger EJ, Pulm W, Knippers R (1983) Chromatin maturation depends on continued DNA-replication. FEBS Lett 156:281–286

Sharp JA, Franco AA, Osley MA, Kaufman PD (2002) Chromatin assembly factor I and Hir proteins contribute to building functional kinetochores in *S. cerevisiae*. Genes Dev 16:85–100

Shelby RD, Vafa O, Sullivan KF (1997) Assembly of CENP-A into centromeric chromatin requires a cooperative array of nucleosomal DNA contact sites. J Cell Biol 136:501–513

Shelby RD, Monier K, Sullivan KF (2000) Chromatin assembly at kinetochores is uncoupled from DNA replication. J Cell Biol 151:1113–1118

Shibahara K, Verreault A, Stillman B (2000) The N-terminal domains of histones H3 and H4 are not necessary for chromatin assembly factor-1-mediated nucleosome assembly onto replicated DNA in vitro. Proc Natl Acad Sci USA 97:7766–7771

Shibahara K, Stillman B (1999) Replication-dependent marking of DNA by PCNA facilitates CAF-1-coupled inheritance of chromatin. Cell 96(4):575–585

Smith JS, Caputo E, Boeke JD (1999) A genetic screen for ribosomal DNA silencing defects identifies multiple DNA replication and chromatin-modulating factors. Mol Cell Biol 19:3184–3197

Smith S, Stillman B (1989) Purification and characterization of CAF-I, a human cell factor required for chromatin assembly during DNA replication in vitro. Cell 58:15–25

Smith S, Stillman B (1991) Stepwise assembly of chromatin during DNA replication in vitro. EMBO J 10:971–980

Sobel RE, Cook RG, Perry CA, Annunziato AT, Allis CD (1995) Conservation of deposition-related acetylation sites in newly synthesized histones H3 and H4. Proc Natl Acad Sci USA 92:1237–1241

Sogo JM, Stahl H, Koller T, Knippers R (1986) Structure of replicating simian virus 40 minichromosomes. The replication fork, core histone segregation and terminal structures. J Mol Biol 189:189–204

Stevenson JB, Gottschling DE (1999) Telomeric chromatin modulates replication timing near chromosome ends. Genes Dev 13:146–151

Stoler S, Keith KC, Curnick KE, Fitzgerald-Hayes M (1995) A mutation in CSE4, an essential gene encoding a novel chromatin-associated protein in yeast, causes chromosome nondisjunction and cell cycle arrest at mitosis. Genes Dev 9:573–586

Strahl BD, Allis CD (2000) The language of covalent histone modifications. Nature 403:41–45

Strohner R, Nemeth A, Jansa P, Hofmann-Rohrer U, Santoro R, Langst G, Grummt I (2001) NoRC – a novel member of mammalian ISWI-containing chromatin remodeling machines. EMBO J 20:4892–4900

Sullivan B, Karpen G (2001) Centromere identity in *Drosophila* is not determined in vivo by replication timing. J Cell Biol 154:683–690

Sullivan KF (2001) A solid foundation: functional specialization of centromeric chromatin. Curr Opin Genet Dev 11:182–188

Taddei A, Roche D, Sibarita JB, Turner BM, Almouzni G (1999) Duplication and maintenance of heterochromatin domains. J Cell Biol 147:1153–1166

Taddei A, Maison C, Roche D, Almouzni G (2001) Reversible disruption of pericentric heterochromatin and centromere function by inhibiting deacetylases. Nat Cell Biol 3:114–120

Tagami H, Ray-Gallet D, Almouzni G, Nakatani Y (2004) Histone h3.1 and h3.3 complexes mediate nucleosome assembly pathways dependent or independent of DNA synthesis. Cell 116:51–61

Tchenio T, Casella JF, Heidmann T (2001) A truncated form of the human CAF-1 p150 subunit impairs the maintenance of transcriptional gene silencing in mammalian cells. Mol Cell Biol 21:1953–1961

Thoma F, Koller T (1977) Influence of histone H1 on chromatin structure. Cell 12:101–107

Tong JK, Hassig CA, Schnitzler GR, Kingston RE, Schreiber SL (1998) Chromatin deacetylation by an ATP-dependent nucleosome remodelling complex. Nature 395:917–921

Tyler JK, Adams CR, Chen SR, Kobayashi R, Kamakaka RT, Kadonaga JT (1999) The RCAF complex mediates chromatin assembly during DNA replication and repair. Nature 402:555–560

Tyler JK, Collins KA, Prasad-Sinha J, Amiott E, Bulger M, Harte PJ, Kobayashi R, Kadonaga JT (2001) Interaction between the *Drosophila* CAF-1 and ASF1 chromatin assembly factors. Mol Cell Biol 21:6574–6584

Van Daal A, Elgin SC (1992) A histone variant, H2AvD, is essential in *Drosophila melanogaster*. Mol Biol Cell 3:593–602

Varga-Weisz PD, Wilm M, Bonte E, Dumas K, Mann M, Becker PB (1997) Chromatin-remodelling factor CHRAC contains the ATPases ISWI and topoisomerase II. Nature 388:598–602

Varga-Weisz PD, Becker PB (1998) Chromatin-remodeling factors: machines that regulate? Curr Opin Cell Biol 10(3):346–353

Verdel A, Jia S, Gerber S, Sugiyama T, Gygi S, Grewal SI, Moazed D (2004) RNAi-mediated targeting of heterochromatin by the RITS complex. Science 303:672–676

Verreault A, Kaufman PD, Kobayashi R, Stillman B (1996) Nucleosome assembly by a complex of CAF-1 and acetylated histones H3/H4. Cell 87:95–104

Vogelauer M, Rubbi L, Lucas I, Brewer BJ, Grunstein M (2002) Histone acetylation regulates the time of replication origin firing. Mol Cell 10:1223–1233

Volpe T, Schramke V, Hamilton GL, White SA, Teng G, Martienssen RA, Allshire RC (2003) RNA interference is required for normal centromere function in fission yeast. Chromosome Res 11:137–146

Wallrath LL (1998) Unfolding the mysteries of heterochromatin. Curr Opin Genet Dev 8:147–153

Xiao H, Sandaltzopoulos R, Wang H, Hamiche A, Ranallo R, Lee K, Fu D, Wu C (2001) Dual functions of largest nurf subunit nurf301 in nucleosome sliding and transcription factor interactions. Mol Cell 8:531–543

Yan Q, Huang J, Fan T, Zhu H, Muegge K (2003) Lsh, a modulator of CpG methylation, is crucial for normal histone methylation. EMBO J 22:5154–5162

Ye X, Franco AA, Santos H, Nelson DM, Kaufman PD, Adams PD (2003) Defective S phase chromatin assembly causes DNA damage, activation of the S phase checkpoint, and S phase arrest. Mol Cell 11:341–351

Yoda K, Ando S, Morishita S, Houmura K, Hashimoto K, Takeyasu K, Okazaki T (2000) Human centromere protein A (CENP-A) can replace histone H3 in nucleosome reconstitution in vitro. Proc Natl Acad Sci USA 97:7266–7271

Zappulla DC, Sternglanz R, Leatherwood J (2002) Control of replication timing by a transcriptional silencer. Curr Biol 12:869–875

Zhang J, Xu F, Hashimshony T, Keshet I, Cedar H (2002) Establishment of transcriptional competence in early and late S phase. Nature 420:198–202

Zhang Z, Shibahara K, Stillman B (2000) PCNA connects DNA replication to epigenetic inheritance in yeast. Nature 408:221–225

Zhou Y, Santoro R, Grummt I (2002) The chromatin remodeling complex NoRC targets HDAC1 to the ribosomal gene promoter and represses RNA polymerase I transcription. EMBO J 21:4632–4640

Epigenetic Inheritance of Chromatin States Mediated by Polycomb and Trithorax Group Proteins in *Drosophila*

Jérôme Déjardin, Giacomo Cavalli

Abstract   Proteins of the Polycomb group (PcG) and of the trithorax group (trxG) are involved in the regulation of key developmental genes, such as homeotic genes. PcG proteins maintain silent states of gene expression, while the trxG of genes counteracts silencing with a chromatin opening function. These factors form multimeric complexes that act on their target chromatin by regulating post-translational modifications of histones as well as ATP-dependent remodelling of nucleosome positions. In *Drosophila*, PcG and trxG complexes are recruited to specific DNA elements named as PcG and trxG response elements (PREs and TREs, respectively). Once recruited, these complexes seem to be able to establish silent or open chromatin states that can be inherited through multiple cell divisions even after decay of the primary silencing or activating signal. In recent years, many components of both groups of factors have been characterized, and the molecular mechanisms underlying their recruitment as well as their mechanism of action on their target genes have been partly elucidated. This chapter summarizes our current knowledge on these aspects and outlines crucial open questions in the field.

1
Introduction

In multicellular organisms, homeostasis is a fundamental process. Regulation of this phenomenon implies that biological relationships among different highly specialized tissues or between tissues and environment must maintain constant responses to dynamic conditions. Homeostasis is essential for viability and it involves multiple layers of regulation, including maintenance of cell

J. Déjardin, G. Cavalli
Institute of Human Genetics, CNRS, 141, rue de la Cardonille, 34396 Montpellier Cedex 5, France, e mail: giacomo.cavalli@igh.cnrs.fr

Progress in Molecular and Subcellular Biology
P. Jeanteur (Ed.)
Epigenetics and Chromatin
© Springer-Verlag Berlin Heidelberg 2005

fate. In many cases, once a cell engages a specific differentiation program, it must remember its identity after each cell cycle. Differentiation and/or cellular fate choices are the result of developmental cues that essentially take place at the transcriptional level. Early stimuli that sculpt gene expression are often transient, but once a cell is committed to a specific differentiation program, gene expression profiles must be maintained. Thus, cellular specialization is the consequence of establishment and maintenance of specific patterns of gene expression. Unscheduled cell death, transformation or developmental abnormalities are often the result of defects in this memory process. One of the best examples of this phenomenon is the establishment and maintenance of specific patterns of homeotic gene expression in *Drosophila*. Homeotic gene expression profiles are responsible for the specification of the antero-posterior axis in all metazoans. In *Drosophila*, a transient cascade of transcription factors provided maternally (i.e. already present as a stock in the egg before fertilization) or produced zygotically by gap, pair-rule and segment polarity genes is responsible for the establishment of such patterns (Ingham and Martinez Arias 1992). Most of these transcription factors act only transiently and disappear at the onset of gastrulation. However, patterns of homeotic gene expression are stable through development and during adult life. Thus, there is a mechanism that relays the action of early transcription factors after their disappearance. This mechanism is able to memorize states of gene expression and to faithfully reproduce these patterns through many rounds of cell division. Failure in the maintenance process leads either to embryonic death if these defects occur early or to developmental abnormalities (named homeotic transformations) if these defects are less profound and/or occur later. Genetic screenings for homeotic transformations in *Drosophila* led to the identification of two counteracting groups of genes. The Polycomb group (PcG) is necessary for the maintenance of the repressed state, while the trithorax group (trxG) is necessary for maintenance of the active state. These genes are normally ubiquitously expressed and encode nuclear factors that assemble in distinct multi-subunit complexes that are conserved from *Drosophila* to mammals (Otte and Kwaks 2003). PcG and trxG proteins act directly on chromatin to regulate their target genes. In *Drosophila*, this is achieved by recruiting PcG and trxG proteins to specific chromosomal elements termed PcG or trxG response elements (PRE/TRE). However, several issues are still poorly understood: first, how these ubiquitous factors are properly targeted to their response elements in tissue- and developmental stage-specific manners; second, how they act on their target genes to maintain a repressed or an active transcriptional state; and third, how they maintain memory of chromatin states through many rounds of cell division. This chapter aims to summarize the current understanding of these issues.

2
Proteins of the Polycomb Group of Genes

Based on genetic screens using deficiencies covering a fraction of the *Drosophila* genome, the PcG might include up to 40 members (Jürgens 1985). To date, only 16 members have been molecularly characterized and these members can be biochemically separated into three main categories depending on their presence or absence in the two complexes described to date: the Polycomb Repressive Complex 1 (PRC1) and the Extra SexCombs/Enhancer of Zeste [ESC/E(Z)] complex and its derivatives. Table 1 lists all the characterized PcG factors, their involvement in distinct biochemically characterized complexes and notable protein domains and their respective functions.

2.1
PcG Complexes

While first members of the PcG of genes were cloned over 10 years ago, their characterization as components of distinct chromatin complexes is quite recent. Early studies suggested that most PcG members were building a single Polycomb complex (Simon et al. 1992). However, detailed genetic and biochemical evidence indicates that this is not the case. In particular, two distinct PcG complexes were characterized (see Table 1). Only 7 of the 16 characterized PcG members are integral components of one of the two complexes. Of the remaining nine, some have been shown to interact directly with components of the two main complexes, suggesting that PcG complex composition may vary in different target genes or in different tissues. Also, some members of the two main complexes were shown to interact with partners that are not known to play a direct role in regulating gene expression. For instance, the PH protein, a main component of PRC1 (see below) was shown to make distinct smaller complexes. Association of a PH isoform with Topoisomerase II as well as the Barren proteins (Lupo et al. 2001) links the PcG to components involved in regulation of chromosomal architecture and condensation. Moreover, PH protein has been recently shown to form a complex containing molecular chaperones (Wang and Brock 2003). These examples suggest that diverse PcG complexes may participate in different types of cellular functions.

2.1.1
The ESC-E(Z) Complex

Since ESC function is required early during development, targeting of the ESC/E(Z) complex is often viewed as the first temporal event in PcG-mediated repression (Ingham 1983; Struhl and Akam 1985). Several variants of this complex have been purified, which all contain the ESC and E(Z) proteins.

Table 1. Characterized proteins of the PcG, their symbol name and functional or biochemical features (see text for details)

Full name	Symbol	Biochemical properties/notable domains
PRC1		
Polycomb	PC	Chromodomain responsible for binding trimethylated K27 of histone H3
Polyhomeotic	PH	Zinc finger, SPM (SAM) domain that binds to SCM[a]
Posterior sex combs	PSC	Zinc finger, HTH domain; interacts with PC and PH
dRing/sex combs extra	–	Ring domain
Sex comb on midleg	SCM	SPM domain that allows interaction with PH[a]
Heat-shock protein cognate 4	HSC70-4	Protein chaperone[b]
ESC-E(Z) complex		
Extra sex combs	ESC	WD40 repeats
Enhancer of zeste	E(Z)	SET domain that methylates histone H3 on lysines 27 and 9
Suppressor of zeste 12	SU(Z)12	Zinc finger, VEFS box
Polycomb-like	PCL	Ring domain, PHD domain
Rpd3	–	*Drosophila* homologue to HDAC1
Other characterized PcG members		
Suppressor of zeste 2	SU(Z)2	'Homology region' (with Psc and Bmi-1) containing a RING finger motif[c]
Pleiohomeotic	PHO	Zinc finger that binds to DNA at 'GCCAT'; a conserved domain (aa 118–172) interacts with PC
Pleiohomeotic-like	PHOL	Zinc finger homologous to that of PHO that binds to the same DNA motif as PHO
Additional sex combs	ASX	Zinc finger, Q-rich domain
Cramped	CRM	Genetically interacts with PCNA, A-rich domain[d]
Enhancer of Polycomb	E(PC)	Q-rich domain, A-rich domain
dMi-2	–	PHD fingers, chromodomains, ATPase domain, HMG-like motif, myb-like motif
Other factors that may show PcG features		
Pipsqueak	PSQ	BTB-POZ domain, Psq domain; binds to 'GAGAG' DNA motif
Corto	–	Chromodomain; reported to interact with E(Z), ESC and GAF[e]
Dorsal switch protein 1	DSP1	HMG box protein[f]

Relevant references that are not discussed in the text are listed here:
[a] Peterson et al. (1997)
[b] Mollaaghababa et al. (2001)
[c] Brunk et al. (1991)
[d] Yamamoto et al. (1997)
[e] Salvaing et al. (2003)
[f] Decoville et al. (2001)

Originally, a 600-kDa complex was isolated (Ng et al. 2000; Tie et al. 2001). This complex was shown to contain the ESC and E(Z) proteins as well as the *Drosophila* HDAC1 homologue Rpd3, the histone binding protein p55 and the Su(Z)12 PcG member (Czermin et al. 2002; Muller et al. 2002). In contrast to the other proteins, Rpd3 may be loosely associated with this complex (Kuzmichev et al. 2002). Later, a larger complex of 1 MDa was purified and shown to contain the PCL protein in addition to the members cited above (Tie et al. 2003). This is consistent with previous two-hybrid data, showing a direct interaction between PCL protein and E(Z) (O'Connell et al. 2001). The apparent discrepancy in composition of the ESC/E(Z) complex probably depends on its dynamic composition during development (Furuyama et al. 2003). In particular, the ESC product is deposited into the egg, and expressed only during embryogenesis. Thus, at larval stages the complex undergoes a major compositional change (Furuyama et al. 2003). How this change affects its regulatory function remains to be studied.

2.1.2
The Polycomb Repressive Complex 1

This large complex, first isolated in the laboratory of Robert Kingston, may take the relay from the early acting ESC/E(Z) complex in order to ensure maintenance of chromatin memory after disappearance of ESC. When isolated from *Drosophila* mid–late embryos, the size of PRC1 is approximately 2 MDa (Shao et al. 1999). PRC1 includes more than 30 identified polypeptides, including Polycomb, Polyhomeotic, Posterior Sex Combs, Sex Comb on Midleg (PC, PH, PSC, SCM) and dRing. dRing has been recently identified as being encoded by the formerly identified *Sex combs extra* (*Sce*) PcG gene (Fritsch et al. 2003). In addition to these main PcG members, PRC1 was shown to contain stoichiometric amounts of several TBP Associated Factors (TAFs) and the Zeste transcriptional activator protein (Saurin et al. 2001). Zeste is a sequence-specific transcription factor that was originally classified as a member of the trxG because of its ability to activate the *Ubx* gene both in vitro and in vivo (Biggin et al. 1988; Laney and Biggin 1992). Using co-immunoprecipitation experiments, it was previously proposed that PC, PH and PSC interact (Strutt and Paro 1997; Kyba and Brock 1998). Moreover, immunostaining of polytene chromosomes using antibodies directed against PC and PH proteins showed a perfect co-localization, again suggesting a possible involvement of these factors in a complex in vivo (Franke et al. 1992). However, the Polycomb-like protein (PCL) also perfectly co-localizes with both PC and PH (Lonie et al. 1994) while it was not found in PRC1, suggesting a loose or a later association of this protein to the complex. The fact that PCL is a component of the ESC-E(Z) complex (Tie et al. 2003) may provide a link between ESC-E(Z) and PRC1 complexes.

2.1.3
Other Identified PcG Proteins and Partners

The other characterized members still remain to be assigned to any biochemical complex, or to a specific function in the framework of the known complexes. The Pleiohomeotic protein (PHO) and, more recently, the PHO-like product (PHOL) are the only members of the PcG shown to bind to DNA at a specific DNA binding motif (Brown et al. 1998, 2003; Fritsch et al. 1999). Consistent with the high degree of conservation in the Zinc finger domain responsible for DNA binding between PHO and PHOL, it was shown that PHOL binds to the same DNA sequences as PHO (Brown et al. 2003), suggesting a redundant function for these two factors.

In the early embryo, at a stage where PcG repression still does not occur, the PC, PH, ESC, E(Z), PHO, Rpd3 and the GAGA Factor (GAF, a member of the trxG) proteins interact (Poux et al. 2001b). This work provides evidence for an early physical interaction between members of the two known PcG complexes and members not yet assigned to any of them and it suggests that contacts between these components may be important for establishment of PcG-mediated silencing. Moreover, the PHO protein was shown to be capable of direct interaction with the PC protein (Mohd-Sarip et al. 2002), providing a link between a sequence-specific PcG member and PRC1. Although YY1, the human PHO homologue was shown to interact with EED, a human ESC homologue (Satijn et al. 2001), no such interaction was found in *Drosophila*, thus leaving open the question whether PHO may link both complexes via a joint recruitment. The function of some other proteins that were originally assigned to the PcG is even less understood. For instance, the two proteins E(PC) and ASX (Table 1) might be classified in the PcG genetically, and on the basis of their partial co-localization with PC in polytene chromosome staining experiments (Sinclair et al. 1998a,b). However, ASX seems to have a dual function, as some mutations in the *Asx* gene also enhance trxG phenotypes, and since this protein has tissue-specific functions via specific cofactors (Dietrich et al. 2001). Moreover, both ASX and E(PC) proteins interfere with the phenomenon of Position Effect Variegation, a gene silencing phenomenon that depends on heterochromatin components and does not involve the other members of the PcG (Sinclair et al. 1998b). Thus, the molecular dissection of the function of these two factors, and perhaps other proteins associated with the PcG, may uncover intriguing links between PcG members and other cellular functions.

2.2
Targeting of PcG-Mediated Repression

2.2.1
PcG Response Elements

PcG targeting requires the presence of a chromosomal element termed PcG Response Element (PRE) at the target locus. PREs are believed to nucleate PcG association to chromatin and they are continuously required during development to anchor PcG proteins to their target genes (Busturia et al. 1997). In *Drosophila*, PREs are DNA sequences ranging from a few hundred to several thousand base pairs. They have been characterized using transgenic approaches where PREs were cloned close to the mini-*white* gene marker (expression of the *white* gene is responsible for the *Drosophila* red eye colour). Flies bearing such constructs usually display variegated eye colour, and homozygous flies for these transgenes often show lower mini-*white* expression than heterozygous individuals. This phenomenon is termed Pairing Sensitive Silencing. Moreover, the PREs present in these constructs induce ectopic binding sites for PcG proteins in polytene chromosomes (Fauvarque and Dura 1993; Chan et al. 1994; Kassis 1994; Zink and Paro 1995; Hagstrom et al. 1997; Shimell et al. 2000; Mishra et al. 2001). Polytene chromosome studies indicate the existence of more than 100 loci bound by PcG proteins, and the low resolution of this technique does not exclude that several PcG binding sites may in fact contain more than one autonomous PRE. For example, there are at least six PREs in the BX-C that are concentrated in only one PC signal in polytene chromosomes. Similarly, the *ph* locus contains two distinct PREs, whereas only one cytological signal can be detected. In vitro studies reveal that many PcG proteins do not bind to specific DNA sequences (Francis et al. 2001). Thus, other factors may probably link PcG proteins to PREs. An attractive hypothesis would be that early acting factors that initiate the pattern of repression recruit PcG proteins. For instance, the early gap repressor Hunchback has binding sites in the *bxd* PRE regulating the *Ubx* expression. Using a two-hybrid assay, Hunchback was shown to interact with dMi-2, an ATP-dependent chromatin remodelling enzyme shown to belong to the PcG (Kehle et al. 1998). However, *bxd*-mediated repression of Ubx was shown to be independent of Hunchback itself, indicating that this may not be the only mechanism of PcG recruitment (Poux et al. 1996).

PRE alignments revealed no common organization at the sequence level but identified three consensus motifs. The first is GCCAT and this motif is specifically bound by PHO (Brown et al. 1998; Mihaly et al. 1998), the fly homologue of the mammalian YY-1 transcription factor (Atchison et al. 2003). This motif was recently shown to be also bound by PHOL (Brown et al. 2003). Mutation of PHO motifs in transgenic PRE has been reported to abolish pairing sensitive silencing effects or maintenance of repression in a number of transgenic assays (Fritsch et al. 1999; Busturia et al. 2001; Mishra et al.

2001). The other conserved motif is the GAGAG sequence, which is a binding site for the GAGA factor (GAF) encoded by the *Trithorax-like* gene (*Trl*). This sequence is also bound by the Pipsqueak protein, another BTB-POZ factor with PcG features (Hodgson et al. 2001; Huang et al. 2002). Mutation of GAGAG motif in transgenic PREs has also been reported to be critical for repression (Busturia et al. 2001; Mishra et al. 2001). It has been suggested recently that GAF association to chromatin may facilitate the binding of PHO at its consensus sites (Mahmoudi et al. 2003). Finally, a third motif, YGAGYG (where Y can be C or T), constitutes a binding site for the Zeste protein and is often found associated to PHO and GAF sites at a subset of PREs. Although Zeste was originally classified as a member of the trxG, mutation of Zeste binding sites at the *Ubx* promoter in a minimal transgenic context led to the hypothesis that Zeste is important for the maintenance of the PcG repressed state of *Ubx* in tissues where *Ubx* is normally silent (Hur et al. 2002). Furthermore, Zeste was found present in stoichiometric amounts in PRC1 (Saurin et al. 2001). Recent in vitro data show that Zeste might favour the PRC1-dependent inhibition of chromatin remodelling mediated by a recombinant human SWI/SNF complex (Mulholland et al. 2003). However, mutation of Zeste sites at a well-characterized PRE does not affect PcG-mediated repression or recruitment in vivo (Déjardin and Cavalli 2004).

As PREs are complex modular elements (Horard et al. 2000), understanding the molecular principles of their action is a challenging task. While PHO and GAF binding motifs are necessary for silencing in transgenic experiments, their presence seems to be insufficient. Indeed, an array containing repeated PHO and GAF binding sites from a minimal PRE of the *engrailed* gene failed to produce detectable pairing-sensitive repression (Americo et al. 2002). This result suggests either that the precise spacing between binding motifs is important, or that there are other recruiter motifs yet to be discovered. Indeed, some PRE fragments containing neither GAF nor PHO binding sites are still able to recruit PC in vitro (Horard et al. 2000). Conversely, mutations in *pho*, *pho-like*, *Trl* or *z* genes do not induce complete loss of PcG proteins from their targets (Brown et al. 2003), strongly suggesting that other factors contribute to recruitment of PcG proteins.

In a chromatin context, the three-dimensional organization of DNA sequence motifs may be critical for anchoring PcG complexes. Thus, favourable configurations may be reached in different PREs, despite different spacings between PHO, GAF and Zeste binding sites. This concept was recently explored using computer-assisted genome-wide prediction of PREs (Ringrose et al. 2003). This analysis showed that, while the presence in a given sequence of PHO, GAF or Zeste binding motifs is not sufficient on its own to predict any PRE, their combination in pairs could distinguish PREs from *Drosophila* genomic sequences. Starting from a pool of known PRE sequences, an algorithm was developed in order to discriminate PRE from non-PRE sequences. This algorithm, tested on the 300 kb of the bithorax complex (BX-C) locus, was able to blindly detect known PREs that were not originally included in the PRE training set. Extension of these studies to the whole genome sequence from *Drosophila* led to the identification of 167 putative PRE sequences, and some of these candidates were functionally validated by genetic and molecular studies. This work represents an important advance in the mapping of new PcG targets but has some limitations. Indeed, PRE sequences selected for the training set are known to be bound by the PC proteins on polytene chromosomes. If the occurrence of PHO and GAF binding sites by pairs was the only parameter to take into account, the algorithm should have predicted a majority of loci that are actually bound by PC on polytene chromosomes. However, about half of PC binding sites escape prediction (for instance, the *ph* locus which is a strong cytological binding site for PC was not predicted). It is likely that, in addition to the known motifs, other DNA sites may be required to attract PcG binding proteins at their target loci. Finding novel functionally important DNA motifs and feeding them into this algorithm may be important to improve its future performance.

2.2.2
Chromatin Determinants Associated with Targeting

The fact that most PcG proteins are ubiquitously expressed while their target genes often display a specific pattern of expression indicates that PcG recruitment might be a highly regulated process. In addition to PRE presence, another requirement for PcG-mediated silencing is that the target gene must be repressed at the time of onset of PcG function (Fig. 1a). This feature suggests that PcG proteins are able to 'sense' whether chromatin of a given gene is favourable to association or not. Thus, together with the presence of a PRE sequence, other features typical of silenced chromatin may provide crucial marks for nucleation and accumulation of PcG complexes.

A few years ago, David Allis proposed that many chromosomal proteins with regulatory roles may be able to interpret the various post-translational modifications of histones in a combinatorial manner, resulting in specific regulatory readouts and in the possibility of propagation of some of these chromatin marks through cell division. This hypothesis, defined as 'the histone code' (Strahl and Allis 2000), has inspired many studies aimed at understanding epigenetic mechanisms of chromosomal regulation. The three-dimensional organization of the chromatin fibre relies on several mechanisms that confer unique features to nucleosomal templates. Chromatin structure, and thus function, can be modulated by covalent modification of histones (Narlikar et al. 2002), by chromatin remodelling mediated by multiprotein complexes that use the energy of ATP in order to displace nucleosomes on the chromatin fibre (Becker and Horz 2002), and, finally, by other non-histone proteins that are able to modify chromatin structure in a manner apparently independent of ATP catalysis. The latter class of factors is less well character-

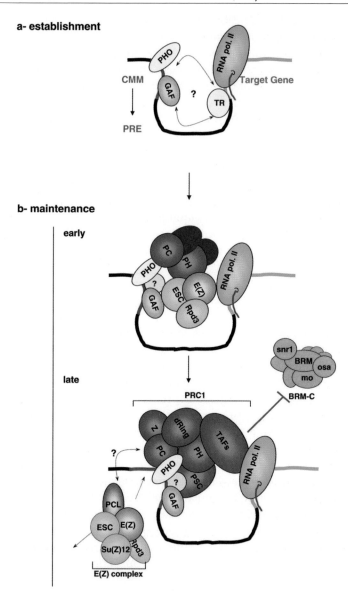

Fig. 1a, b. PcG-dependent memory of the repressed state. **a** Establishment of PcG silencing at homeotic genes requires that the target gene is silenced during early embryogenesis. In the particular case depicted here, the target gene (*green*) is silenced by a transcriptional repressor (*TR*) at the promoter. Transcriptional competence is, via an unknown mechanism (*arrows*), sensed by distantly located cellular memory modules (*CMM, red*), which may favour the binding of PHO and GAF, and possibly other early DNA binding activities (not shown here). Thus, this step may be critical for switching an undetermined *CMM* towards a *PRE*. **b** The maintenance process may occur in two phases: an early one, where *PC, PH, ESC, E(Z), PHO, GAF* and *Rpd3* may build a single complex. This early complex possibly marks chromatin by Rpd3-mediated histone deacetylation and E(Z)-dependent histone methylation. Later, the maintenance phase occurs through the action of two distinct complexes. The

ized, but seems to include GAF (Tsukiyama et al. 1994; Leibovitch et al. 2002) and HMG box-containing factors (Travers 2003).

Histone features can be dynamically modified by several covalent modifications, such as acetylation, phosphorylation, methylation, ADP-ribosylation and ubiquitination. Among these different marks, histone methylation is the best candidate to account for the relative stability of PRE-mediated silencing through development. Indeed, in contrast to other modifications, methylation is thought to be the most stable, as no demethylase activity has been isolated to date.

Examination of PRC1 components does not suggest any enzymatic function that could account for the transcriptional silencing mechanisms described above. In contrast, the ESC/E(Z) complex is endowed with histone-modifying activities. First, the histone deacetylase Rpd3 was identified as a component of an ESC/E(Z) complex, suggesting that this complex might deacetylate its target chromatin in order to favour stable PcG recruitment (Tie et al. 2001; Kuzmichev et al. 2002).

Moreover, the identification of SET [for Su(Var)3–9 E(Z) and TRX] domain-containing proteins as histone methyl transferase enzymes (Rea et al. 2000) and the presence of such domain in the PcG member E(Z) suggested a role for histone methylation in recruitment of PcG complexes to their target elements. Indeed, the ESC/E(Z) complex was shown to methylate histone H3 at lysine 27 and, to a lesser extent, at lysine 9 (Cao et al. 2002; Czermin et al. 2002; Kuzmichev et al. 2002; Muller et al. 2002). Methylated H3 lysines can be specifically recognized by chromodomain proteins. For instance, the heterochromatin protein Su(Var)3–9 [Su(Var) stands for Suppressor of Variegation] methylates histone H3 at lysine 9 via its SET domain, and this mark is specifically recognized by the chromodomain of the Heterochromatin Protein 1 (HP1) (Lachner et al. 2001). A similar but distinct pathway applies to PcG regulators: the chromodomain of PC binds stably to the histone H3 trimethylated on lysine 27. PC binds to this mark with much higher affinity than to trimethylated lysine 9 of histone H3, and, conversely, HP1 prefers this latter mark to trimethylated lysine 27 (Fischle et al. 2003; Min et al. 2003). Moreover, patterns of trimethylated lysine 27 and 9 correlate well with binding patterns

ESC/E(Z) complex may undergo dynamic changes. PCL may become an integral component of this complex and ESC is downregulated at the end of embryogenesis (marked by an *arrow*). Again, E(Z) complexes mark chromatin by histone methylation. The PCL protein may somehow interact with the Polycomb Repressive Complex 1 (*PRC1*). The latter may interfere with the cross-talk between TAFs and RNA polymerase II. PRC1 may also inhibit chromatin remodelling by inhibiting BRM Complex (*BRM-C*) action. PRC1 may be anchored to the PRE by recruitment of PHO and other unknown factors (the '?' *labelled circle*). This recruitment may be stabilized by an interaction between the chromodomain of PC and methylated lysine residues on histone H3. For simplicity, nucleosomes, modified histones or higher order chromatin structures are not represented in this scheme, and the possible role of non-coding RNA is not considered because of lack of detailed information

of PC and HP1 at polytene chromosomes, respectively (Fischle et al. 2003). Thus, the specific chromodomains of PC and HP1 proteins are responsible for their chromatin targeting specificity to a large extent. This is consistent with the pioneering work of Platero and colleagues, in which the chromodomain of HP1 was swapped with that of PC protein, with a re-localization of the chimeric protein in polytene chromosomes as a consequence (Platero et al. 1995). Although these experiments suggest a role of E(Z) and histone methylation in providing an anchor for PC and thus PRC1, it remains to be demonstrated in vivo that trimethylation of histone H3 is sufficient to attract PcG complexes to specific loci in chromatin.

How does the transition between the ESC/E(Z) and the PRC1 complexes occur? In one view, GAF might favour PHO binding at the target PRE. PHO might in turn recruit the ESC/E(Z) complex, which would methylate histone H3 at lysines 9 and 27 at the site of recruitment (i.e. PRE containing accessible PHO site). This modification might stably anchor PRC1 and maintain it at the target chromatin. This rather simple view does not explain a certain number of observations, such as the direct interaction between PHO and PC, and the lack of interaction between PHO and components of the ESC/E(Z) complex in *Drosophila*. On the other hand, ESC, E(Z), PC, PH, PHO, GAF and Rpd3 were reported to interact in early embryos and, in particular, a transient interaction between PC and ESC was observed (Poux et al. 2001b). This may provide a functional link between the two apparently distinct activities observed later in development and suggests an alternative scenario (Fig. 1b). Once PHO is recruited at its target chromatin, it might directly contact the PC protein (Mohd-Sarip et al. 2002). PC might then recruit the ESC/E(Z) complex through a direct interaction with ESC. E(Z) would then methylate histone H3 and stabilize binding by PRC1. In this speculative view, specific DNA sequences and histone modification marks would be required in order to maintain silencing on chromatin. This might also explain the strong interplay between PcG members that was originally suggested by genetic analysis. However, even this scenario might be too simplistic, and in particular it does not account for the fact that GAF and PHO may not suffice to recruit PcG proteins and are likely to be assisted by additional factors.

In addition to recognition of histone methylation marks, chromodomains were also shown to be able to interact with RNA. In *Drosophila*, one chromodomain protein involved in the regulation of the process of dosage compensation was shown to bind to a non-coding RNA, and this binding seems to be important for its chromatin targeting (Akhtar et al. 2000). As the Polycomb protein does contain a chromodomain that is necessary for its correct targeting to chromatin, it is tempting to speculate that an RNA moiety may be involved in PcG silencing (or targeting). Indeed, there is a connection between silencing mediated by the RNA interference (RNAi) machinery and by Polycomb (Pal-Bhadra et al. 2002). Thus, it would be interesting to explore in molecular terms this interplay: is the putative RNA moiety a product of RNAi? Are PcG/trxG genes themselves regulated by small RNAs?

Finally, some pieces of the puzzle may be missing. For instance, among the known members of chromatin remodelling machineries, dMi-2 is important for early stages of PcG function in silencing (Kehle et al. 1998), but molecular evidence of its mechanism of action is still lacking. However, dMi-2 also contains a chromodomain, suggesting that this protein may also be tethered to PRE via recognition of methylated histones and/or RNA moieties.

2.3
Mechanisms of Repression

While PcG proteins are known and well characterized, their mechanisms of action leading to stable repression remain elusive. Silencing of gene expression might involve multiple non-exclusive mechanisms. PREs might interfere with activators by competing for binding to overlapping DNA sequences. PRE silencing might also be achieved by altering chromatin architecture (for instance by changing specific positions or the packing density of nucleosomes) in order to prevent access of specific regulatory elements to transcription factors. Another mechanism may be to interfere with the RNA polymerase II machinery. This can be achieved at any of the multiple steps involved in the processes of transcription initiation and elongation. Finally, a way to prevent transcription may be to segregate target genes in nuclear compartments that are depleted from RNA polymerases.

2.3.1
Spreading or Looping?

The best characterized PREs are those from the BX-C homeotic locus. These elements are located very far from the promoter of their target genes, an organization similar to that found for enhancers. How can PREs work from such a distance? Two models for PRE action can be hypothesized. The first one involves spreading from a nucleation site (i.e. the PRE) towards the regulated promoter, coating the chromatin fibre in order to render it inaccessible. This model is similar to the one proposed to account for telomeric silencing mediated by SIR proteins in yeast, or for heterochromatin repression in *Drosophila* and vertebrates. Consistent with this model, several studies have indicated that chromatin accessibility is reduced at loci regulated by PREs (Schlossherr et al. 1994; McCall and Bender 1996; Boivin and Dura 1998; Fitzgerald and Bender 2001). On the other hand, other works failed to detect differences in chromatin accessibility (Schlossherr et al. 1994). Moreover, core PRE sequences display hypersensitivity to nuclease digestion, which argues for chromatin accessibility at least locally, at the level of the PRE itself (Galloni et al. 1993; Karch et al. 1994; Dellino et al. 2002). To date, there is still no evidence arguing for an extensive spreading of PcG factors from the PRE to the promoter. Indeed, chromatin immunoprecipitation experiments suggest that PcG proteins do not coat large chromatin domains uniformly, but may bind preferentially to the promoter and the PRE, whereas the intervening chromatin between these elements may be less enriched (Strutt et al. 1997; Orlando et al. 1998).

The second model for long-distance action of PREs proposes an interaction with the promoter region through a looping mechanism, similar to that proposed for enhancers in transcriptional activation (Bulger and Groudine 1999). This model is supported by the discovery that several TAFs and TBP are present in PRC1 in stoichiometric amounts with Polycomb (Saurin et al. 2001), and by chromatin immunoprecipitation of PcG proteins, TBP and TAFs at promoters of PcG target genes (Breiling et al. 2001). Thus, a complex comprising PcG proteins and general transcription factors may bind PREs and promoters at the same time. This would be consistent with a model proposing that PREs may contact their cognate promoters via a looping mechanism. A novel approach named Capturing Chromosome Conformation (Dekker et al. 2002) was used recently to demonstrate such a physical interaction between enhancers and specific genes of the β-globin locus in mouse (Tolhuis et al. 2002). It might be of interest to apply this technique for PREs and promoters in order to directly demonstrate a physical proximity between them.

2.3.2
Proposed Silencing Mechanisms

Artificial targeting of different PcG proteins to DNA leads to stable embryonic repression of reporter genes (Müller 1995; Poux et al. 2001a). This suggests that single PcG components can reconstitute a silencing machinery in vivo if efficiently recruited. However, silencing is lost during larval stages, suggesting that in addition to PcG recruitment, other events are required for maintenance of the memory of chromatin states. One of these events may be represented by histone methylation. The histone methyl transferase activity from the ESC/E(Z) complex has not been demonstrated to have any repressive activity per se on chromatin. This complex may function primarily by depositing a histone H3 methylation mark that may stabilize recruitment of PRC1 on its target chromatin.

However, how does PRC1 repress chromatin? There is no evidence for enzymatic activities associated with PRC1. In vitro approaches have suggested that a possible mechanism of action for PRC1 may involve inhibition of chromatin remodelling (Fig. 1). Pre-incubation of a reconstituted nucleosomal array with PRC1 inhibited remodelling mediated by a human SWI/SNF complex (Shao et al. 1999). A Polycomb Core Complex (PCC) comprising only four PcG members from PRC1 (PC, PH, PSC and dRing) and even the PSC protein alone were also shown to be able to prevent hSWI/SNF-mediated chromatin remodelling in vitro, albeit to a lesser extent than PRC1 (Francis et al. 2001). The addition of Zeste into PCC strongly stimulates the ability of the resulting PCC-Z to inhibit SWI/SNF action (Mulholland et al. 2003). These experiments suggest that at least a part of the function of PcG proteins is to repress genes via a modulation of chromatin at the level of nucleosome structure or positioning.

In addition, PcG proteins may interact directly with the promoter of their target genes (Breiling et al. 2001; Saurin et al. 2001). This may occur in two different ways. First, as discussed above, via looping of the PRE. Mapping of specific interactions of some of the TAFs binding at promoters with components of the PcG and identification of genetic interactions between the interacting partners would be an important advance in this field. Second, if PREs are located at regions overlapping the promoter of some target genes, PcG proteins may physically interfere with the assembly of functional transcription initiation complexes. In this context it is interesting to note that many of the putative PREs predicted in silico by Ringrose and colleagues are indeed located at or close to gene promoters (Ringrose et al. 2003).

PcG proteins may interfere with not only transcription initiation, but also transcription elongation and, conversely, trxG members may be supposed to stimulate the elongation process. This last point is supported by recent work on the function of the TRX and GAF proteins. Transcriptional silencing may thus be achieved by blocking transcriptional elongation. Again, while there is no direct evidence for PcG-mediated inhibition of transcriptional elongation, the presence of putative PREs in intronic sequences of the majority of target genes (Ringrose et al. 2003) makes this possibility attractive.

Finally, regulation of nuclear compartmentalization of PcG target genes may contribute to silencing. While there is no evidence for strong homologous interactions in mammals, homologous chromosomes are paired in *Drosophila* (Hiraoka et al. 1993). As a consequence, PRE sequences are also paired at the homozygous state, implying that such elements are closely located in the nuclear space. For many transgenic PREs, silencing is more robust when present at the homozygous state, a phenomenon known as the pairing sensitive (or dependent) silencing effect (Kassis 2002). Moreover, even in *cis*, multimerization of weak PRE sequences leads to a more robust maintenance of silencing (Horard et al. 2000). In addition to homologous interactions, long-distance association of PREs was also identified. In particular, a transgenic PRE from the *Fab-7* regulatory region of the homeotic gene *Abdominal-B* can exert its full repressive effect only if the endogenous copy of the *Fab-7* is also present (Bantignies et al. 2003). The intensity/stability of repression was correlated with a physical interchromosomal interaction between the transgene containing the *Fab-7* fragment and the endogenous *Fab-7*. This observation suggests that natural PREs might co-localize in specific PcG nuclear compartments in order to stabilize repression. Altogether, these observations suggest that PcG-dependent maintenance of silencing occurs in a highly cooperative way: the more PRE sequences co-localize in a given region, the more favourable is PcG nucleation and the more efficient is silencing. Foci of PcG factors can be observed in nuclei from embryos or SL2 *Drosophila*-cultured cells (Messmer et al. 1992). Thus, localization of target genes in nuclear regions with a high concentration of PcG proteins may be an important aspect of PcG-mediated gene regulation, although it is not known whether PcG foci per se are able to exclude interaction of DNA with the transcriptional machinery.

3
Proteins of the Trithorax Group

Identification of a group of genes whose products counteract the action of the PcG proteins at homeotic genes originated the dichotomic view of the PcG and the trxG displaying opposing effects at PREs/TREs. However, except for the *trithorax* (*trx*) (Ingham 1983) and the *absent, small*, and *homeotic discs 1* and *2* (*ash1* and *ash2*) genes (LaJeunesse and Shearn 1995), whose products seem to counteract specifically PcG-mediated repression and to share with PcG proteins many of the targets, other members of the activating group appear to be general transcriptional activators or co-activators. For such members, counteracting the action of PcG at some of their target genes may represent only one of the roles that these proteins have in genome function regulation. In contrast to PcG genes, mutations of trxG genes (except for *trx, ash1* and *ash2*) do not lead to homeotic transformation by themselves, although they suppress PcG phenotypes.

In general, when a given gene mutation produces homeotic gain-of-function or loss-of-function phenotypes, and when it interacts genetically with known PcG/trxG genes, the corresponding gene is assigned to the trxG or the PcG respectively. Mutations of members of a group can complement mutations in members of the other one. However, by using genetic approaches, Gildea and co-workers found mutations in some PcG genes (mainly supposed to disrupt homeotic gene repression on their own) that did not complement mutations in some trxG genes (Gildea et al. 2000). Moreover, some formerly identified members of the trxG like Zeste or GAF are also able to mediate some aspects of PcG-dependent silencing. Thus, it was proposed that genes with ambiguous behaviour should be classified as 'Enhancers of Trithorax and Polycomb' (ETP). Members of such a group would thus be required for maintenance of silencing or activation, respectively, depending on the gene, the tissue and the developmental stage under study. This classification is useful for interpreting puzzling genetic evidence concerning phenotypes of several genes, but it does not attempt to provide a molecular basis for the explanation of their function. For this reason, we maintain in this chapter the original classification of *E(z)* in the PcG, and of *Trl* and *z* in the trxG. Table 2 lists all characterized members of the trxG, their notable protein domains and their involvement in distinct complexes.

Table 2. Characterized proteins of the trxG, their symbol name and functional or biochemical features (see text for details)

Full name	Symbol	Biochemical properties/notable domains
TAC1		
Trithorax	TRX	SET domain, PHD domain
CREB-binding protein	dCBP	Histone acetyl transferase
SBF1	–	Protein phosphatase
BRM complex		
Brahma	BRM	DNA-dependent ATPase, helicase domains; SWI2/SNF2 homologue
Moira	MOR	Yeast SWI3 homologue
Osa	–	ARID DNA binding domain[a]
Snf5 related 1	SNR1	Interacts with TRX
GAF-FACT complex		
GAGA factor	GAF	Zinc finger, BTB-POZ domain; interacts with TRX
SPT16	–	DNA unwinding
SSRP1	–	HMG box domain
trxG members that build distinct complexes		
Absent, small or homeotic discs 1	ASH1	SET domain; histone H3 and H4 methyl transferase; interacts with dCBP
Absent, small, or homeotic discs 2	ASH2	–
Other characterized trxG members		
Kismet	KIS	Chromodomain, SNF2-related domain, helicase domain, Myb-like domain
Modifier of mdg4	MOD	BTB-POZ domain; participates in Gypsy insulator function[b]
Little imaginal discs	LID	ARID, PHD domain[c]
Zeste	Z	HTH and Leucine Zipper domains. Zeste protein forms oligomers

Relevant references that are not discussed in the text are listed here:
[a] Vazquez et al. (1999)
[b] Gerasimova and Corces (1998)
[c] Gildea et al. (2000)

Like PcG factors, trxG proteins assemble into distinct complexes. Five main complexes were described in *Drosophila*: the TAC1 complex containing TRX, the Brahma complex that includes BRM and other members of the trxG, ASH1 and ASH2 complexes, which were not studied in detail, and lastly the FACT complex containing GAF. As in the case of PcG complexes, distinct trxG complexes contain proteins that are excluded from others. This suggests that, at least at homeotic genes, distinct mechanisms that depend on different trxG factors may cooperate to maintain the memory of active gene expression. In contrast to PcG proteins of PRC1, most of the trxG members do contain protein domains with known enzymatic activity, suggesting that trxG-mediated maintenance of transcription may occur through 'active' mechanisms. Similar to the PcG, several genes classified as members of the trxG encode factors that are not contained in any of the above complexes, and the lack of detailed studies does not allow us to conclude whether they act separately or in conjunction with any of the known complexes.

3.1
trxG Complexes

3.1.1
The Trithorax Acetylation Complex (TAC1)

The 1-MDa size complex TAC1 was purified from *Drosophila* embryos. This complex contains three components: TRX, the *Drosophila* CREB-binding protein (dCBP) histone acetyl-transferase and the SBF1 anti-phosphatase protein (Petruk et al. 2001). While most of the available TRX and SBF1 enter into the formation of TAC1, only 5% of the total dCBP is involved, consistent with the fact that dCBP has many other roles besides regulation of trxG target genes. Components of the TAC1 complex were shown to bind to a PRE/TRE and to be required for activation of the *Ubx* gene in vivo (Petruk et al. 2001). Moreover, an indirect connection between TRX and histone acetylation was also suggested by the finding that maintenance of the active state mediated by *Fab-7* was dependent on *trx* and correlated with hyperacetylation of histone H4 (Cavalli and Paro 1999).

3.1.2
The Brahma Complex

The BRM-containing complex is well characterized and its composition is conserved from yeast to mammals. BRM is the fly homologue of the yeast ATP-dependent chromatin remodelling enzyme SNF2, and most of its partners are also found in the related mammalian BRM/BRG1 complexes. The BRM complex seems to act predominantly, although perhaps not exclusively, in transcriptional activation (Armstrong et al. 2002; Zraly et al. 2003). In the fly complex, ten subunits have been identified (Kal et al. 2000). Besides BRM, other characterized members are the SNR1, MOR and OSA proteins, which are integral components of the complex and interact genetically with mutations in the *brm* gene. At least in the case of SNR1, this protein may be specifically required in some of the tissues and cell types, suggesting that more than one BRM complex may exist in *Drosophila* (Zraly et al. 2003).

3.1.3
The GAF–FACT Complex

GAF was suggested for a long time to act in maintaining the active state of homeotic genes as well as of other genes, but its mechanism of action was unknown. Recently, however, a complex named FACT (Facilitates Chromatin Transcription) was purified as a partner of GAF (Shimojima et al. 2003), suggesting that GAF may stimulate transcriptional elongation at least at some of its target genes. FACT was previously shown to facilitate elongation by RNA polymerase II in mammals (Orphanides et al. 1998) via its SPT16 and SSRP1 subunits. The interaction between GAF and FACT subunits was substantiated by genetic analysis and they were shown to cooperate in vivo for the proper activation of homeotic genes. The role of GAF in regulating transcriptional elongation is consistent with earlier findings on the localization of this protein mapped by UV cross-linking followed by chromatin immunoprecipitation on heat-shock inducible genes. Before heat-shock, GAF was present at the promoter, while after heat-dependent induction of transcription it was found throughout the transcribed region (O'Brien et al. 1995).

3.1.4
Other trxG Complexes and Partners

Using gel-filtration analysis, it was shown that ASH1 and ASH2 proteins form two biochemically distinct complexes, but other members of these complexes have not been identified (Papoulas et al. 1998). ASH1 interacts genetically and physically with the histone acetyl transferase dCBP (Bantignies et al. 2000). Moreover, ASH1 is itself a histone methyltransferase (Beisel et al. 2002; Byrd and Shearn 2003), suggesting that at least in the case of the ASH1 complex, it may have redundant composition and function with TAC1. Clearly, more work is required on these complexes in order to understand their function.

3.2
Targeting of trxG Complexes at TREs

TREs are much less well characterized than PREs. Genetic experiments suggest that PREs and TREs are in fact constituted of overlapping DNA sequences (Tillib et al. 1999). This is further supported by the observation that transgenic PREs are also responsive to trxG mutations (Fauvarque et al. 1995; Cavalli et al. 1999; Horard et al. 2000). In addition, insertion of a *bxd* PRE-containing transgene at some genomic loci induces an active chromatin state in the construct that requires a wild type *trx* gene (Poux et al. 2002). Finally, PcG and some trxG proteins co-localize in the same chromosomal regions. About half of the TRX and ASH1 sites in polytene chromosomes are bound by PC

(Chinwalla et al. 1995; Tripoulas et al. 1996). Chromatin immunoprecipitation experiments have also shown that the same sequences in the BX-C are bound by PC as well as TRX (Orlando et al. 1998).

This proximity of PREs and TREs is reflected by their dual ability to maintain active as well as silent chromatin states. For instance, the *Fab-7* element from the BX-C is able to maintain a PcG-dependent repressed chromatin state in transgenic assays. However, the same element can be switched to a heritable active state upon a transient embryonic pulse of transactivation (Cavalli and Paro 1998). Because *Fab-7* can integrate and maintain both chromatin states depending on early and transient regulatory cues, it was termed Cellular Memory Module (CMM). Memory of the active state mediated by *Fab-7* was shown to be dependent on *trx* (Cavalli and Paro 1999). Good candidate regulatory factors responsible for recruitment of trxG proteins at TRE are GAF and Zeste and their DNA binding sites.

GAF may participate in this function, since it co-immunoprecipitates with TRX from embryonic extracts and GAF binding motifs are able to recruit TRX in vitro at DNA sequences from the *bxd* PRE (Poux et al. 2002). Mutations in the *Trl* gene can result in anterior homeotic transformations, a phenotype reminiscent of trxG mutations (Farkas et al. 1994). GAF regulates nucleosome positioning at the promoter of heat-shock genes (Lu et al. 1993; Leibovitch et al. 2002). Whether GAF plays a similar role at PRE/TRE sequences remains to be evaluated, but these elements were found to be DNase hypersensitive (Galloni et al. 1993; Karch et al. 1994), and these hypersensitive sites correspond to a region bound by GAF (Karch et al. 1994; Cavalli and Paro 1998). Chromatin opening by GAF might thus be important for recruitment of some PcG and trxG proteins at PRE/TRE elements.

Zeste binding sites are also found at many PRE/TRE, but they are not critical for PcG recruitment at a minimal *Fab-7* element (Déjardin and Cavalli 2004). Rather, they are involved in the stable recruitment of the BRM chromatin remodelling complex. Mutations of these sites selectively affect maintenance of the active state, suggesting that binding of Zeste at core PRE/TRE is more likely to be involved in the trxG rather than the PcG response (Déjardin and Cavalli 2004). This is consistent with previous analysis of Zeste binding sites at the *bxd* PRE (Horard et al. 2000), and with the fact that Zeste can contact the BRM complex in vitro via direct interactions with its MOR and OSA subunits (Kal et al. 2000). TRX may cooperate with Zeste at two levels in order to recruit the BRM complex. First, an interaction was detected between TRX and the SNR1 protein, an integral component of the *Drosophila* BRM complex (Rozenblatt-Rosen et al. 1998; Zraly et al. 2003). Second, the protein domain named bromodomain, which is present in many factors including the trxG protein BRM, was shown to recognize specifically nucleosomes carrying acetylated histones (Hassan et al. 2002). The dCBP protein present in the TRX-containing TAC1 complex may acetylate histones, creating a binding site for the bromodomain of BRM. The BRM complex might thus be recruited by three distinct activities: direct interaction between TRX and SNR1, TAC1-dependent histone acetylation and Zeste-mediated recruitment.

How TRX itself is recruited is still unclear. Except for Zeste and GAF, no other trxG members were shown to bind directly to DNA in a sequence-specific manner. Although previous work identified the AACAA motif in the *bxd* element as important for the genetic response to a *trx* allele in a transgenic context, no protein binding to this motif was identified (Tillib et al. 1999). Further research will be required in order to understand the interplay between activating/repressing sequences at PRE/TRE. Identification of sequence mutations that selectively affect trxG or PcG response in these elements would provide a good starting point for future investigations.

3.3
Mechanisms of Action

Three chromatin activities identified in trxG complexes may contribute to maintain chromatin open. The first activity is deposition of histone modifications, such as histone acetylation and methylation, in particular of histone H3 at lysine 4. The second activity is chromatin remodelling by the ATP-dependent BRM machinery. A third activity may be to remodel chromatin structure in an ATP-independent manner (Fig. 2).

Histone acetylation may be carried by the TAC1 complex as well as the ASH1 complex, both involving the histone acetyl transferase dCBP (Bantignies et al. 2000; Petruk et al. 2001). Histone acetylation may open chromatin in two ways. First, it may loosen histone to DNA associations by neutralization of positive charges present at unmodified lysines at neutral pH, and this may increase chromatin accessibility (Nightingale et al. 1998). This chromatin opening may occur at promoters and in the coding regions. Indeed, an intriguing finding shows that TAC1 is involved in regulation of heat-shock genes via stimulation of transcriptional elongation (Smith et al. 2004). Second, histone acetylation may provide signals to target other chromatin remodelling components like the BRM complex which is able to loosen nucleosome positions on the DNA (see Sect. 3.2). Methylation of lysine 4 of histone H3 is another hallmark of actively transcribed genes (Bernstein et al. 2002; Zegerman et al. 2002; Schneider et al. 2004) and such activity was found in a TRX-containing complex (Czermin et al. 2002) as well as in the trxG member ASH1, which also contains a SET domain (Beisel et al. 2002; Byrd and Shearn 2003). How this mark may be translated into physical opening of chromatin is not understood. One possibility is that some chromatin remodelling complexes may associate to this mark through binding mediated by chromodomains, similarly to recognition of methylated lysine 9 and 27 of histone H3 by silencing proteins such as HP1 and PC. To date, only one member of the trxG, Kismet (KIS), possesses a chromodomain. However, no complexes containing KIS have been purified, leaving the possibility of trxG protein recruit-

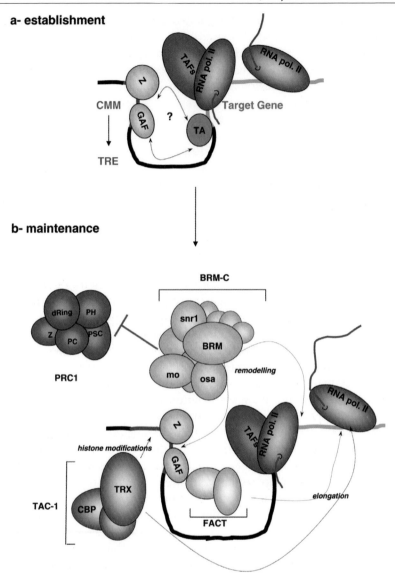

Fig. 2a, b. trxG-dependent memory of the active state. **a** This figure represents the same CMM and target gene as in Fig. 1, but in a population of cells in which the gene is activated by an early transcriptional activator (*TA*). This active status is sensed by the CMM, which may favour recruitment of DNA binding factors such as Zeste and GAF. Therefore, the CMM switches towards a TRE. **b** Maintenance of the active state is achieved through the action of distinct trxG complexes, which act at multiple levels to maintain an active transcription. TAC1 acts by modifying histone tails through acetylation (mediated by dCBP) and possibly by methylation (mediated by TRX). TAC1 also stimulates the elongation of transcription. GAF and FACT association also favours target gene expression by stimulating transcriptional elongation. The BRM complex (*BRM-C*) may be recruited by Zeste and acetylated histone tails. BRM-C may act at two levels: it may stimulate transcription by locally remodelling chromatin and at the same time it may prevent PRC1 reassociation to the CMM

ment by histone methylation open. If KIS is not involved, trxG proteins might be anchored to methylated lysine residues via other protein domains (Santos-Rosa et al. 2003), or via trxG members yet to be discovered.

SWI/SNF-dependent remodelling is an important function encoded in TREs. The molecular action of this complex is primarily to mobilize nucleosomes. This may be required to displace PcG complexes at PREs and to allow access to the transcription machinery at promoters. In yeast, SWI/SNF was shown to be dispensable for gene expression in the majority of cases. However, if these genes are located in a 'closed' chromatin conformation, SWI/SNF becomes absolutely necessary for gene expression. In *Drosophila*, BRM has been shown to act as a member of the trxG because it may be required to counteract silencing effects due to PRE proximity. It is interesting to note that, in contrast to the TRX protein, BRM and PC binding patterns on polytene chromosomes are mutually exclusive (Armstrong et al. 2002). This may reflect a direct competition between these two activities in vivo (Fig. 2), as was suggested to be the case in vitro (Shao et al. 1999).

ATP-independent chromatin remodelling mediated by the GAF proteins may cooperate with the above activities to open chromatin. The recent identification of FACT as a partner of GAF suggests that maintenance of an active transcription may involve facilitating the elongation step of transcription. Moreover, addition of FACT facilitated GAF-mediated chromatin remodelling in vitro. Since this activity does not require ATP, it seems that different steps may be taken to increase chromatin accessibility at trxG target genes.

The current view of trxG protein action is the maintenance of an active state of transcription established by an early acting activator (Fig. 2). However, recent data suggest that TRX protein might also be required to directly modulate/attenuate PcG-mediated silencing (Poux et al. 2002). Thus, multiple trxG-dependent chromatin activities may be used differently at specific target genes in order to achieve gene-specific regulation.

In addition to chromatin modification dependent on the interplay between DNA binding proteins and chromatin complexes, non-coding RNAs may also have important regulatory inputs in trxG pathways. In particular, as well as *Fab-7* other CMM were identified (Maurange and Paro 2002; Rank et al. 2002). In the case of *Fab-7*, chromatin inheritance of activated states was correlated with the appearance of a non-coding *Fab-7* transcript. Other non-coding RNAs were described all along the *Drosophila* BX-C (Bae et al. 2002; Rank et al. 2002). A striking observation is that such RNAs are expressed co-linearly to homeotic genes, and their expression may occur before homeotic gene expression starts (Bae et al. 2002). Thus, it was proposed that tissue-specific non-coding transcription in the BX-C may be able to prevent PREs from silencing while stimulating TRE function. Indeed, when transcription is forced through a PRE, it perturbs PcG silencing (Bender and Fitzgerald 2002; Hogga and Karch 2002). Thus, a model was suggested in which transcription through PREs would leave an imprint, preventing further PcG protein nucleation (Drewell et al. 2002; Rank et al. 2002;

Orlando 2003). Whether such transcription is absolutely required to inactivate PREs outside their domains of action, or whether non-coding RNA are just a by-product of activation of homeotic enhancers (i.e. cryptic transcription may start while the enhancers begin to search for their cognate promoters) remains to be elucidated.

4
Modes of Inheritance

PcG and trxG proteins constitute a memory system that is able to faithfully reproduce patterns of target gene expression through many cell divisions. Although recent studies have significantly advanced our knowledge on the molecular function and biochemical features of individual members or protein complexes of the PcG and the trxG, the molecular mechanism of maintenance of chromatin structure through cell division is still unknown. An obvious hypothesis was that PcG or trxG proteins might remain bound to their targets during S phase and mitosis. However, the current evidence does not support this hypothesis, at least concerning mitosis. Only a small fraction of PcG proteins, if at all, might be retained on mitotic chromatin (Buchenau et al. 1998; Dietzel et al. 1999). Thus, alternative mechanisms were proposed (Francis and Kingston 2001): (1) a component, yet to be identified, may stay on target chromatin during mitosis and specify re-assembly of an activating or a repressing complex in daughter cells; or (2) covalent histone modifications and in particular histone methylation could ensure inheritance. This might be particularly attractive if, once specific patterns of histone modification are established on chromatin, they could be maintained upon DNA replication and mitosis. Semi-conservative segregation followed by restoration of original patterns is known to function in the case of DNA methylation, but this modification is unlikely to be crucial for maintenance of PcG-mediated silencing in *Drosophila* (Lyko et al. 1999; Kunert et al. 2003). On the other hand, recent work suggests that histone dimers in nucleosomes might also be replicated in a semi-conservative manner during DNA replication (Tagami et al. 2004). If this was true, and if ways of restoring chromatin modifications by maintenance activities similar to maintenance DNA methyl transferases were identified, this may be a key to explain memory of chromatin states.

In this context, it is also important to note that re-establishment of specific chromatin patterns may be facilitated by the regulation of the timing of replication of PcG target loci. There is good evidence that chromatin replication during the late S-phase is regulated in specific manners that may facilitate the memory process (Collins et al. 2002). It has been recently shown that PRE-regulated loci were late replicating (Zhimulev and Belyaeva 2003). Thus, a late timing of DNA replication, semi-conservative segregation of nucleosomal modifications, and re-establishment of the marks after the DNA replication process may provide a framework for maintenance of memory of chromatin states dependent on PcG and trxG proteins during DNA replication. For instance, one might envisage that PcG proteins may compact their target chromatin and make it late-replicating during the S-phase. Upon replication of DNA, some PcG complexes might be displaced from chromatin. In theory this would give a window of opportunity for transcription factors to bind chromatin and counteract silencing. However, if PcG target chromatin is replicated late, these factors might be already titrated away from the already replicated chromatin, lowering the probability of competing with PcG factors (Wolffe 1994). Moreover, if histone modifications can segregate semi-conservatively and be faithfully re-established by maintenance machineries, PcG proteins may be re-attracted to the replicated target chromatin via its recruiting marks soon after the passage of the replication fork. Conversely, trxG proteins may shift the timing of DNA replication of their target loci in addition to deposing specific chromatin marks. Early replication during the S-phase and copying the appropriate trxG type marks would attract trxG proteins back to their target while making it unlikely for PcG proteins to be able to compete with them.

If memory is based on transmission of histone marks, maintenance of determined states through mitosis could be explained even if most PcG and trxG proteins are lost, since the state of histones at their target elements is likely to be conserved. Therefore, PcG and trxG proteins could bind to their appropriately marked targets at the onset of the next G1 phase.

5
Concluding Remarks

The identification and analysis of the mechanisms leading to inheritance of chromatin states through the two 'bottlenecks' imposed by DNA replication and mitosis is certainly one of the most fascinating areas of PcG and trxG science that remains to be explored. This field will certainly profit from advances in the understanding of the tethering of PcG/trxG proteins as well as of the action of these proteins at chromatin templates. However, understanding molecular memory might require novel approaches and techniques. One poorly understood issue is the involvement of non-coding RNA components in addition to proteins and DNA sequences. These components clearly play a crucial role in setting up heterochromatin, and they may be central in PcG-mediated silencing as well. Another crucial area of investigation is the comparison of PcG- and trxG-mediated gene regulation between *Drosophila* and vertebrates. For instance, no sequences similar to *Drosophila* PRE/TRE could be identified in vertebrates to date, although general mechanisms of homeotic gene regulation are conserved. While there is a mammalian PHO counterpart (YY-1), there are no identified mammalian homologues for GAF or Zeste. In the only documented cases of PcG target elements in mammals, PcG recruitment is dependent on sequence-specific binding factors that are not PcG or trxG members (Dahiya et al. 2001; Ogawa et al. 2002). Thus, if mammalian PREs/TREs exist, it is not clear whether the sequence determinants and DNA binding proteins will turn out to be similar to *Drosophila*.

Moreover, some basic molecular rules suggested by data obtained in *Drosophila* may not apply to mammals. For instance, mammalian X chromosome inactivation is mediated by the action of EED/E(Z)2 [the mammalian counterpart of the ESC/E(Z) complex], leading to trimethylation of lysine 27 of histone H3. However, this methylation is not followed by recruitment of PcG proteins. Therefore, it will be important to study how the PcG/trxG gene regulatory machinery has evolved in order to cope with the increasingly complex gene regulatory pathways that have emerged during the evolution of eukaryotes.

Acknowledgements. We wish to thank Florence Maschat, Olivier Cuvier and the Cavalli laboratory members for critical reading of the manuscript. We thank Julia Roche for editing the manuscript. J.D. was supported by a grant from the Ministère de l'Enseignement Supérieur et de la Recherche and by a grant from La Ligue Nationale Contre le Cancer. G.C. was supported by grants from the Centre National de la Recherche Scientifique, the Human Frontier Science Program Organization, the Fondation pour la Recherche Médicale and the Association pour la Recherche sur le Cancer'.

References

Akhtar A, Zink D, Becker PB (2000) Chromodomains are protein-RNA interaction modules. Nature 407:405–409

Americo J, Whiteley M, Brown JL, Fujioka M, Jaynes JB, Kassis JA (2002) A complex array of DNA-binding proteins required for pairing-sensitive silencing by a polycomb group response element from the *Drosophila* engrailed gene. Genetics 160:1561–1571

Armstrong JA, Papoulas O, Daubresse G, Sperling AS, Lis JT, Scott MP, Tamkun JW (2002) The *Drosophila* BRM complex facilitates global transcription by RNA polymerase II. EMBO J 21:5245–5254

Atchison L, Ghias A, Wilkinson F, Bonini N, Atchison ML (2003) Transcription factor YY1 functions as a PcG protein in vivo. EMBO J 22:1347–1358

Bae E, Calhoun VC, Levine M, Lewis EB, Drewell RA (2002) Characterization of the intergenic RNA profile at abdominal-A and abdominal-B in the *Drosophila* bithorax complex. Proc Natl Acad Sci USA 99:16847–16852

Bantignies F, Goodman RH, Smolik SM (2000) Functional interaction between the coactivator *Drosophila* CREB-binding protein and ASH1, a member of the trithorax group of chromatin modifiers. Mol Cell Biol 20:9317–9330

Bantignies F, Grimaud C, Lavrov S, Gabut M, Cavalli G (2003) Inheritance of Polycomb-dependent chromosomal interactions in *Drosophila*. Genes Dev 17:2406–2420

Becker PB, Horz W (2002) ATP-dependent nucleosome remodeling. Annu Rev Biochem 71:247–273

Beisel C, Imhof A, Greene J, Kremmer E, Sauer F (2002) Histone methylation by the *Drosophila* epigenetic transcriptional regulator Ash1. Nature 419:857–862

Bender W, Fitzgerald DP (2002) Transcription activates repressed domains in the *Drosophila* bithorax complex. Development 129:4923–4930

Bernstein BE, Humphrey EL, Erlich RL, Schneider R, Bouman P, Liu JS, Kouzarides T, Schreiber SL (2002) Methylation of histone H3 Lys 4 in coding regions of active genes. Proc Natl Acad Sci USA 99:8695–8700

Biggin MD, Bickel S, Benson M, Pirrotta V, Tjian R (1988) Zeste encodes a sequence-specific transcription factor that activates the Ultrabithorax promoter in vitro. Cell 53:713–722

Boivin A, Dura JM (1998) In vivo chromatin accessibility correlates with gene silencing in Drosophila. Genetics 150:1539–1549

Breiling A, Turner BM, Bianchi ME, Orlando V (2001) General transcription factors bind promoters repressed by Polycomb group proteins. Nature 412:651–655

Brown JL, Mucci D, Whiteley M, Dirksen M-L, Kassis JA (1998) The Drosophila Polycomb group gene pleiohomeotic encodes a DNA binding protein with homology to the transcription factor YY1. Mol Cell 1:1057–1064

Brown JL, Fritsch C, Mueller J, Kassis JA (2003) The Drosophila pho-like gene encodes a YY1-related DNA binding protein that is redundant with pleiohomeotic in homeotic gene silencing. Development 130:285–294

Brunk BP, Martin EC, Adler PN (1991) Drosophila genes Posterior sex combs and Suppressor two of zeste encode proteins with homology to the murine bmi-1 oncogene. Nature 353:351–353

Buchenau P, Hodgson J, Strutt H, Arndt-Jovin DJ (1998) The distribution of Polycomb-group proteins during cell division and development in Drosophila embryos: impact on models for silencing. J Cell Biol 141:469–481

Bulger M, Groudine M (1999) Looping versus linking: toward a model for long-distance gene activation. Genes Dev 13:2465–2477

Busturia A, Wightman CD, Sakonju S (1997) A silencer is required for maintenance of transcriptional repression throughout Drosophila development. Development 124:4343–4350

Busturia A, Lloyd A, Bejarano F, Zavortink M, Xin H, Sakonju S (2001) The MCP silencer of the Drosophila Abd-B gene requires both pleiohomeotic and GAGA factor for the maintenance of repression. Development 128:2163–2173

Byrd KN, Shearn A (2003) ASH1, a Drosophila trithorax group protein, is required for methylation of lysine 4 residues on histone H3. Proc Natl Acad Sci USA 100:11535–11540

Cao R, Wang L, Wang H, Xia L, Erdjument-Bromage H, Tempst P, Jones RS, Zhang Y (2002) Role of histone H3 lysine 27 methylation in Polycomb-group silencing. Science 298:1039–1043

Cavalli G, Paro R (1998) The Drosophila Fab-7 chromosomal element conveys epigenetic inheritance during mitosis and meiosis. Cell 93:505–518

Cavalli G, Paro R (1999) Epigenetic inheritance of active chromatin after removal of the main transactivator. Science 286:955–958

Cavalli G, Orlando V, Paro R (1999) Mapping DNA target sites of chromatin-associated proteins by formaldehyde cross-linking in Drosophila embryos. In: Bickmore WA (ed) Chromosome structural analysis: a practical approach. Oxford University Press, Oxford, pp 20–37

Chan CS, Rastelli L, Pirrotta V (1994) A Polycomb response element in the Ubx gene that determines an epigenetically inherited state of repression. EMBO J 13:2553–2564

Chinwalla V, Jane EP, Harte PJ (1995) The Drosophila Trithorax protein binds to specific chromosomal sites and is co-localized with Polycomb at many sites. EMBO J 14:2056–2065

Collins N, Poot RA, Kukimoto I, Garcia-Jimenez C, Dellaire G, Varga-Weisz PD (2002) An ACF1-ISWI chromatin-remodeling complex is required for DNA replication through heterochromatin. Nat Genet 32:627–632

Czermin B, Melfi R, McCabe D, Seitz V, Imhof A, Pirrotta V (2002) *Drosophila* enhancer of Zeste/ESC complexes have a histone H3 methyltransferase activity that marks chromosomal Polycomb sites. Cell 111:185–196

Dahiya A, Wong S, Gonzalo S, Gavin M, Dean DC (2001) Linking the Rb and polycomb pathways. Mol Cell 8:557–569

Decoville M, Giacomello E, Leng M, Locker D (2001) DSP1, an HMG-like protein, is involved in the regulation of homeotic genes. Genetics 157:237–244

Déjardin J, Cavalli G (2004) Chromatin inheritance upon Zeste mediated Brahma recruitment at a minimal cellular memory module. EMBO J 23:857–868

Dekker J, Rippe K, Dekker M, Kleckner N (2002) Capturing chromosome conformation. Science 295:1306–1311

Dellino GI, Tatout C, Pirrotta V (2002) Extensive conservation of sequences and chromatin structures in the bxd polycomb response element among Drosophilid species. Int J Dev Biol 46:133–141

Dietrich BH, Moore J, Kyba M, dosSantos G, McCloskey F, Milne TA, Brock HW, Krause HM (2001) Tantalus, a novel ASX-interacting protein with tissue-specific functions. Dev Biol 234:441–453

Dietzel S, Niemann H, Bruckner B, Maurange C, Paro R (1999) The nuclear distribution of Polycomb during *Drosophila melanogaster* development shown with a GFP fusion protein. Chromosoma 108:83–94

Drewell RA, Bae E, Burr J, Lewis EB (2002) Transcription defines the embryonic domains of cis-regulatory activity at the *Drosophila* bithorax complex. Proc Natl Acad Sci USA 99:16853–16858

Farkas G, Gausz J, Galloni M, Reuter G, Gyurkovics H, Karch F (1994) The trithorax-like gene encodes the *Drosophila* GAGA factor. Nature 371:806–808

Fauvarque M-O, Dura J-M (1993) *polyhomeotic* regulatory sequences induce developmental regulator-dependent variegation and targeted P-element insertions in *Drosophila*. Genes Dev 7:1508–1520

Fauvarque M-O, Zuber V, Dura J-M (1995) Regulation of *polyhomeotic* transcription may involve local changes in chromatin activity in *Drosophila*. Mech Dev 52:343–355

Fischle W, Wang Y, Jacobs SA, Kim Y, Allis CD, Khorasanizadeh S (2003) Molecular basis for the discrimination of repressive methyl-lysine marks in histone H3 by Polycomb and HP1 chromodomains. Genes Dev 17:1870–1881

Fitzgerald DP, Bender W (2001) Polycomb group repression reduces DNA accessibility. Mol Cell Biol 21:6585–6597

Francis NJ, Kingston RE (2001) Mechanisms of transcriptional memory. Nat Rev Mol Cell Biol 2:409–421

Francis NJ, Saurin AJ, Shao Z, Kingston RE (2001) Reconstitution of a functional core polycomb repressive complex. Mol Cell 8:545–556

Franke A, Decamillis M, Zink D, Cheng NS, Brock HW, Paro R (1992) Polycomb and polyhomeotic are constituents of a multimeric protein complex in chromatin of *Drosophila melanogaster*. EMBO J 11:2941–2950

Fritsch C, Beuchle D, Muller J (2003) Molecular and genetic analysis of the Polycomb group gene Sex combs extra/Ring in *Drosophila*. Mech Dev 120:949–954

Fritsch C, Brown JL, Kassis JA, Muller J (1999) The DNA-binding polycomb group protein pleiohomeotic mediates silencing of a *Drosophila* homeotic gene. Development 126:3905–3913

Furuyama T, Tie F, Harte PJ (2003) Polycomb group proteins ESC and E(Z) are present in multiple distinct complexes that undergo dynamic changes during development. Genesis 35:114–124

Galloni M, Gyurkovics H, Schedl P, Karch F (1993) The bluetail transposon: evidence for independent *cis*-regulatory domains and domain boundaries in the bithorax complex. EMBO J 12:1087–1097

Gerasimova TI, Corces VG (1998) Polycomb and trithorax group proteins mediate the function of a chromatin insulator. Cell 92:511–521

Gildea JJ, Lopez R, Shearn A (2000) A screen for new trithorax group genes identified little imaginal discs, the *Drosophila melanogaster* homologue of human retinoblastoma binding protein 2. Genetics 156:645–663

Hagstrom K, Muller M, Schedl P (1997) A *Polycomb* and GAGA dependent silencer adjoins the *Fab-7* boundary in the *Drosophila* bithorax complex. Genetics 146:1365–1380

Hassan AH, Prochasson P, Neely KE, Galasinski SC, Chandy M, Carrozza MJ, Workman JL (2002) Function and selectivity of bromodomains in anchoring chromatin-modifying complexes to promoter nucleosomes. Cell 111:369–379

Hiraoka Y, Dernburg AF, Parmelee SJ, Rykowski MC, Agard DA, Sedat JW (1993) The onset of homologous chromosome pairing during *Drosophila melanogaster* embryogenesis. J Cell Biol 120:591–600

Hodgson JW, Argiropoulos B, Brock HW (2001) Site-specific recognition of a 70-base-pair element containing d(GA)(n) repeats mediates bithoraxoid polycomb group response element-dependent silencing. Mol Cell Biol 21:4528–4543

Hogga I, Karch F (2002) Transcription through the iab-7 cis-regulatory domain of the bithorax complex interferes with maintenance of Polycomb-mediated silencing. Development 129:4915–4922

Horard B, Tatout C, Poux S, Pirrotta V (2000) Structure of a polycomb response element and in vitro binding of polycomb group complexes containing GAGA factor. Mol Cell Biol 20:3187–3197

Huang DH, Chang YL, Yang CC, Pan IC, King B (2002) Pipsqueak encodes a factor essential for sequence-specific targeting of a polycomb group protein complex. Mol Cell Biol 22:6261–6271

Hur MW, Laney JD, Jeon SH, Ali J, Biggin MD (2002) Zeste maintains repression of Ubx transgenes: support for a new model of Polycomb repression. Development 129:1339–1343

Ingham PW (1983) Differential expression of bithorax complex genes in absence of the *extra sex combs* and *trithorax* genes. Nature 306:591–593

Ingham PW, Martinez Arias A (1992) Boundaries and fields in early embryos. Cell 68:221–235

Jürgens G (1985) A group of genes controlling the spatial expression of the bithorax complex in *Drosophila*. Nature 316:153–155

Kal AJ, Mahmoudi T, Zak NB, Verrijzer CP (2000) The *Drosophila* brahma complex is an essential coactivator for the trithorax group protein zeste. Genes Dev 14:1058–1071

Karch F, Galloni M, Sipos L, Gausz J, Gyurkovics H, Schedl P (1994) *Mcp* and *Fab-7*: molecular analysis of putative boundaries of *cis*-regulatory domains in the bithorax complex of *Drosophila melanogaster*. Nucleic Acids Res 22:3138–3146

Kassis JA (1994) Unusual properties of regulatory DNA from the *Drosophila engrailed* gene: three "pairing-sensitive" sites within a 1.6-kb region. Genetics 136:1025–1038

Kassis JA (2002) Pairing-sensitive silencing, polycomb group response elements, and transposon homing in *Drosophila*. Adv Genet 46:421–438

Kehle J, Beuchle D, Treuheit S, Christen B, Kennison JA, Bienz M, Müller J (1998) dMi-2, a Hunchback-interacting protein that functions in *Polycomb* repression. Science 282:1897–1900

Kunert N, Marhold J, Stanke J, Stach D, Lyko F (2003) A Dnmt2-like protein mediates DNA methylation in *Drosophila*. Development 130:5083–5090

Kuzmichev A, Nishioka K, Erdjument-Bromage H, Tempst P, Reinberg D (2002) Histone methyltransferase activity associated with a human multiprotein complex containing the Enhancer of Zeste protein. Genes Dev 16:2893–2905

Kyba M, Brock HW (1998) The *Drosophila* polycomb group protein Psc contacts ph and Pc through specific conserved domains. Mol Cell Biol 18:2712–2720

Lachner M, O'Carroll D, Rea S, Mechtler K, Jenuwein T (2001) Methylation of histone H3 lysine 9 creates a binding site for HP1 proteins. Nature 410:116–120

LaJeunesse D, Shearn A (1995) Trans-regulation of thoracic homeotic selector genes of the Antennapedia and bithorax complexes by the trithorax group genes: *absent, small*, and *homeotic discs 1* and *2*. Mech Dev 53:123–139

Laney JD, Biggin MD (1992) *zeste*, a nonessential gene, potently activates Ultrabithorax transcription in the *Drosophila* embryo. Genes Dev 6:1531–1541

Leibovitch BA, Lu Q, Benjamin LR, Liu Y, Gilmour DS, Elgin SC (2002) GAGA factor and the TFIID complex collaborate in generating an open chromatin structure at the *Drosophila melanogaster* hsp26 promoter. Mol Cell Biol 22:6148–6157

Lonie A, Dandrea R, Paro R, Saint R (1994) Molecular characterisation of the Polycomblike gene of *Drosophila melanogaster*, a trans-acting negative regulator of homeotic gene expression. Development 120:2629–2636

Lu Q, Wallrath LL, Granok H, Elgin SCR (1993) (CT)n.(GA)n repeats and heat shock elements have distinct roles in chromatin structure and transcriptional activation of the *Drosophila*-hsp26 gene. Mol Cell Biol 13:2802–2814

Lupo R, Breiling A, Bianchi ME, Orlando V (2001) *Drosophila* chromosome condensation proteins Topoisomerase II and Barren colocalize with Polycomb and maintain Fab-7 PRE silencing. Mol Cell 7:127–136

Lyko F, Ramsahoye BH, Kashevsky H, Tudor M, Mastrangelo MA, Orr-Weaver TL, Jaenisch R (1999) Mammalian (cytosine-5) methyltransferases cause genomic DNA methylation and lethality in *Drosophila*. Nat Genet 23:363–366

Mahmoudi T, Zuijderduijn LM, Mohd-Sarip A, Verrijzer CP (2003) GAGA facilitates binding of pleiohomeotic to a chromatinized polycomb response element. Nucleic Acids Res 31:4147–4156

Maurange C, Paro R (2002) A cellular memory module conveys epigenetic inheritance of hedgehog expression during *Drosophila* wing imaginal disc development. Genes Dev 16:2672–2683

McCall K, Bender W (1996) Probes for chromatin accessibility in the *Drosophila* bithorax complex respond differently to *Polycomb*-mediated repression. EMBO J 15:569–580

Messmer S, Franke A, Paro R (1992) Analysis of the functional role of the Polycomb chromo domain in *Drosophila melanogaster*. Genes Dev 6:1241–1254

Mihaly J, Mishra RK, Karch F (1998) A conserved sequence motif in *Polycomb*-response elements. Mol Cell 1:1065–1066

Min J, Zhang Y, Xu RM (2003) Structural basis for specific binding of Polycomb chromodomain to histone H3 methylated at Lys 27. Genes Dev 17:1823–1828

Mishra RK, Mihaly J, Barges S, Spierer A, Karch F, Hagstrom K, Schweinsberg SE, Schedl P (2001) The iab-7 polycomb response element maps to a nucleosome-free region of chromatin and requires both GAGA and pleiohomeotic for silencing activity. Mol Cell Biol 21:1311–1318

Mohd-Sarip A, Venturini F, Chalkley GE, Verrijzer CP (2002) Pleiohomeotic can link polycomb to DNA and mediate transcriptional repression. Mol Cell Biol 22:7473–7483

Mollaaghababa R, Sipos L, Tiong SY, Papoulas O, Armstrong JA, Tamkun JW, Bender W (2001) Mutations in *Drosophila* heat shock cognate 4 are enhancers of Polycomb. Proc Natl Acad Sci USA 98:3958–3963

Mulholland NM, King IF, Kingston RE (2003) Regulation of Polycomb group complexes by the sequence-specific DNA binding proteins Zeste and GAGA. Genes Dev 17:2741–2746

Muller J, Hart CM, Francis NJ, Vargas ML, Sengupta A, Wild B, Miller EL, O'Connor MB, Kingston RE, Simon JA (2002) Histone methyltransferase activity of a *Drosophila* Polycomb group repressor complex. Cell 111:197–208

Müller J (1995) Transcriptional silencing by the Polycomb protein in *Drosophila* embryos. EMBO J 14:1209–1220

Narlikar GJ, Fan HY, Kingston RE (2002) Cooperation between complexes that regulate chromatin structure and transcription. Cell 108:475–487

Ng J, Hart CM, Morgan K, Simon JA (2000) A *Drosophila* ESC-E(Z) protein complex is distinct from other polycomb group complexes and contains covalently modified ESC. Mol Cell Biol 20:3069–3078

Nightingale KP, Wellinger RE, Sogo JM, Becker PB (1998) Histone acetylation facilitates RNA polymerase II transcription of the *Drosophila* hsp26 gene in chromatin. EMBO J 17:2865–2876

O'Brien T, Wilkins RC, Giardina C, Lis JT (1995) Distribution of GAGA protein on *Drosophila* genes in vivo. Genes Dev 9:1098–1110

O'Connell S, Wang L, Robert S, Jones CA, Saint R, Jones RS (2001) Polycomblike PHD fingers mediate conserved interaction with enhancer of zeste protein. J Biol Chem 276:43065–43073

Ogawa H, Ishiguro K, Gaubatz S, Livingston DM, Nakatani Y (2002) A complex with chromatin modifiers that occupies E2F- and Myc-responsive genes in G0 cells. Science 296:1132–1136

Orlando V (2003) Polycomb, epigenomes, and control of cell identity. Cell 112:599–606

Orlando V, Jane EP, Chinwalla V, Harte PJ, Paro R (1998) Binding of trithorax and Polycomb proteins to the bithorax complex: dynamic changes during early *Drosophila* embryogenesis. EMBO J 17:5141–5150

Orphanides G, LeRoy G, Chang CH, Luse DS, Reinberg D (1998) FACT, a factor that facilitates transcript elongation through nucleosomes. Cell 92:105–116

Otte AP, Kwaks TH (2003) Gene repression by Polycomb group protein complexes: a distinct complex for every occasion? Curr Opin Genet Dev 13:448–454

Pal-Bhadra M, Bhadra U, Birchler JA (2002) RNAi related mechanisms affect both transcriptional and posttranscriptional transgene silencing in *Drosophila*. Mol Cell 9:315–327

Papoulas O, Beek SJ, Moseley SL, McCallum CM, Sarte M, Shearn A, Tamkun JW (1998) The *Drosophila* trithorax group proteins BRM, ASH1 and ASH2 are subunits of distinct protein complexes. Development 125:3955–3966

Peterson AJ, Kyba M, Bornemann D, Morgan K, Brock HW, Simon J (1997) A domain shared by the Polycomb group proteins Scm and ph mediates heterotypic and homotypic interactions. Mol Cell Biol 17:6683–6692

Petruk S, Sedkov Y, Smith S, Tillib S, Kraevski V, Nakamura T, Canaani E, Croce CM, Mazo A (2001) Trithorax and dCBP acting in a complex to maintain expression of a homeotic gene. Science 294:1331–1334

Platero JS, Hartnett T, Eissenberg JC (1995) Functional analysis of the chromodomain of HP1. EMBO J 14:3977–3986

Poux S, Kostic C, Pirrotta V (1996) Hunchback-independent silencing of late *Ubx* enhancers by a *Polycomb* group response element. EMBO J 15:4713–4722

Poux S, McCabe D, Pirrotta V (2001a) Recruitment of components of Polycomb group chromatin complexes in *Drosophila*. Development 128:75–85

Poux S, Melfi R, Pirrotta V (2001b) Establishment of polycomb silencing requires a transient interaction between PC and ESC. Genes Dev 15:2509–2514

Poux S, Horard B, Sigrist CJ, Pirrotta V (2002) The *Drosophila* trithorax protein is a coactivator required to prevent re-establishment of polycomb silencing. Development 129:2483–2493

Rank G, Prestel M, Paro R (2002) Transcription through intergenic chromosomal memory elements of the *Drosophila* bithorax complex correlates with an epigenetic switch. Mol Cell Biol 22:8026–8034

Rea S, Eisenhaber F, O'Carroll D, Strahl BD, Sun ZW, Schmid M, Opravil S, Mechtler K, Ponting CP, Allis CD, Jenuwein T (2000) Regulation of chromatin structure by site-specific histone H3 methyltransferases. Nature 406:593–599

Ringrose L, Rehmsmeier M, Dura JM, Paro R (2003) Genome-wide prediction of Polycomb/trithorax response elements in *Drosophila melanogaster*. Dev Cell 5:759–771

Rozenblatt-Rosen O, Rozovskaia T, Burakov D, Sedkov Y, Tillib S, Blechman J, Nakamura T, Croce CM, Mazo A, Canaani E (1998) The C-terminal SET domains of ALL-1 and TRITHORAX interact with the INI1 and SNR1 proteins, components of the SWI/SNF complex. Proc Natl Acad Sci USA 95:4152–4157

Salvaing J, Lopez A, Boivin A, Deutsch JS, Peronnet F (2003) The *Drosophila* Corto protein interacts with Polycomb-group proteins and the GAGA factor. Nucleic Acid Res 31:2873–2882

Santos-Rosa H, Schneider R, Bernstein BE, Karabetsou N, Morillon A, Weise C, Schreiber SL, Mellor J, Kouzarides T (2003) Methylation of histone H3 K4 mediates association of the Isw1p ATPase with chromatin. Mol Cell 12:1325–1332

Satijn DP, Hamer KM, den Blaauwen J, Otte AP (2001) The polycomb group protein EED interacts with YY1, and both proteins induce neural tissue in *Xenopus* embryos. Mol Cell Biol 21:1360–1369

Saurin AJ, Shao Z, Erdjument-Bromage H, Tempst P, Kingston RE (2001) A *Drosophila* Polycomb group complex includes Zeste and dTAFII proteins. Nature 412:655–660

Schlossherr J, Eggert H, Paro R, Cremer S, Jack RS (1994) Gene inactivation in *Drosophila* mediated by the Polycomb gene product or by position-effect variegation does not involve major changes in the accessibility of the chromatin fibre. Mol Gen Genet 243:453–462

Schneider R, Bannister AJ, Myers FA, Thorne AW, Crane-Robinson C, Kouzarides T (2004) Histone H3 lysine 4 methylation patterns in higher eukaryotic genes. Nat Cell Biol 6:73–77

Shao Z, Raible F, Mollaaghababa R, Guyon JR, Wu CT, Bender W, Kingston RE (1999) Stabilization of chromatin structure by PRC1, a Polycomb complex. Cell 98:37–46

Shimell MJ, Peterson AJ, Burr J, Simon JA, O'Connor MB (2000) Functional analysis of repressor binding sites in the iab-2 regulatory region of the abdominal-A homeotic gene. Dev Biol 218:38–52

Shimojima T, Okada M, Nakayama T, Ueda H, Okawa K, Iwamatsu A, Handa H, Hirose S (2003) *Drosophila* FACT contributes to Hox gene expression through physical and functional interactions with GAGA factor. Genes Dev 17:1605–1616

Simon J, Chiang A, Bender W (1992) Ten different *Polycomb* group genes are required for spatial control of abdA and AbdB homeotic products. Development 114:493–505

Sinclair DA, Milne TA, Hodgson JW, Shellard J, Salinas CA, Kyba M, Randazzo F, Brock HW (1998a) The additional sex combs gene of *Drosophila* encodes a chromatin protein that binds to shared and unique Polycomb group sites on polytene chromosomes. Development 125:1207–1216

Sinclair DA, Clegg NJ, Antonchuk J, Milne TA, Stankunas K, Ruse C, Grigliatti TA, Kassis JA, Brock HW (1998b) Enhancer of Polycomb is a suppressor of position-effect variegation in *Drosophila melanogaster*. Genetics 148:211–220

Smith ST, Petruk S, Sedkov Y, Cho E, Tillib S, Canaani E, Mazo A, (2004) Modulation of heat shock gene expression by the TAC1 chromatin-modifying complex. Nat Cell Biol 6:162–167

Strahl BD, Allis CD (2000) The language of covalent histone modifications. Nature 403:41–45

Struhl G, Akam M (1985) Altered distributions of *Ultrabithorax* transcripts in *extra sex combs* mutant embryos of *Drosophila*. EMBO J 4:3259–3264

Strutt H, Paro R (1997) The *Polycomb* group protein complex of *Drosophila melanogaster* has different compositions at different target genes. Mol Cell Biol 17:6773–6783

Strutt H, Cavalli G, Paro R (1997) Co-localization of *Polycomb* protein and GAGA factor on regulatory elements responsible for the maintenance of homeotic gene expression. EMBO J 16:3621–3632

Tagami H, Ray-Gallet D, Almouzni G, Nakatani Y (2004) Histone H3.1 and H3.3 complexes mediate nucleosome assembly pathways dependent or independent of DNA synthesis. Cell 116:51–61

Tie F, Furuyama T, Prasad-Sinha J, Jane E, Harte PJ (2001) The *Drosophila* Polycomb group proteins ESC and E(Z) are present in a complex containing the histone-binding protein p55 and the histone deacetylase RPD3. Development 128:275–286

Tie F, Prasad-Sinha J, Birve A, Rasmuson-Lestander A, Harte PJ (2003) A 1-megadalton ESC/E(Z) complex from *Drosophila* that contains polycomblike and RPD3. Mol Cell Biol 23:3352–3362

Tillib S, Petruk S, Sedkov Y, Kuzin A, Fujioka M, Goto T, Mazo A (1999) Trithorax- and Polycomb-group response elements within an Ultrabithorax transcription maintenance unit consist of closely situated but separable sequences. Mol Cell Biol 19:5189–5202

Tolhuis B, Palstra RJ, Splinter E, Grosveld F, de Laat W (2002) Looping and interaction between hypersensitive sites in the active beta-globin locus. Mol Cell 10:1453–1465

Travers AA (2003) Priming the nucleosome: a role for HMGB proteins? EMBO Rep 4:131–136

Tripoulas N, LaJeunesse D, Gildea J, Shearn A (1996) The *Drosophila* ash1 gene product, which is localized at specific sites on polytene chromosomes, contains a SET domain and a PHD finger. Genetics 143:913–928

Tsukiyama T, Becker PB, Wu C (1994) ATP-dependent nucleosome disruption at a heatshock promoter mediated by binding of GAGA transcription factor. Nature 367:525–532

Vazquez M, Moore L, Kennison JA (1999) The trithorax group gene osa encodes an ARID-domain protein that genetically interacts with the brahma chromatin-remodeling factor to regulate transcription. Development 126:733–742

Wang YJ, Brock HW (2003) Polyhomeotic stably associates with molecular chaperones Hsc4 and Droj2 in *Drosophila* Kc1 cells. Dev Biol 262:350–360

Wolffe AP (1994) Inheritance of chromatin states. Dev Genet 15:463–470

Yamamoto Y, Girard F, Bello B, Affolter M, Gehring WJ (1997) The cramped gene of *Drosophila* is a member of the Polycomb-group, and interacts with mus209, the gene encoding proliferating cell nuclear antigen. Development 124:3385–3394

Zegerman P, Canas B, Pappin D, Kouzarides T (2002) Histone H3 lysine 4 methylation disrupts binding of nucleosome remodeling and deacetylase (NuRD) repressor complex. J Biol Chem 277:11621–11624

Zhimulev IF, Belyaeva ES (2003) Intercalary heterochromatin and genetic silencing. Bioessays 25:1040–1051

Zink D, Paro R (1995) *Drosophila Polycomb*-group regulated chromatin inhibits the accessibility of a *trans*-activator to its target DNA. EMBO J 14:5660–5671

Zraly CB, Marenda DR, Nanchal R, Cavalli G, Muchardt C, Dingwall AK (2003) SNR1 is an essential subunit in a subset of *Drosophila* brm complexes, targeting specific functions during development. Dev Biol 253:291–308

How to Pack the Genome for a Safe Trip

Cécile Caron, Jérôme Govin, Sophie Rousseaux, Saadi Khochbin

Abstract The transformation of the somatic chromatin into a unique and highly compact structure occurring during the post-meiotic phase of spermatogenesis is one of the most dramatic known processes of chromatin remodeling. Paradoxically, no information is available on the mechanisms controlling this specific reorganization of the haploid cell genome. The only existing hints suggest a role for histone variants, as well as for stage-specific post-translational histone modifications, before and during the incorporation of testis-specific basic nuclear proteins. Moreover, the exact functions of the latter remain obscure. This chapter summarizes the major chromatin-associated events taking place during the post-meiotic differentiation of male haploid cells in mammals and discusses some of the basic issues that remain to be solved to finally understand chromatin remodeling during spermatogenesis.

1
Introduction

The basic unit of chromatin, the nucleosome, is a nucleoprotein octameric complex formed by the association of two copies of each of the four core histones, H2A, H2B, H3 and H4, around which 146 bp of DNA is wrapped in a 1.65-superhelical turn. A fifth histone, H1, is believed to bind to the DNA as it enters and exits the nucleosome, and the approximately DNA 160-bp unit containing the nucleosome and H1 has been termed the chromatosome (Wolffe 1995).

For any process requiring access to DNA, such as transcription or replication, the chromatin has to undergo a very complex and regulated alteration of its structure. Understanding this process, known as remodeling, is today a

C. Caron, J. Govin, S. Rousseaux, S. Khochbin
Laboratoire de Biologie Moléculaire et Cellulaire de la Différenciation – INSERM U309, Equipe "Chromatine et Expression des Gènes", Institut Albert Bonniot, Faculté de Médecine, Domaine de la Merci, 38706 La Tronche Cedex, France, e-mail: khochbin@ujf-grenoble.fr

Progress in Molecular and Subcellular Biology
P. Jeanteur (Ed.)
Epigenetics and Chromatin
© Springer-Verlag Berlin Heidelberg 2005

challenging but crucial step in many areas of biology including cell differentiation. Two major classes of factors have been found to be involved in chromatin remodeling. The first is composed of enzymes capable of modifying chromatin structure in an ATP-dependent fashion (Lusser and Kadonaga 2003). The second group of enzymes are involved in histone post-translational modifications, such as acetylation, methylation, phosphorylation or ubiquitination (Berger 2002). A third mechanism involving histones with different primary sequences (histone variants) has recently emerged as another powerful way to specifically alter chromatin structure and function. Indeed, in many organisms, conventional histones can be replaced by histones bearing divergent sequences. Together with other nucleosome modification pathways, histone variants participate in the functional specialization of nucleosomes and chromatin domains (Malik and Henikoff 2003).

All three known mechanisms of chromatin remodeling seem to actively participate in one of the most dramatic chromatin remodeling ever observed, accompanying the differentiation of male germ cells or spermatogenesis, where, in mammals, precursor germinal cells, spermatogonia, differentiate into spermatozoa (Hess 1999). Spermatogonia are mitotically dividing somatic cells, which eventually enter meiosis and form primary spermatocytes. The latter undergo the preleptotene stage, during which they replicate DNA, and subsequently go through the leptotene, zygotene, pachytene and diplotene stages of the first meiotic division prophase. In pachytene spermatocytes, homologous chromosomes are paired and exchange DNA segments through a process of homologous recombination (or meiotic crossing-overs). This is helped by a number of proteins, which are localized in the sites of recombination along the paired chromosomes, in structures called synaptonemal complexes. Meiotic I division yields secondary spermatocytes, which then rapidly go through meiotic II division, generating haploid round spermatids. During its post-meiotic maturation or spermiogenesis, the spermatid undergoes a dramatic reorganization of its nucleus, which elongates and compacts into a very unique structure in spermatozoa. During this process, the core histones are replaced by small basic testis-specific proteins, the "transition proteins", and then by sperm-specific nucleoproteins, the protamines (Lewis et al. 2003a; Meistrich et al. 2003). This chromatin remodeling, which has been observed in most studied species, is accompanied by a hyperacetylation of the core histones prior to their replacement (Meistrich et al. 1992; Hazzouri et al. 2000). It is also associated with the extinction of gene transcription. Transcription is active in spermatogonia, pachytene spermatocytes and round spermatids. It is thought to stop shortly after, in elongating spermatids (Sassone-Corsi 2002). Spermiogenesis is also preceded by the synthesis and assembly of particular types of nucleosomes bearing variants of the core and linker histones.

The remodeling of the haploid cell nucleus is therefore a unique process where histone variants, histone post-translational modifications and many non-histone factors act in concert to package the genome in a new histone-less structure. Here, we summarize our knowledge of the different steps of this dramatic structural transition and discuss the many basic questions that remain unanswered regarding events controlling this global reorganization of the haploid genome during spermiogenesis.

2
Synthesis of Histone Variants

Histone variants are non-allelic isoforms of major core or linker histones, which can replace them within the nucleosome. In contrast to the conventional histones, which are synthesized and assembled into chromatin during the S phase of the cell cycle, most of the variants are produced throughout the cell cycle. Meiosis is a privileged period when most of the testis-specific histone variants are synthesized and assembled into nucleosomes. However, during spermatogenesis, many non-tissue-specific histone variants are also incorporated into chromatin (Fig. 1).

Here, these variants are described and their potential function discussed.

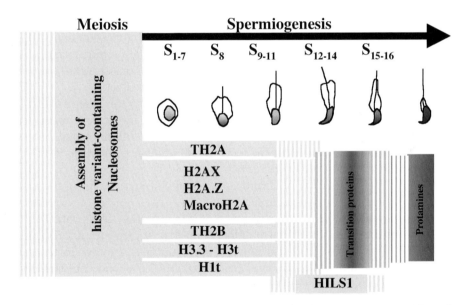

Fig. 1. Histone variants create a spermatogenesis-specific chromatin. Testis-specific and non-testis-specific histone variants are mainly synthesized and assembled in nucleosomes before and during meiosis. An already testis-specific and differentiated chromatin serves therefore as a template for the remodeling processes occurring during spermiogenesis and leading to histone replacement by transition proteins and protamines. *Arrow* shows the major stages of mouse spermiogenesis

2.1
Non-Testis-Specific Core Histone Variants

The most studied core histone variants are the H3 variants H3.3 and CenpA, as well as the H2A variants H2AX, H2AZ and macroH2A. Although their role is not yet fully understood, these variants have been associated in somatic cells with specific chromatin structure and functions (Malik and Henikoff 2003). Most of these variants are also found in male germ cells.

The H2AX variant, which is involved in DNA double-strand break (DSB) surveillance and repair, also operates during meiotic recombination (Lewis et al. 2003a). H2AX disruption in mice induces infertility in the male and an absence of DNA cleavage and alignment in synapsis during zygotene and early pachytene (Celeste et al. 2002).

MacroH2A is a high molecular weight variant of H2A, which contains a large C-terminal non-histone portion and is expressed as two non-allelic variants, macroH2A.1 and macroH2A.2, which contain 80 % identity (Chadwick et al. 2001). A high concentration of histone macroH2A, more specifically of macroH2A1.2, has been found in mice testis (Pehrson et al. 1997; Rasmussen et al. 1999). It has been observed in the nuclei of germ cells, with a localization that is largely, if not exclusively, to the developing XY body in early pachytene spermatocytes (HoyerFender et al. 2000a; Turner et al. 2001). The XY body, or sex vesicle, is a densely stained region of the chromatin corresponding to the pairing of the sex chromosomes during meiosis in male mammals and it is associated with ongoing heterochromatinization and X inactivation. Its precise function during male meiosis has not been unraveled yet.

The centromeric H3 variant CENP-A, which harbors an N-terminal tail completely divergent from that of H3, is also present in the centromere of germ cells. In somatic cells, CENP-A is deposited on newly duplicated sister centromeres and is required for the recruitment of other proteins to the centromere and kinetochore (Smith 2002). Similarly, in germ cells, it could be involved in the segregation of chromosomes and chromatids during the first and second meiotic divisions, respectively.

H3.3, a variant of H3 highly conserved during evolution, is also abundant in male germ cells. H3.3 sequence differs from that of conventional H3 only in four or five positions depending on H3.3 subtypes. Indeed, in several organisms, including mammals and *Drosophila*, H3.3 is encoded by two different genes, H3.3A and H3.3B. H3.3 mRNAs have been detected in human and mouse testis (Albig et al. 1995; Bramlage et al. 1997). On mouse testis sections, H3.3A mRNA was shown to be present in pre- and post-meiotic cells, whereas expression of the H3.3B gene was found to be essentially restricted to cells of the meiotic prophase (Bramlage et al. 1997). The localization of H3.3-containing chromatin has not been determined in mammalian germ cells, but in *Drosophila*, H3.3 is incorporated in chromatin during first meiotic prophase (Akhmanova et al. 1997). It remains concentrated in specific regions (compared to H3, which is evenly distributed) in round and elongating spermatids, and disappears in condensed spermatids like other histones. In somatic cells, H3.3 is known to be incorporated in the absence of DNA replication, and has been shown to replace H3 in transcriptionally active chromatin (Ahmad and Henikoff 2002). It has been proposed that the replacement of H3 by H3.3 in spermatocytes could correspond to the very active transcription that takes place during meiosis (Hennig 2003; Lewis et al. 2003a).

2.2
Testis-Specific Histone Variants

2.2.1
Linker Histones

In mammals, the linker histone family includes seven H1 subtypes: H1.1–H1.5, H1^0 and H1t (Khochbin 2001). The predominant subtype in spermatogonia is the H1.1 isoform, which is produced at high level at early stages of spermatogenesis and then decreases upon further development during mitotic and meiotic cell divisions (Meistrich et al. 1985; Franke et al. 1998). It is noteworthy that mice lacking the H1.1 gene were fertile and showed normal spermatogenesis and testicular morphology (Rabini et al. 2000).

A testis-specific linker histone subtype H1t is also expressed in mammalian spermatogenic cells. This protein, although presenting the tri-partite structure of linker histones, is highly divergent in its primary sequence from the five other members (H1.1–H1.5) of this family. It is also unique in that it exhibits a truly tissue-specific pattern of expression. H1t is indeed found only in the testis, and more precisely from pachytene spermatocytes until round to elongated spermatid stages, where it constitutes up to 55 % of the total linker histones in chromatin (Drabent et al. 1996; Steger et al. 1998). In vitro experimental data suggest that H1t is less tightly associated to oligonucleosomes and has a lower DNA condensing capacity than the other rat H1 subtypes (de Lucia et al. 1994; Khadake and Rao 1995). This property has been proposed to help maintain chromatin in a relatively open state during meiosis, facilitating meiotic events such as recombination (Oko et al. 1996). However, H1t-deficient mice show no specific phenotype and are as fertile as wild-type mice. Several studies propose two opposite explanations for this result. Some studies show that other H1-subtypes, including H1.1, H1.2 and H1.4, fully compensate for the absence of this very specific linker histone (Drabent et al. 2000; Lin et al. 2000), whereas another independent study reports that the other linker histones only partially compensate for H1t in spermatocytes and spermatids (Fantz et al. 2001). In the latter case, H1-deficient chromatin, containing less linker histones, would then be, like H1t-containing chromatin, less tightly compacted, allowing spermatogenesis to proceed.

A new spermatid-specific H1 variant, HILS1 (H1-like protein in spermatids 1), has been recently found in mouse and human (Iguchi et al. 2003; Yan et al. 2003). In contrast to H1t, mHILS1 is exclusively detected in the nuclei of elongating and condensing spermatids, whereas H1t is essentially detected until the round/elongating stages. This expression pattern highly suggests that HILS1 could replace H1t in elongating spermatids and play a role in the chromatin reorganization occurring in these cells.

2.2.2
Core Histones

Testis-specific subtypes have been described as TH2A (Trostle-Weige et al. 1982), TH2B (Hwang and Chae 1989) and TH3 (Trostle-Weige et al. 1984) in the rat. A mouse and a human TH2B were also cloned and were strikingly similar to the rat TH2B (Choi et al. 1996; Zalensky et al. 2002). The main differences in sequence between H2B and TH2B are located in the N-terminal tail of the histone, and to a lesser extent in the globular domain. The C-terminal parts are completely conserved. Interestingly, most of the differences in the N-ter tail (and also globular domain) are conserved between the three species, and could be used in a spermatogenesis-specific signalization process (see Sect. 5.5). TH2A differs from H2A in several residues located in its histone fold domain and in its C-terminal part but presents only two divergent residues in the N-terminal tail.

In the rat, TH2A and TH2B are actively synthesized in early primary spermatocytes (around the preleptotene stage) and their synthesis continues through mid- or late pachytene (Meistrich et al. 1985). Although no further synthesis occurs, TH2B remains the major form of H2B in round and elongating spermatids. In human testis, TH2B immunostaining is first apparent in spermatogonia and reaches an intense signal in round spermatids. During condensation of the spermatid nucleus, the immunodetectability of TH2B disappears gradually, from the anterior region of the nucleus onwards (van Roijen et al. 1998).

No gene or sequence data exist on the TH3 variant, but this protein, purified from rat testis extracts, has an amino acid composition and a mobility on triton/acid/urea gels which differ from all other H3 subtypes (Trostle-Weige et al. 1984). In contrast to the other testis-specific histones, actively synthesized and incorporated in spermatocytes, high amounts of TH3 are present in spermatogonia (Trostle-Weige et al. 1984; Meistrich et al. 1985). TH3 is then maintained in similar or slightly higher amounts in spermatocytes and round spermatids.

A human testis-specific H3 histone gene has also been isolated and characterized (Albig et al. 1996; Witt et al. 1996), but the protein encoded by this gene apparently is not a homologue of rat TH3. The human H3 variant gene encodes for a testis-specific protein, which is nearly identical to human H3 (only four residues are replaced with similar residues), and the corresponding transcripts are found only in pachytene spermatocytes. This human-specific testis variant of H3 was named H3t.

3
Histone Modifications

The post-translational modifications of histones emerge now as crucial elements of a powerful signalization system leading to the establishment of differentiated nucleosomes and chromatin domains. Indeed, the histone code hypothesis proposes that specific histone modifications or combinations of modifications create signals for the docking of specific cellular factors, themselves mediating particular chromatin-related functions (Strahl and Allis 2000).

The histone code is very likely in action during spermatogenesis, since several particular and stage-specific histone modifications have already been reported (Fig. 2). Our current knowledge on these histone modifications is summarized below.

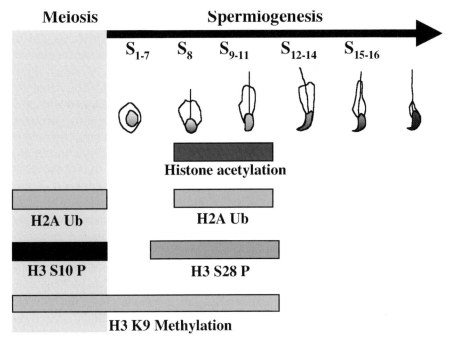

Fig. 2. Chromatin remodeling during spermiogenesis is associated with a specific set of histone modifications. Several stage-specific histone modifications, mainly acetylation, ubiquitination and methylation, have been reported to occur during spermatogenesis in several species. Among these modifications, histone acetylation and ubiquitination are known to be associated with the start of a global chromatin condensation and histone replacement. *Bars* indicate timing of histone modifications. Data showing H3 S10 and S28 phosphorylation as well as H3 K9 methylation are from our unpublished results

3.1
Acetylation

A detailed analysis of the waves of histone acetylation that occur throughout spermatogenesis in mouse (Hazzouri et al. 2000) has shown that spermatogonia and preleptotene spermatocytes contain acetylated core histones H2A, H2B and H4, whereas histones are globally underacetylated during meiosis in leptotene or pachytene spermatocytes. Spermatogonia are cycling cells, which divide through mitosis, and preleptotene spermatocytes also undergo DNA replication, before meiosis. In both these stages, acetylated H4, as well as acetylated H2A and H2B, could have a role in histone deposition during DNA replication (Verreault 2000).

As spermatocytes enter meiosis, a general deacetylation of the core histones occurs and all core histones remain globally underacetylated throughout the long prophase period, including the pachytene stage, as well as during the early stages of post-meiotic maturation in round spermatids. This is somewhat surprising because spermatocytes and round spermatids are known to be actively transcribing cells (Sassone-Corsi 2002). However, a core histones hyperacetylation restricted to regions containing promoters of actively transcribed genes could take place but remain below the detection threshold of the current methods. In agreement with this hypothesis, Moens (1995) has shown that hyperacetylated H4 histones accumulated in euchromatic regions of pachytene spermatocytes.

In mouse spermatids, the spatial distribution pattern of acetylated H4 within the nuclei was examined by immunofluorescence combined with confocal microscopy, showing a spatial sequence of events tightly associated with chromatin condensation (Hazzouri et al. 2000). Indeed, following hypoacetylation of chromatin in round spermatids, a global hyperacetylation of the nucleus is observed in early elongating spermatids. Later, the acetylated chromatin becomes heterogeneously distributed in the nucleus, with areas of condensed acetylated chromatin localized in the center of the nucleus. At the late stages of elongation the acetylated histones progressively disappear from the anterior portion of the nucleus onwards, following a similar pattern to that of nuclear condensation. No histone is then detected in the condensed spermatids. A hyperacetylation of histone H4 was also observed in situ in the rat testis by Meistrich et al. (1992), which appeared in spermatids beginning elongation and showed a maximum in late elongating spermatids. In the latter, an intense acetylated histone H4 staining was observed at the caudal area of the spermatid nucleus. A hyperacetylation of histones in post-meiotic cells was also observed in the testis of various other animal species including trout (Christensen and Dixon 1982; Christensen et al. 1984) and rooster (Oliva and Mezquita 1982).

3.2
Ubiquitination

Among the four histones – H1, H3, H2A and H2B – known to be ubiquitinated in vivo, H2A and H2B have been the most studied (Jason et al. 2002). Five to 15 % of H2A and 1–2 % of H2B are ubiquitinated in higher eukaryotes, on a unique target lysine located in the C-terminal part of each histone.

Histone ubiquitination during spermatogenesis has been investigated in rat and mouse (Chen et al. 1998; Baarends et al. 1999). Two-dimensional gel electrophoresis performed on mouse testis extracts established that the major ubiquitinated histone in mouse testis is H2A (Baarends et al. 1999). Immuno-histochemistry on mouse testis sections showed that uH2A, present in Sertoli cells and spermatogonia, becomes abundant in pachytene spermatocytes. It first (at early pachytene stage) co-localizes with the XY body, and then (at mid-pachytene stage) extends and covers the whole nucleus, before being again concentrated in the XY body, and finally completely disappears (at late pachytene stage). Although the role of H2A ubiquitination is still not clear, uH2A is deubiquitinated in heterochromatin regions and several data argue that it could participate in maintaining an open chromatin conformation (Jason et al. 2002; Moore et al. 2002). An hypothesis is that ubiquitination of H2A in pachytene cells could facilitate the replacement of somatic histones by testis-specific histone variants.

After meiosis, the deubiquitination of H2A occurring in late spermatocytes is maintained until round spermatid stages (Chen et al. 1998; Baarends et al. 1999). In the mouse, it was shown that H2A is then again ubiquitinated in elongating spermatids (Baarends et al. 1999). Although never detected in mouse or in other organisms or cell lines, ubiquitinated H3 was detected in elongating spermatids of rat, which could be involved in post-meiotic chromatin reorganization (Chen et al. 1998).

3.3
Phosphorylation

Histone phosphorylation has also been shown to occur during meiosis. For instance, a transient phosphorylation of H2AX on S139 accompanies the double strand break damage repair, as well as DNA cleavage events such as those associated with meiotic recombination (Mahadevaiah et al. 2001). Phosphorylation of histone H3 on S10 is associated with chromosome condensation in mitosis, and has been observed also during meiosis in several organisms (Prigent and Dimitrov 2003). However, histone H3 and H2AX phosphorylation are not specific features of the male germ cell differentiation, and will therefore not be further developed here.

3.4
Methylation

The isolation of an H3 methyltransferase specifically expressed in the adult
testis suggested a particular role of this histone modification in the male germ
line (O'Carroll et al. 2000). This methyltransferase, Suv39h2, is a homologue of
Suv39h1, an enzyme responsible for methylation of K9 of H3, and associated
with heterochromatin in somatic cells (Sims et al. 2003). Suv39h2 has been
found to be enriched in heterochromatin regions from the leptotene spermato-
cyte to round spermatid stages. A specific antibody against dimethylated
H3-K9 decorates heterochromatin foci in spermatogonia and preleptotene
spermatocytes, stains a larger region in zygotene and early pachytene spermatocytes, before being again concentrated in heterochromatin from late
pachytene to elongating spermatid stages (Peters et al. 2001). A double knock-
out of Suv39h1/Suv39h2 in mice leads to a severe impairment of spermatogenesis at mid- to late pachytene stages, characterized by non-homologous
interactions during chromosome pairing, and delayed synapsis (Peters et al.
2001). H3-K9 methylation and HP1 colocalization on heterochromatin are
completely disrupted in spermatogonia and early spermatocytes of these
mice. In contrast, normal H3-K9 methylation as well as a correct colocalization of the H3-K9 interacting factor HP1 occur during mid- and late meiosis,
as well as in the very rare spermatids which are present in the testes of these
KO mice. These data define an early and stage-specific role for Suv39h
HMTases, and suggest that other HMTases can methylate H3-K9 in mid- and
late meiosis and in spermatids.

4
Transition Proteins

During spermatid differentiation in mammals, transition nuclear proteins
(TPs) constitute 90% of the chromatin basic proteins accumulating in cells
after histone removal and before the deposition of the protamines (Meistrich
et al. 2003). They are thought to be involved in the disruption of the nucleosomal organization or in the deposition of protamines, or both. However, their
precise role remains largely unknown.

Transition proteins are quite variable with regards to size and amino-acid
composition. They are generally more basic than histones and less basic than
protamines. In boar, bull, man, mouse, ram and rat, this family consists of four
proteins, TP1–4, of which TP1 and TP2 are the best characterized. TP1 is a low
molecular size basic protein (54 residues, 6,200 Da), rich in arginine (20%),
lysine (19%) and serine (14%). All the known mammalian TP1s exhibit a
high degree of similarity, particularly in the very basic region 29–42 which
contains a tyrosine which might be important for the protein to destabilize
the chromatin structure (Singh and Rao 1987). TP2 is about the molecular

size of a core histone (13,000 Da) and is characterized by a large amount of basic residues (32 %), serine (22 %) and proline (13 %) and by the presence of cysteine (5 %), arginine (10 %) and lysine (10 %) (Grimes et al. 1975). In contrast to TP1, the TP2 sequence is poorly conserved.

A role of TP in single-strand break (SSB) repairs has been proposed. Indeed, TP1 can stimulate the repair of SSB in vitro. In vivo, it enhances the repair of UV-induced DNA lesions in mammalian cells (Caron et al. 2001). The authors suggest that this major transition protein may contribute to the yet unidentified enzymatic activity responsible for the repair of SSB at mid-spermiogenesis steps.

To investigate the role of TPs in vivo, KO mice for TP1 or TP2 have been generated. However, these mice are fertile and present no major defect in spermatogenesis. Thus, mice lacking TP1 manage to produce relatively normal sperm, although fertility is reduced (60 % of TP1 null mice were infertile) and chromatin condensation is abnormal (Yu et al. 2000). Spermatogenesis in TP2-null mice was also almost normal, with testis weights and epididymal sperm counts being unaffected and a subnormal fertility (the mice were fertile but produced small litters) (Adham et al. 2001; Zhao et al. 2001). Most of the changes in chromatin during spermiogenesis were normal in TP-/- mice, with histones being completely removed and the protamine 2:protamine 1 ratio close to that of the wild-type animals. However, the sperm chromatin of these mutants was less compacted. It was also shown that the processing of protamine 2 was incomplete, in both TP1-/- and TP2-/- mice, with cauda epididymal sperm containing high proportions of intermediate partially processed forms of protamine 2. TP1 or TP2 are thus not critical for histone displacement and initiation of chromatin condensation, but are necessary for maintaining the normal processing of P2 and, consequently, the completion of chromatin condensation. However, these works suggest a likely compensatory effect of TP1 in the TP2 null mice, or TP2 in the TP1 null mice, and more information about the possible role of TPs in chromatin reorganization should be obtained from the double KO mice.

5
Final Components of the Sperm Chromatin

At the end of spermiogenesis the mammalian haploid genome is packaged in a highly compact structure, containing protamines and some remaining histones. A challenging issue is to understand how this structure conveys epigenetic information and how it controls early embryonic events.

5.1
Protamines

Protamines are generally defined as low molecular mass and highly basic proteins associated to nuclear DNA in the spermatozoon. Despite obvious structural similarities and their ability to keep DNA in a highly compact structure, their primary sequence appears less conserved than that of histones (Lewis et al. 2003b).

In most mammals, the sperm nuclei contain only one protamine called P1, an arginine- and cystein-rich polypeptide of about 50 residues, which appears to be well conserved from species to species. The polypeptide chain of protamine P1 can be divided into three different structural domains: (1) the amino-terminal region (residues 1–12) is characterized by an ARYRCC motif (residues 1–6) highly conserved in all known mammalian protamines P1, as well as by the presence of a serine or a threonine at position 8 and a serine at positions 10 and 12; (2) the central region (residues 13–27) is very arginine-rich and is highly conserved from species to species and might be the primary DNA-binding site of the protein (Fita et al. 1983); and (3) the carboxyterminal region (residues 28–50), more variable in sequence, contains most of the bulky hydrophobic amino acid residues present in the protamine P1.

In a few mammals, including man and mouse, a second class of protamines called P2 are also present in sperm nuclei. P2 protamines differ from P1 by a large amount of histidines and a slightly higher size, 54–63 amino acid residues in man and mouse, respectively. In protamine P2, the arginine residues are mostly clustered in the central part of the molecule (residues 30–45) and the histidine and cysteine residues do not appear to be randomly distributed along the polypeptide chain.

Protamines undergo a variety of chemical modifications during nucleoprotamine assembly, including phosphorylation and dephosphorylation (Marushige and Marushige 1975), disulfide bond formation (during mammalian sperm maturation in the epididymis; Calvin and Bedford 1971) and in certain species (like in mouse) proteolytic processing of protamine 2 (Yelick et al. 1987). Protamine modification appears to be necessary for the production of stable, highly condensed mammalian sperm nuclei. Although their function has not been clearly assessed, it has been shown that in mice both protamines are necessary for post-meiotic chromatin condensation.

It was indeed shown that premature translation of P1 mRNA caused precocious condensation of spermatid nuclear DNA, abnormal head morphogenesis and incomplete processing of P2 protein (Lee et al. 1995), and in mice hemizygous for the transgene it caused dominant male sterility, which in some cases was accompanied by a complete arrest in spermatid differentiation. Therefore it was suggested that correct temporal synthesis of P1 is necessary for the transition from nucleohistones to nucleoprotamines.

Using targeted disruptions of one allele of either P1 or P2 in mice, Cho et al. (2001) showed that both proteins are essential for normal sperm function, and that haplo-insufficiency of protamines prevents genetic transmission of both mutant and wild-type alleles. Recently, this effect has been shown to be due to alterations in the organization and integrity of sperm DNA of these mice, arguing for the crucial role of P2 in compaction and protection of DNA from damage (Cho et al. 2003).

5.2
Histones

In many species, a variable proportion of the sperm genome remains associated with histones. For instance, in human spermatozoa, about 15% of the genomic DNA is bound by histones (Gatewood et al. 1990). It has been postulated that these histone-containing regions encompass a specific subset of early embryonic expressed genes (Gatewood et al. 1987). However, it has recently been shown that in human sperm, regions containing the genes of the two protamines and the transition protein TP2 are also enriched in histones (Wykes and Krawetz 2003). The labeling of specific gene-containing regions could not be the only function of the histone-containing genomic domains, since in both human and mouse spermatozoa, LINE/L1 elements (Pittoggi et al. 1999), as well as telomeric sequences (Zalenskaya et al. 2000), were found associated with histones.

Among histone variants, only CENP-A has been reported to survive histone displacement in mammals, but its possible role in late spermiogenesis or in post-fertilization events is not known (Palmer et al. 1990).

6
Mechanisms Controlling Post-Meiotic Chromatin Reorganization: A General Discussion

Despite the accumulation of data concerning the successive events accompanying the transition from the round spermatid somatic-like chromatin to the highly specialized and unique sperm chromatin structure, the molecular mechanisms involved are entirely unknown. Some of the existing data can, however, help to address several basic questions relative to the molecular basis of chromatin remodeling during the spermiogenesis. These questions include:

- Which mechanisms mediate the general transcriptional repression occurring in early elongating spermatids?
- Why and how do some genes escape this early transcriptional repression?
- How do histones become hyperacetylated in elongating spermatids?
- Why and how do histones become modified (acetylated) in elongating spermatids and is there a spermatid-specific histone code?
- What is the role of testis-specific histone variants?

- How are histones replaced, removed and degraded/recycled?
- What is the link between histone displacement and the assembly of new nucleoprotein complexes?

Although these questions cannot be directly answered yet, some recent data allow speculation on possible solutions to these issues.

6.1
Active Transcription Followed by Repression in Round Spermatids

The molecular basis of transcriptional repression in round spermatids has not yet been established. Nevertheless, recent observations suggest that some of the somatic cell-type mechanisms could be involved in this general transcriptional repression. These events include histone deacetylation and histone H3-K9 methylation.

Histone deacetylases (HDACs) seem to be highly active in round spermatids and highly involved in the global histone underacetylation observed in these cells. Indeed, the treatment of round spermatids with HDAC inhibitor trichostatin A (TSA) very efficiently induces chromatin hyperacetylation (Hazzouri et al. 2000). These data are in good agreement with classical models of transcriptional repression, where class I HDACs are recruited on promoters by transcriptional repressors to deacetylate histones. Until recently, no factor had been identified capable of large-scale recruitment of HDACs to the chromatin of spermatids. We have recently shown that a chromodomain protein, CDYL, known to be overexpressed in the testis (Lahn and Page 1999), is a transcriptional corepressor and is very probably involved in histone deacetylation in round spermatids. Indeed, this protein is specifically overexpressed in post-meiotic germ cells, it can repress transcription when targeted to a promoter in somatic cells, and immunoprecipitation assays have shown that it forms a complex with HDAC1 and HDAC2 in spermatids (Caron et al. 2003).

Data in somatic cells suggest that the deacetylation of histone H3-K9 would allow its subsequent methylation by methyltransferases, and the recruitment of chromodomain proteins like HP1, involved in the spreading of heterochromatin-like structures. This model, often proposed for heterochromatinization in somatic cells, could also stand for round spermatids. Indeed, HP1-alpha and -beta are found associated to heterochromatin in round spermatids (HoyerFender et al. 2000b; Martianov et al. 2002). Furthermore, Western blots on extracts of pachytene spermatocytes and round spermatids have revealed an increase in H3-K9 di-methylation in round spermatids (our unpubl. data). The protein Suv39h2, found enriched in the heterochromatin regions of round spermatid (O'Carroll et al. 2000), could be one of the methyltransferases modifying H3K9 in these cells. However, H3K9 remains methylated in the spermatids of Suv39h1/h2 KO mice

(Peters et al. 2001), which means that other, yet unknown, HMTases are also involved.

Interestingly, Baarends and coworkers have also detected a strong deubiquitination of H2A in round spermatids, which was unexpected because such a hypoubiquitination had never been described in other mammalian cell types (except in metaphase somatic cells). Like HDACs, deubiquitinating enzymes should therefore be very active in round spermatids. Global deubiquitination in round spermatids is thus in good agreement with data reporting that uH2A and uH2B are globally associated to transcriptionally active regions (Jason et al. 2002).

Global hypoacetylation and hypoubiquitination could therefore constitute key events in the whole transcriptional repression that takes place as round spermatids differentiate.

6.2
Functional Link Between Histone Acetylation and Chromatin Condensation and Histone Replacement

Post-meiotic histone hyperacetylation is a global event, observed in many species where histones are displaced. This specific histone modification seems to be tightly associated with histone replacement, since it has not been observed in species where somatic histones are completely retained in spermatozoa, such as in winter flounder and carp (Kennedy and Davies 1980, 1981). Moreover, acetylated H4 disappearance, in mice spermatids, follows an antero-caudal pattern, which is similar to that of chromatin condensation, suggesting a tight link between histone replacement and nucleus condensation (Hazzouri et al. 2000).

According to the "histone code hypothesis", histone hyperacetylation would likely serve as a signal for the recruitment of a machinery mediating the histone replacement. In this respect, bromodomain-containing proteins appear to be excellent candidates to mediate and control the events following histone acetylation during spermiogenesis. Indeed, bromodomains are acetyl-lysine binding modules present in many ATP-dependent chromatin remodeling factors as well as in many HATs and some other nuclear proteins (Zeng and Zhou 2002).

Recently, a search for testis-specific bromodomain-containing proteins has led to the identification of a double bromodomain testis-specific protein of unknown function, BRDT (Pivot-Pajot et al. 2003). The murine BRDT gene encodes two RNAs transcripts, which are exclusively present in the testis and, more precisely, in the germinal cells from the early meiotic until late spermatid stages. The functional study of the product of the shortest RNA, mBRDT2, has established that the bromodomains of BRDT indeed bind histone H4 in its hyperacetylated form, and that BRDT is able to induce a dramatic compaction of acetylated chromatin in vivo as well as in vitro (Pivot-Pajot et al. 2003).

Hence, the testis-specific protein BRDT is the first identified factor that can induce a condensation of acetylated chromatin, suggesting that histone acetylation during spermiogenesis could primarily be a signal for chromatin condensation. In support of this hypothesis, our in situ data show the compaction of hyperacetylated chromatin occurring before histone replacement in elongating spermatids (unpubl. results).

The mechanism responsible for the induction of a global histone acetylation during chromatin elongation also remains unknown. However, recent work has shown that this hyperacetylation is associated with the degradation of major cellular HDACs (Caron et al. 2003), which could certainly play an important role in inducing histone hyperacetylation by disrupting the cellular acetylation equilibrium.

6.3
Does Histone Ubiquitination Play a Role in Spermatid-Specific Chromatin Remodeling?

The human HR6, an E2 ligase homologous to yeast RAD6 known to ubiquitinate H2A and H2B in vitro (Sung et al. 1988), is strongly expressed in the testis (Koken et al. 1996). It is specifically abundant in the nucleus of elongating spermatids, at the step of histones/protamines replacement. In mice, the KO of HR6B induces a severe defect in spermiogenesis, with an arrest of cell maturation at the round/elongating spermatids step, confirming an important requirement of HR6 at this stage of spermiogenesis (Roest et al. 1996). However, H2A is probably ubiquitinated by other enzymes since in testis sections from HR6B KO mice, uH2A is normally detected in pachytene spermatocytes, and is also observed in the rare round/elongating spermatids present. These data suggest that the defect in spermiogenesis in HR6B KO mice is due either to an insufficient (but not completely null) level of uH2A ubiquitination or to a defect in ubiquitination of other substrates of HR6B, like, for instance, H2B.

Like acetylation, histone ubiquitination is likely to play a role in histone replacement, but the mechanism involved is not known. Considering that ubiquitin represents 60 % of the size of H2A, this modification has been suggested to affect the global structure of the nucleosome, facilitating its disruption. However, experiments performed on nucleosomes reconstituted with and without uH2A argue against a structural impact of ubiquitinated K119 of H2A on nucleosomal structure and stability or chromatin folding (Moore et al. 2002).

The most probable hypothesis is that histone mono-ubiquitination could serve as a signal for the recruitment of specific factors involved in chromatin reorganization or histone displacement. Indeed, mono-ubiquitination is a protein modification involved in several cell signaling pathways (such as endocytosis, transcription, etc.), which would serve as a tag recognized by the ubiquitin-binding domains such as UIM, ZnF-UBP, UEV or UBA (Schnell and Hicke 2003). The identification of such factors is a future challenge.

6.4
Is There a Spermiogenesis-Specific Histone Code?

Although histone modifications associated with spermiogenesis are not fully characterized, it already appears that they could be very different from those occurring during meiosis. In meiosis, for instance, a global H4 hypoacetylation is associated with ubiquitination of H2A. After meiosis, in contrast, H2A ubiquitination and acetylation are underrepresented in the nuclei of round spermatids, and increase together in elongated spermatids. Moreover, phosphorylation of S10 and S28 of H3, which is normally associated with mitotic chromatin condensation, occurs differently during meiosis and in elongating spermatids. Indeed, in these cells while S10 phosphorylation dramatically decreases, that of S28 remains, and even increases as the differentiation proceeds (our unpubl. data). This confirms that different combinations of modifications are associated with specific chromatin remodeling events, and suggests that there could be a code specific for histone replacement.

An important question to answer would be whether the events controlled by this particular histone code are strictly specific to spermiogenesis. Indeed, several nucleosome modifications observed during spermiogenesis before histone displacement – such as histone acetylation, ubiquitination of H2A, methylation of K4 of H3 (our unpubl. data) and enrichment in the histone variant H3.3 – are also present on transcriptionally active chromatin in somatic cells. Very interestingly, a recent paper has shown that, upon activation of the yeast promoter of the *PHO5* gene, the histones of a few nucleosomes of the promoter were eliminated just after their acetylation (Reinke and Horz 2003). This result suggests that the global histone modifications (at least acetylation) and displacement observed in spermiogenesis could correspond to specific events also occurring locally in somatic cells, for instance on transcriptionally active promoters.

On the other hand, the replacement of histones by TP and protamines is very unique to spermatids. Therefore, if nucleosome modifications play a role in the assembly of these small basic proteins, some combinations of modifications could only take place in male germ cells, and define a spermatid-specific histone code. For instance, dimethylation of K4 and K9 of H3 increases post-meiotically (our unpubl. data). It is so far unclear whether both modifications coexist or not on the same histones or within the same nucleosomes. This question is of great interest, since the association of methylated K4 and K9 in the same chromatin regions could strongly support the existence of a new and spermatid-specific histone code.

6.5
Do Histone Variants Play a Role in Spermatid-Specific Chromatin Remodeling?

Another strong argument for a spermatid-specific histone code is the presence of core histone variants in late spermiogenesis. Most, if not all, of the post-translationally modified residues of N- and C-terminal parts of H3, H2A and H2B are conserved in H3.3, TH2A and TH2B. These residues have therefore the potential to be modified in the same way as in somatic histones. However, the phosphorylable S14 of H2B is replaced by phenylalanine in TH2B. This poorly characterized modification has recently been reported to target H2B in apoptotic chromatin (Cheung et al. 2003). TH2B will therefore escape this particular signalization.

Very interestingly, S19 of H2A is also replaced by a phenylalanine in TH2A. This suggests, first, that H2A S19 could also be a site of phosphorylation, and, second, that the hydrophobic phenylalanine residue could create a special signal in the N-terminal part of TH2A and TH2B for the interaction with specific factors in spermatids. Furthermore, the important variations in the N-terminal tail part of TH2B could modify the recognition of these amino acids by histone-modifying enzymes. For instance, it could accelerate or inhibit the modification of a specific residue.

Finally, sequence divergence observed in the histone fold regions of testis-specific histones compared to the somatic types may have dramatic effects on the nucleosome stability and facilitate histone replacement in elongating spermatids (Fan et al. 2002).

7
Concluding Remarks

Although recent investigations allow speculations on the mechanisms controlling several aspects of the global chromatin remodeling that takes place during spermiogenesis, many questions remain unanswered. It is, however, possible to propose a working model as follows (Fig. 3).

The chromatin of post-meiotic cells has a significantly different composition because of the presence of specific variants of H2A, H2B, H3 and H1. Moreover, some spermatid-specific factors such as CDYL could recruit HDACs to massively deacetylate histones in round spermatids. We speculate that deubiquitinating enzymes would also be recruited in the same way. These events, in association with a massive methylation of H3K9, would lead to a gradual repression of transcription as the round spermatids differentiate. The degradation of the major nuclear HDACs would then induce a sudden and massive histone hyperacetylation, which would in turn constitute a signal for chromatin condensation, through the action of bromodomain-containing factors such as BRDT. Since this chromatin condensation is associated with

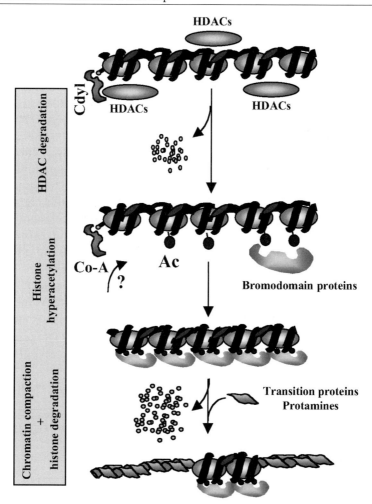

Fig. 3. A working hypothesis for the molecular mechanisms of chromatin condensation and histone replacement during mouse spermiogenesis. Post-meiotic transcriptional repressors such as CDYL would recruit class I HDACs to chromatin in round spermatids to initiate a general histone deacetylation and a progressive transcriptional repression. A stage-specific degradation of class I HDACs would then induce a global hyperacetylation of histones, leading to incorporation of bromodomain-containing proteins such as BRDT, which would in turn induce compaction of the chromatin associated with removal and degradation of the acetylated histones. CDYL has a co-enzyme A binding pocket, which, after degradation of HDACs, might be involved in HAT activity (Lahn et al. 2002) or alternatively participate in a yet unknown enzymatic activity involving Co-A derivatives. (Caron et al. 2003)

histone replacement, these bromodomain-containing factors could also recruit a machinery involved in histone removal/degradation as well as TP assembling.

Understanding the mechanisms controlling histone degradation not only would allow a better understanding of a crucial step of the spermiogenesis, but also may convey new information on an important chromatin remodeling process operating in specific loci in somatic cells (Reinke and Horz 2003).

It is also very important to understand the nature of the epigenetic information contained in the nucleus of spermatozoa. An interesting hint is the presence of genomic islands that survive the dramatic transition in the organization of the genome during spermiogenesis and maintain a somatic-like chromatin structure (see Sect. 5.2). One possibility would be that these chromatin-like structures provide an essential mark for the establishment of adequate epigenetic information in the offspring. In agreement with this hypothesis, the paternally imprinted IGF2 gene was found preferentially associated with histones in the human sperm nucleus (Banerjee and Smallwood 1998), a phenomenon that might be important for the imprinting of this gene.

In summary, considering all the data available today, it appears that understanding of the molecular basis of genome reorganization during spermiogenesis remains one of the major challenges in the years to come.

References

Adham IM, Nayernia K, Burkhardt-Gottges E, Topaloglu O, Dixkens C, Holstein AF, Engel W (2001) Teratozoospermia in mice lacking the transition protein 2 (Tnp2). Mol Hum Reprod 7:513–520

Ahmad K, Henikoff S (2002) The histone variant h3.3 marks active chromatin by replication-independent nucleosome assembly. Mol Cell 9:1191–1200

Akhmanova A, Miedema K, Wang Y, van Bruggen M, Berden JH, Moudrianakis EN, Hennig W (1997) The localization of histone H3.3 in germ line chromatin of Drosophila males as established with a histone H3.3-specific antiserum. Chromosoma 106:335–347

Albig W, Bramlage B, Gruber K, Klobeck HG, Kunz J, Doenecke D (1995) The human replacement histone H3.3B gene (H3F3B). Genomics 30:264–272

Albig W, Ebentheuer J, Klobeck G, Kunz J, Doenecke D (1996) A solitary human H3 histone gene on chromosome 1. Hum Genet 97:486–491

Baarends WM, Hoogerbrugge TW, Roest HP, Ooms M, Vreeburg J, Hoeijmakers JHJ, Grootegoed JA (1999) Histone ubiquitination and chromatin remodeling in mouse spermatogenesis. Dev Biol 207:322–333

Banerjee S, Smallwood A (1998) Chromatin modification of imprinted H19 gene in mammalian spermatozoa. Mol Reprod Dev 50:474–484

Berger SL (2002) Histone modifications in transcriptional regulation. Curr Opin Genet Dev 12:142–148

Bramlage B, Kosciessa U, Doenecke D (1997) Differential expression of the murine histone genes H3.3A and H3.3B. Differentiation 62:13–20

Calvin HI, Bedford JM (1971) Formation of disulphide bonds in the nucleus and accessory structures of mammalian spermatozoa during maturation in the epididymis. J Reprod Fertil Suppl 13:65–75

Caron C, Pivot-Pajot C, van Grunsven LA, Col E, Lestrat C, Rousseaux S, Khochbin S (2003) Cdyl: a new transcriptional co-repressor. EMBO Rep 4:877–882

Caron N, Veilleux S, Boissonneault G (2001) Stimulation of DNA repair by the spermatidal TP1 protein. Mol Reprod Dev 58:437–443

Celeste A, Petersen S, Romanienko PJ, Fernandez-Capetillo O, Chen HT, Sedelnikova OA, Reina-San-Martin B, Coppola V, Meffre E, Difilippantonio MJ, Redon C, Pilch DR, Olaru A, Eckhaus M, Camerini-Otero RD, Tessarollo L, Livak F, Manova K, Bonner WM, Nussenzweig MC, Nussenzweig A (2002) Genomic instability in mice lacking histone H2AX. Science 296:922–927

Chadwick BP, Valley CM, Willard HF (2001) Histone variant macroH2A contains two distinct macrochromatin domains capable of directing macroH2A to the inactive X chromosome. Nucleic Acids Res 29:2699–2705

Chen HY, Sun JM, Zhang Y, Davie JR, Meistrich ML (1998) Ubiquitination of histone H3 in elongating spermatids of rat testes. J Biol Chem 273:13165–13169

Cheung WL, Ajiro K, Samejima K, Kloc M, Cheung P, Mizzen CA, Beeser A, Etkin LD, Chernoff J, Earnshaw WC, Allis CD (2003) Apoptotic phosphorylation of histone H2B is mediated by mammalian sterile twenty kinase. Cell 113:507–517

Cho C, Willis WD, Goulding EH, Jung-Ha H, Choi YC, Hecht NB, Eddy EM (2001) Haploinsufficiency of protamine-1 or -2 causes infertility in mice. Nat Genet 28:82–86

Cho C, Jung-Ha H, Willis WD, Goulding EH, Stein P, Xu Z, Schultz RM, Hecht NB, Eddy EM (2003) Protamine-2 deficiency leads to sperm DNA damage and embryo death in mice. Biol Reprod 69:211–217

Choi YC, Gu W, Hecht NB, Feinberg AP, Chae CB (1996) Molecular cloning of mouse somatic and testis-specific H2B histone genes containing a methylated CpG island. DNA Cell Biol 15:495–504

Christensen ME, Dixon GH (1982) Hyperacetylation of histone H4 correlates with the terminal, transcriptionally inactive stages of spermatogenesis in rainbow trout. Dev Biol 93:404–415

Christensen ME, Rattner JB, Dixon GH (1984) Hyperacetylation of histone H4 promotes chromatin decondensation prior to histone replacement by protamines during spermatogenesis in rainbow trout. Nucleic Acids Res 12:4575–4592

De Lucia F, Faraone-Mennella MR, D'Erme M, Quesada P, Caiafa P, Farina B (1994) Histone-induced condensation of rat testis chromatin: testis-specific H1 t versus somatic H1 variants. Biochem Biophys Res Commun 198:32–39

Drabent B, Bode C, Bramlage B, Doenecke D (1996) Expression of the mouse testicular histone gene H1 t during spermatogenesis. Histochem Cell Biol 106:247–251

Drabent B, Saftig P, Bode C, Doenecke D (2000) Spermatogenesis proceeds normally in mice without linker histone H1 t. Histochem Cell Biol 113:433–442

Fan JY, Gordon F, Luger K, Hansen JC, Tremethick DJ (2002) The essential histone variant H2A.Z regulates the equilibrium between different chromatin conformational states. Nat Struct Biol 19:19

Fantz DA, Hatfield WR, Horvath G, Kistler MK, Kistler WS (2001) Mice with a targeted disruption of the H1 t gene are fertile and undergo normal changes in structural chromosomal proteins during spermiogenesis. Biol Reprod 64:425–431

Fita I, Campos JL, Puigjaner LC, Subirana JA (1983) X-ray diffraction study of DNA complexes with arginine peptides and their relation to nucleoprotamine structure. J Mol Biol 167:157–177

Franke K, Drabent B, Doenecke D (1998) Testicular expression of the mouse histone H1.1 gene. Histochem Cell Biol 109:383–390

Gatewood JM, Cook GR, Balhorn R, Bradbury EM, Schmid CW (1987) Sequence-specific packaging of DNA in human sperm chromatin. Science 236:962–964

Gatewood JM, Cook GR, Balhorn R, Schmid CW, Bradbury EM (1990) Isolation of four core histones from human sperm chromatin representing a minor subset of somatic histones. J Biol Chem 265:20662–20666

Grimes SR Jr, Platz RD, Meistrich ML, Hnilica LS (1975) Partial characterization of a new basic nuclear protein from rat testis elongated spermatids. Biochem Biophys Res Commun 67:182–189

Hazzouri M, Pivot-Pajot C, Faure AK, Usson Y, Pelletier R, Sele B, Khochbin S, Rousseaux S (2000) Regulated hyperacetylation of core histones during mouse spermatogenesis: involvement of histone deacetylases. Eur J Cell Biol 79:950–960

Hennig W (2003) Chromosomal proteins in the spermatogenesis of *Drosophila*. Chromosoma 111:489–494

Hess RA (1999) Spermatogenesis, overview. In: Nobil E, Neill JD (eds) Encyclopedia of reproduction. Academic Press, San Diego

HoyerFender S, Costanzi C, Pehrson JR (2000a) Histone MacroH2A1.2 is concentrated in the XY-body by the early pachytene stage of spermatogenesis. Exp Cell Res 258:254–260

HoyerFender S, Singh PB, Motzkus D (2000b) The murine heterochromatin protein M31 is associated with the chromocenter in round spermatids and is a component of mature spermatozoa. Exp Cell Res 254:72–79

Hwang I, Chae CB (1989) S-phase-specific transcription regulatory elements are present in a replication-independent testis-specific H2B histone gene. Mol Cell Biol 9:1005–1013

Iguchi N, Tanaka H, Yomogida K, Nishimune Y (2003) Isolation and characterization of a novel cDNA encoding a DNA-binding protein (Hils1) specifically expressed in testicular haploid germ cells. Int J Androl 26:354–365

Jason LJ, Moore SC, Lewis JD, Lindsey G, Ausio J (2002) Histone ubiquitination: a tagging tail unfolds? Bioessays 24:166–174

Kennedy BP, Davies PL (1980) Acid-soluble nuclear proteins of the testis during spermatogenesis in the winter flounder. Loss of the high mobility group proteins. J Biol Chem 255:2533–2539

Kennedy BP, Davies PL (1981) Phosphorylation of a group of high molecular weight basic nuclear proteins during spermatogenesis in the winter flounder. J Biol Chem 256:9254–9259

Khadake JR, Rao MR (1995) DNA- and chromatin-condensing properties of rat testes H1a and H1 t compared to those of rat liver H1bdec; H1 t is a poor condenser of chromatin. Biochemistry 34:15792–15801

Khochbin S (2001) Histone H1 diversity: bridging regulatory signals to linker histone function. Gene 271:1–12

Koken MHM, Hoogerbrugge JW, Jaspersdekker I, Dewit J, Willemsen R, Roest HP, Grootegoed JA, Hoeijmakers JHJ (1996) Expression of the ubiquitin-conjugating DNA-repair enzymes hhr6a and hhr6b suggests a role in spermatogenesis and chromatin modification. Dev Biol 173:119–132

Lahn BT, Page DC (1999) Retroposition of autosomal mRNA yielded testis-specific gene family on human Y chromosome. Nat Genet 21:429–433

Lahn BT, Tang ZL, Zhou J, Barndt RJ, Parvinen M, Allis CD, Page DC (2002) Previously uncharacterized histone acetyltransferases implicated in mammalian spermatogenesis. Proc Natl Acad Sci USA 99:8707–8712

Lee K, Haugen HS, Clegg CH, Braun RE (1995) Premature translation of protamine 1 mRNA causes precocious nuclear condensation and arrests spermatid differentiation in mice. Proc Natl Acad Sci USA 92:12451–12455

Lewis JD, Abbott DW, Ausio J (2003a) A haploid affair: core histone transitions during spermatogenesis. Biochem Cell Biol 81:131–140

Lewis JD, Song Y, de Jong ME, Bagha SM, Ausio J (2003b) A walk through vertebrate and invertebrate protamines. Chromosoma 111:473–482

Lin Q, Sirotkin A, Skoultchi AI (2000) Normal spermatogenesis in mice lacking the testis-specific linker histone H1 t. Mol Cell Biol 20:2122–2128

Lusser A, Kadonaga JT (2003) Chromatin remodeling by ATP-dependent molecular machines. Bioessays 25:1192–1200

Mahadevaiah SK, Turner JM, Baudat F, Rogakou EP, de Boer P, Blanco-Rodriguez J, Jasin M, Keeney S, Bonner WM, Burgoyne PS (2001) Recombinational DNA double-strand breaks in mice precede synapsis. Nat Genet 27:271–276

Malik HS, Henikoff S (2003) Phylogenomics of the nucleosome. Nat Struct Biol 10:882–891

Martianov I, Brancorsini S, Gansmuller A, Parvinen M, Davidson I, Sassone-Corsi P (2002) Distinct functions of TBP and TLF/TRF2 during spermatogenesis: requirement of TLF for heterochromatic chromocenter formation in haploid round spermatids. Development 129:945–955

Marushige Y, Marushige K (1975) Transformation of sperm histone during formation and maturation of rat spermatozoa. J Biol Chem 250:39–45

Meistrich ML, Bucci LR, Trostle-Weige PK, Brock WA (1985) Histone variants in rat spermatogonia and primary spermatocytes. Dev Biol 112:230–340

Meistrich ML, Trostle-Weige PK, Lin R, Bhatnagar YM, Allis CD (1992) Highly acetylated H4 is associated with histone displacement in rat spermatids. Mol Reprod Dev 31:170–181

Meistrich ML, Mohapatra B, Shirley CR, Zhao M (2003) Roles of transition nuclear proteins in spermiogenesis. Chromosoma 111:483–488

Moens PB (1995) Histones H1 and H4 of surface-spread meiotic chromosomes. Chromosoma 104:169–174

Moore SC, Jason L, Ausio J (2002) The elusive structural role of ubiquitinated histones. Biochem Cell Biol 80:311–319

O'Carroll D, Scherthan H, Peters AH, Opravil S, Haynes AR, Laible G, Rea S, Schmid M, Lebersorger A, Jerratsch M, Sattler L, Mattei MG, Denny P, Brown SD, Schweizer D, Jenuwein T (2000) Isolation and characterization of Suv39h2, a second histone H3 methyltransferase gene that displays testis-specific expression. Mol Cell Biol 20:9423–9433

Oko RJ, Jando V, Wagner CL, Kistler WS, Hermo LS (1996) Chromatin reorganization in rat spermatids during the disappearance of testis-specific histone, H1 t, and the appearance of transition proteins TP1 and TP2. Biol Reprod 54:1141–1157

Oliva R, Mezquita C (1982) Histone H4 hyperacetylation and rapid turnover of its acetyl groups in transcriptionally inactive rooster testis spermatids. Nucleic Acids Res 10:8049–8059

Palmer DK, O'Day K, Margolis RL (1990) The centromere specific histone CENP-A is selectively retained in discrete foci in mammalian sperm nuclei. Chromosoma 100:32–36

Pehrson JR, Costanzi C, Dharia C (1997) Developmental and tissue expression patterns of histone macroH2A1 subtypes. J Cell Biochem 65:107–113

Peters AH, O'Carroll D, Scherthan H, Mechtler K, Sauer S, Schofer C, Weipoltshammer K, Pagani M, Lachner M, Kohlmaier A, Opravil S, Doyle M, Sibilia M, Jenuwein T (2001) Loss of the suv39 h histone methyltransferases impairs mammalian heterochromatin and genome stability. Cell 107:323–337

Pittoggi C, Renzi L, Zaccagnini G, Cimini D, Degrassi F, Giordano R, Magnano AR, Lorenzini R, Lavia P, Spadafora C (1999) A fraction of mouse sperm chromatin is organized in nucleosomal hypersensitive domains enriched in retroposon DNA. J Cell Sci 112:3537–3548

Pivot-Pajot C, Caron C, Govin J, Vion A, Rousseaux S, Khochbin S (2003) Acetylation-dependent chromatin reorganization by BRDT, a testis-specific bromodomain-containing protein. Mol Cell Biol 23:5354–5365

Prigent C, Dimitrov S (2003) Phosphorylation of serine 10 in histone H3, what for? J Cell Sci 116:3677–3685

Rabini S, Franke K, Saftig P, Bode C, Doenecke D, Drabent B (2000) Spermatogenesis in mice is not affected by histone H1.1 deficiency. Exp Cell Res 255:114–124

Rasmussen TP, Huang T, Mastrangelo MA, Loring J, Panning B, Jaenisch R (1999) Messenger RNAs encoding mouse histone macroH2A1 isoforms are expressed at similar levels in male and female cells and result from alternative splicing. Nucleic Acids Res 27:3685–3689

Reinke H, Horz W (2003) Histones are first hyperacetylated and then lose contact with the activated PHO5 promoter. Mol Cell 11:1599–1607

Roest HP, van Klaveren J, de Wit J, van Gurp CG, Koken MH, Vermey M, van Roijen JH, Hoogerbrugge JW, Vreeburg JT, Baarends WM, Bootsma D, Grootegoed JA, Hoeijmakers JH (1996) Inactivation of the HR6B ubiquitin-conjugating DNA repair enzyme in mice causes male sterility associated with chromatin modification. Cell 86:799–810

Sassone-Corsi P (2002) Unique chromatin remodeling and transcriptional regulation in spermatogenesis. Science 296:2176–2178

Schnell JD, Hicke L (2003) Non-traditional functions of ubiquitin and ubiquitin-binding proteins. J Biol Chem 278:35857–35860

Sims RJ, Nishioka K, Reinberg D (2003) Histone lysine methylation: a signature for chromatin function. Trends Genet 19:629–639

Singh J, Rao MR (1987) Interaction of rat testis protein, TP, with nucleic acids in vitro. Fluorescence quenching, UV absorption, and thermal denaturation studies. J Biol Chem 262:734–740

Smith MM (2002) Centromeres and variant histones: what, where, when and why? Curr Opin Cell Biol 14:279–285

Steger K, Klonisch T, Gavenis K, Drabent B, Doenecke D, Bergmann M (1998) Expression of mRNA and protein of nucleoproteins during human spermiogenesis. Mol Hum Reprod 4:939–945

Strahl BD, Allis CD (2000) The language of covalent histone modifications. Nature 403:41–45

Sung P, Prakash S, Prakash L (1988) The RAD6 protein of *Saccharomyces cerevisiae* polyubiquitinates histones, and its acidic domain mediates this activity. Genes Dev 2:1476–1485

Trostle-Weige PK, Meistrich ML, Brock WA, Nishioka K, Bremer JW (1982) Isolation and characterization of TH2A, a germ cell-specific variant of histone 2A in rat testis. J Biol Chem 257:5560–5567

Trostle-Weige PK, Meistrich ML, Brock WA, Nishioka K (1984) Isolation and characterization of TH3, a germ cell-specific variant of histone 3 in rat testis. J Biol Chem 259:8769–8776

Turner JM, Burgoyne PS, Singh PB (2001) M31 and macroH2A1.2 colocalise at the pseudoautosomal region during mouse meiosis. J Cell Sci 114:3367–3375

Van Roijen HJ, Ooms MP, Spaargaren MC, Baarends WM, Weber RF, Grootegoed JA, Vreeburg JT (1998) Immunoexpression of testis-specific histone 2B in human spermatozoa and testis tissue. Hum Reprod 13:1559–1566

Verreault A (2000) De novo nucleosome assembly: new pieces in an old puzzle. Genes Dev 14:1430–1438

Witt O, Albig W, Doenecke D (1996) Testis-specific expression of a novel human H3 histone gene. Exp Cell Res 229:301–306

Wolffe A (1995) Chromatin – structure and function. Academic Press, London

Wykes SM, Krawetz SA (2003) The structural organization of sperm chromatin. J Biol Chem 278:29471–29477

Yan W, Ma L, Burns KH, Matzuk MM (2003) HILS1 is a spermatid-specific linker histone H1-like protein implicated in chromatin remodeling during mammalian spermiogenesis. Proc Natl Acad Sci USA 100:10546–10551

Yelick PC, Balhorn R, Johnson PA, Corzett M, Mazrimas JA, Kleene KC, Hecht NB (1987) Mouse protamine 2 is synthesized as a precursor whereas mouse protamine 1 is not. Mol Cell Biol 7:2173–2179

Yu YE, Zhang Y, Unni E, Shirley CR, Deng JM, Russell LD, Weil MM, Behringer RR, Meistrich ML (2000) Abnormal spermatogenesis and reduced fertility in transition nuclear protein 1-deficient mice. Proc Nat Acad Sci USA 97:4683–4688

Zalenskaya IA, Bradbury EM, Zalensky AO (2000) Chromatin structure of telomere domain in human sperm. Biochem Biophys Res Commun 279:213–218

Zalensky AO, Siino JS, Gineitis AA, Zalenskaya IA, Tomilin NV, Yau P, Bradbury EM (2002) Human testis/sperm-specific histone H2B (hTSH2B). Molecular cloning and characterization. J Biol Chem 277:43474–43480

Zeng L, Zhou MM (2002) Bromodomain: an acetyl-lysine binding domain. FEBS Lett 513:124–128

Zhao M, Shirley CR, Yu YE, Mohapatra B, Zhang Y, Unni E, Deng JM, Arango NA, Terry NH, Weil MM, Russell LD, Behringer RR, Meistrich ML (2001) Targeted disruption of the transition protein 2 gene affects sperm chromatin structure and reduces fertility in mice. Mol Cell Biol 21:7243–7255

Chromatin Modifications on the Inactive X Chromosome

Hannah R. Cohen, Morgan E. Royce-Tolland, Kathleen A. Worringer,
Barbara Panning

Abstract In female mammals, one X chromosome is transcriptionally silenced to achieve dosage compensation between XX females and XY males. This process, known as X-inactivation, occurs early in development, such that one X chromosome is silenced in every cell. Once X-inactivation has occurred, the inactive X chromosome is marked by a unique set of epigenetic features that distinguishes it from the active X chromosome and autosomes. These modifications appear sequentially during the transition from a transcriptionally active to an inactive state and, once established, act redundantly to maintain transcriptional silencing. In this review, we survey the unique epigenetic features that characterize the inactive X chromosome, describe the mechanisms by which these marks are established and maintained, and discuss how each contributes to silencing the inactive X chromosome.

1
Introduction

Equalization of X-linked gene dosage between XY male and XX female mammals occurs by X-inactivation, the transcriptional silencing of one X chromosome in female cells (Lyon 1961). The term X-inactivation encompasses two processes: the initial transition from a transcriptionally active to an inactive state and the subsequent stable maintenance of the silent state. Early in female embryogenesis, at approximately the time when pluripotent cells differentiate into more developmentally restricted lineages, one X chromosome is silenced in each cell. X chromosome silencing is initiated at the *X-inactivation center*, a *cis*-element that is necessary and sufficient to nucleate chromosome-wide silencing. This process occurs at random, such that there is an equal probability that the X chromosome inherited from either parent is transcriptionally

H.R. Cohen, M.E. Royce-Tolland, K.A. Worringer, B. Panning
Department of Biochemistry and Biophysics, University of California San Francisco,
San Francisco, California 94143, USA, e-mail: bpanning@biochem.ucsf.edu

Progress in Molecular and Subcellular Biology
P. Jeanteur (Ed.)
Epigenetics and Chromatin
© Springer-Verlag Berlin Heidelberg 2005

inactivated. Once silencing has occurred, it is stably maintained throughout all ensuing cell divisions. As a result, females are mosaic, with the paternally derived X chromosome silenced in 50 % of differentiated cells and the maternally derived X chromosome silenced in the remaining half.

The inactive X chromosome (Xi) differs from the active X chromosome (Xa) and autosomes in differentiated cells, as it is characterized by a unique combination of epigenetic features including histone modifications and DNA methylation. These modifications are acquired sequentially during the onset of X-inactivation and act redundantly to maintain X chromosome silencing. In this chapter, we discuss the mechanisms by which the unique chromatin structure of the Xi is established and maintained, and the role of epigenetic modifications in regulating transcriptional silencing.

2
Features of Xi Chromatin

In the nuclei of all eukaryotic cells, DNA is highly compacted into chromatin. The basic unit of chromatin is the nucleosome, which consists of DNA wrapped around an octamer of core histones, H2A, H2B, H3 and H4. The amino terminal tails of the core histones protrude from the surface of the nucleosome and are subject to covalent modifications such as acetylation, methylation, phosphorylation and ubiquitination. Different combinations of histone modifications are thought to establish transcriptionally active euchromatin and transcriptionally silent heterochromatin (Strahl and Allis 2000; Turner 2000; Jenuwein and Allis 2001). According to this 'histone code' hypothesis, epigenetic marks on the histone tails provide binding sites for proteins that regulate gene expression. Replacement of core histones with variant histones is another chromatin alteration that is employed to modulate gene expression. In addition to histone modifications, CpG methylation is a covalent DNA modification that is implicated in chromatin structure and transcription. Given that much transcriptional regulation is achieved via changes in chromatin structure, it is not surprising that the Xi shows a distinct signature of chromatin marks when compared to the Xa and autosomes. Below we describe the features of chromatin that distinguish the Xi from the Xa and autosomes (Table 1).

2.1
Histone H3 Lysine 9 Methylation

Methylation can occur on lysine and arginine residues in the amino-terminal tails of core histones. Lysine residues are found in mono-, di- or tri-methylated forms, while arginine can be mono- or di-methylated (Bannister et al. 2002). Enrichment for histone H3 methylated at lysine 9 (H3-K9) is one of the

Table 1. Histone modifications and histone variants on the Xi and Xa. Relative levels of histone modifications that show unique densities on the Xi in somatic cells are shown for three classes of genes: X-linked (genes that are expressed from the Xa and silenced on the Xi), Xist (expressed from the Xi and silenced on the Xa) and escapees (genes that are expressed from both the Xa and Xi). + indicates modification enriched at that locus relative to autosomal levels; – indicates modification is underrepresented at that locus relative to autosomal levels; = indicates modification is found at autosomal levels; *question marks* appear where the enrichment at a particular location is unknown

Histone modification	Xi			Xa		
	X-linked	Xist	Escapees	X-linked	Xist	Escapees
H3–2 mK9	+	=	=	=	?	=
H3–3 mK27	+	?	?	=	?	?
H3–2 mK4	–	=	=	=	–	=
H3–2 mR17	–	?	?	=	?	?
H3–2 mK36	–	?	?	=	?	?
H3 acetylation	–	=	=	=	–	=
H4 acetylation	–	=	=	=	–	=
macroH2A	+	?	?	=	?	?
H2-Bbd	–	?	?	=	?	?
H2AZ	–	?	?	=	?	?

hallmarks of heterochromatin (Jenuwein and Allis 2001; Bannister et al. 2002). The di-methylated form of H3-K9 (H3–2 mK9) is enriched on the Xi, specifically at the promoters of silenced genes, such that H3-K9 methylation at the promoter correlates with transcriptional inactivity on the Xi (Heard et al. 2001).

There are two types of heterochromatin: facultative, in which silencing is reversible, and constitutive, in which silencing is irreversible. The heterochromatin of the Xi is facultative and pericentric heterochromatin is constitutive. Both the Xi and pericentric heterochromatin show enrichment for methylated H3-K9 (Heard et al. 2001). Two antibodies have been used to study H3–2 mK9 distribution in the nucleus; one antibody was raised against a branched peptide and the other against a linear peptide. Depending on which antibody is used, different staining patterns are observed. In studies using branched peptide antibodies, staining was observed on both the Xi and pericentric regions (Heard et al. 2001; Maison et al. 2002), while linear peptide antibodies recognized the Xi exclusively (Heard et al. 2001). When antibodies were raised against linear peptides for tri-methylated form of H3-K9 (H3–3 mK9), these antibodies recognized pericentric regions but not the Xi (Peters et al. 2003; Plath et al. 2003; Rice et al. 2003). These results in combination with peptide competition studies suggest that the branched peptide antibodies recognize both H3–2 mK9 and H3–3 mK9 epitopes (Plath et al. 2003). Therefore, the facultative heterochromatin of the Xi and the constitutive heterochromatin of pericentric regions differ in that they are enriched for H3–2 mK9 and H3–3 mK9, respectively.

The SET domain family of histone methyltransferases (HMTase) catalyzes the methylation of lysine residues (Bannister et al. 2002). Five mammalian SET domain proteins, Suv39h1, Suv39h2, G9a, Eset/SETDB1 and EZH2, have HMTase activity on H3-K9 in vitro (Lachner et al. 2003) and are candidates for catalyzing H3-K9 methylation on the Xi. H3-K9 methylation is detected on the Xi in cells bearing a deletion of Ezh2 and in cells deleted for both Suv39h1 and Suv39h2 (Peters et al. 2002; Erhardt et al. 2003; Silva et al. 2003). Eset/SETDB1 mediates formation of H3–3 mK9 (Wang et al. 2003), but this modification is not found on the Xi (Plath et al. 2003). As G9a catalyzes production of the H3–2 mK9 in vitro and in vivo (Rice et al. 2003), this enzyme is the most likely candidate to serve as the HMTase that mediates H3–2 mK9 enrichment on the Xi. Alternatively, multiple HMTases may act redundantly to di-methylate H3-K9 on the Xi.

The significance of H3–2 mK9 enrichment on the Xi is likely to involve recruitment of H3–2 mK9 binding proteins that further regulate chromatin structure. Heterochromatin protein 1 (HP1) binds methylated H3-K9 in vitro, and this protein is required for formation of pericentric heterochromatin (Bannister et al. 2001; Lachner et al. 2001; Fischle et al. 2003). All three forms of human HP1 appear to be enriched on the Xi as well as on pericentric regions (Chadwick and Willard 2003), suggesting that H3–2 mK9 contributes to heterochromatin formation by recruiting HP1 to the Xi. None of the three forms of Hp1 is enriched on the Xi in mouse cells (Peters et al. 2002), suggesting that HP1 enrichment on the Xi may be specific to human cells. Alternatively, as adult human cells and embryonic mouse cells were examined, it is possible that HP1 enrichment on the Xi is specific to a particular developmental stage.

2.2
Histone H3 Lysine 27 Methylation

Regulated silencing of homeotic genes during *Drosophila* development requires methylation of H3 at lysine 27 (H3-K27), mediated by the ESC-E(Z) complex (Cao et al. 2002; Muller et al. 2002). Thus, this HMTase complex is involved in the formation of facultative heterochromatin. Loss of function of Eed, the murine ESC homologue, results in reactivation of X-linked genes, suggesting a role for H3-K27 methylation in regulating X-inactivation in female mammals (Wang et al. 2001). Further investigation revealed that Eed is present in a complex with Ezh2, the mouse homologue of E(Z) (Denisenko et al. 1998), and that both of these proteins are enriched on the Xi (Mak et al. 2002; Plath et al. 2003; Silva et al. 2003). Ezh2 is capable of methylating H3-K27 in vitro (Cao et al. 2002; Czermin et al. 2002; Kuzmichev et al. 2002; Muller et al. 2002), and there is enrichment of the trimethylated form of H3-K27

A. Barr Body B. H3-3mK27 C. H3-2mK4

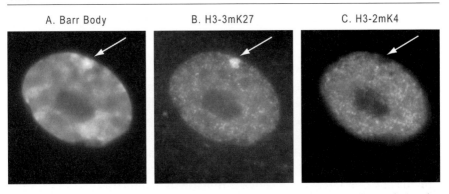

Fig. 1. Histone modifications enriched and deficient on the Xi. **A** A DNA-intercalating dye stains the Xi intensely, marking the classical Barr body. **B** Immunofluorescence, on the same cell, shows that H3-3 mK27 is enriched on the Xi. **C** Immunofluorescence against H3-2 mK4 shows that this modification is excluded from the Xi

(H3–3 mK27) on the Xi in some cell types (Fig. 1; Gilbert et al. 2003). Cells from *Ezh2* and *eed* mutant mice have no detectable staining with antibodies directed against the tri-methylated form of H3-K27 (H3–3 mK27) (Erhardt et al. 2003; Silva et al. 2003). In combination, these data argue that the Eed/Ezh2 complex mediates the accumulation of H3–3 mK27 on the Xi.

A model for the role of methylated H3-K27 in transcriptional silencing is based on findings in *Drosophila*. In flies, H3–2 mK27 provides a binding site for the chromo-domain protein Polycomb (PC) (Cao et al. 2002; Czermin et al. 2002; Fischle et al. 2003; Kuzmichev et al. 2002), a component of the Polycomb repressive complex (PRC1), which is essential for maintaining homeotic gene silencing (Francis and Kingston 2001; Simon and Tamkun 2002). PRC1 can block SWI/SNF-mediated chromatin remodeling in vitro (Shao et al. 1999). Alterations in chromatin structure mediated by H3–3 mK27-bound mammalian PRC1 provide an attractive model for the role of this histone modification in mediating transcriptional repression on the Xi. However, PRC1 components were not detected on the Xi (Mak et al. 2002), suggesting that alternative models for the contribution of H3–3 mK27 to transcriptional silencing of the Xi should also be considered.

2.3
Methylation at Other Histone H3 Residues

In contrast to H3-K9 and H3-K27 methylation, which correlate with transcriptional silencing, methylation of histone H3 at lysine 4 (H3-K4) or arginine 17 (H3-R17) shows a strong correlation with gene activity (Strahl et al. 1999). Immunofluorescence and ChIP with antibodies directed against the dimethylated form of H3-K4 (H3-2 mK4) shows that H3-2 mK4 is underrepresented on the Xi (Fig. 1; Boggs et al. 2002; Rougeulle et al. 2003). A similar result is obtained using antibodies raised against dimethylated H3-R17 (H3–2 mR17) (Chaumeil et al. 2002). The absence of these two methylation marks on the Xi is consistent with its silent state.

It has been suggested that histone H3 dimethylated on lysine 36 (H3–2 mK36) causes gene repression in yeast (Strahl et al. 2002; Landry et al. 2003). However, there is also evidence linking this modification to gene expression in yeast and *Tetrahymena* (Strahl et al. 2002; Krogan et al. 2003; Li et al. 2003; Schaft et al. 2003; Xiao et al. 2003). H3–2 mK36 is underrepresented on the Xi (Chaumeil et al. 2002), indicating that enrichment of this modification may be characteristic of active chromatin in mammals.

2.4
Histone Acetylation

Hypoacetylation of histones H3 and H4 at lysine residues is commonly associated with heterochromatin and transcriptional inactivity (Richards and Elgin 2002). In agreement with this observation, the Xi appears devoid of histone acetyl modifications. When interphase cells are immunostained with an antibody raised against H3 acetylated on lysine 9, the Xi is understained, appearing as a hole (Belyaev et al. 1996; Boggs et al. 1996; Chaumeil et al. 2002). The same result is observed using antibodies raised against H4 acetylated at lysines 5, 8, 12 and 16 (Jeppesen and Turner 1993; Keohane et al. 1996; Heard et al. 2001; Chaumeil et al. 2002; Goto et al. 2002).

Histone acetylation is regulated by a combination of histone acetyltransferase and histone deacetylase (HDAC) activities that add and remove acetyl groups, respectively. Human EED and EZH2 interact with HDAC1 and HDAC2 in vitro and in vivo (van der Vlag and Otte 1999), suggesting that this HMTase complex may regulate acetylation on the Xi by recruiting deacetylases. However, Hdac1 and Hdac2 are not enriched on the Xi in mouse cells with Xi-enrichment of Eed and Ezh2 (Mak et al. 2002), indicating that enrichment of the Eed/Ezh2 complex is insufficient to cause an enrichment of these Hdacs. HDAC1 and HDAC2 form corepressor complexes with SIN3A and SIN3B (Yang et al. 2003). SIN3A and SIN3B complexes are excluded from the Xi (Chadwick and Willard 2003), further suggesting that these deacetylases do not regulate levels of acetylation on the Xi. There are at least ten HDAC family members in mammals, providing a number of candidates for the HDACs that might mediate the decrease in histone H3 and H4 acetylation on the Xi. Alternatively, the underacetylation of the Xi may be achieved by the exclusion of histone acetyltransferase activities from the Xi.

2.5
Histone MacroH2A

In addition to exhibiting a unique combination of post-translational modifications on core histones, the Xi contains a high proportion of nucleosomes in which canonical H2A is replaced by the variant histones macroH2A1.2, macroH2A1.2 or macroH2A2 (Mermoud et al. 1999; Chadwick and Willard 2001a; Changolkar and Pehrson 2002). The N terminus of each variant is homologous to canonical H2A, while the C termini or non-histone regions (NHRs) of macroH2A proteins show no H2A homology (Pehrson and Fried 1992; Pehrson and Fuji 1998; Chadwick and Willard 2001a; Costanzi and Pehrson 2001). Most cell types display Xi-enrichment of all three macroH2A isoforms (Costanzi and Pehrson 1998, 2001; Chadwick and Willard 2001a). Both the H2A-like domain and the NHR may be involved in proper localization of macroH2A to the Xi, as a truncated protein consisting of the H2A-like domain of macroH2A1 or macroH2A2 localizes to the Xi, and a fusion protein consisting of canonical H2A fused to the macroH2A1.2 NHR also localizes to the Xi (Chadwick et al. 2001).

The enrichment of macroH2A on the Xi suggests that this protein may contribute to gene silencing. Indeed, ectopic macroH2A can downregulate gene expression in vivo. Using Gal4 to tether the macroH2A NHR to the promoter of the luciferase gene reduced luciferase activity more than twofold (Perche et al. 2000). MacroH2A NHR was not assembled into nucleosomes in this assay, indicating that the NHR may be sufficient for silencing outside the context of the nucleosome.

A recent study proposes two mechanisms for transcriptional repression by macroH2A: interference with transcription factor binding and resistance to nucleosome remodeling (Angelov et al. 2003). The transcription factor NF-κB binds chromatin assembled with conventional histones, but not chromatin assembled with histone octamers containing macroH2A1.2. The NHR is necessary to prevent NF-κB binding. In addition, the DNA near the NF-κB binding site has different DNase accessibility in H2A- and macroH2A-containing nucleosomes. In combination, these results suggest that the NHR sterically blocks transcription factor access to its target DNA sequence. Nucleosome remodeling by SWI/SNF complexes usually promotes gene expression; however, nucleosomes containing macroH2A1.2 are resistant to SWI/SNF activity, suggesting that macroH2A may also regulate gene expression by inhibiting nucleosome remodeling (Angelov et al. 2003). The H2A-like domain is responsible for resistance to SWI/SNF activity. Thus, both the H2A-like domain and the NHR may contribute to transcriptional regulation by macroH2A.

2.6
Other Variant Histones

Two variant histones are less abundant on the Xi than on the Xa and autosomes: H2A-Barr body deficient (H2A-Bbd) and H2AZ (Chadwick and Willard 2001b, 2003). While little is known about the function of H2A-Bbd, H2AZ has been shown to antagonize silencing in *Saccharomyces cerevisiae* (Meneghini et al. 2003). H2AZ may be underrepresented on the Xi because most genes on this chromosome are repressed. Localization of H2AZ to pericentric heterochromatin has been observed in mouse extraembryonic tissue (Leach et al. 2000; Rangasamy et al. 2003), indicating that this modification is not excluded from all heterochromatin. It therefore seems likely that variant histones are used in combination with other epigenetic marks to establish diverse forms of chromatin.

2.7
Nucleosome Position

Another chromatin-based mechanism of regulating transcription involves the position of nucleosomes at promoters. On the Xa, nucleosomes occupy defined positions at the HPRT promoter, leaving transcription factor binding sites and the transcription initiation region accessible to transcription machinery. In contrast, the same promoter on the Xi is blocked by nucleosomes that are more randomly positioned (Chen and Yang 2001). These results can be explained in two ways: either transcription forces nucleosomes away from the promoter or nucleosome positioning acts upstream of transcription to control gene expression. In support of the latter model, chromatin remodeling precedes occupation of transcription factor binding sites during reactivation of X-linked genes (Litt et al. 1997). The factors that determine nucleosome positioning are unknown, but may include histone modifications and histone variants. Perhaps nucleosome positioning is an additional level of transcriptional control.

2.8
Shape of the Xi

The unique combination of chromatin modifications on the Xi may also be responsible for the cytological characteristics of this chromosome. The region of the nucleus occupied by the Xi can be detected as a region of intense staining with DNA intercalating dyes (Fig. 1; Barr 1961). This brightly staining region, known as the Barr body, suggests that the Xi is condensed relative to other chromosomes. This condensation could result in transcriptional silencing by excluding transcriptional machinery. When assessed using chromosome paints, the volume of the Xi appeared to be the same as that of the Xa (Bischoff et al. 1993; Clemson et al. 1996; Eils et al. 1996). These data indicate that overall the Xi is not highly condensed relative to the Xa, but do not rule out that local regions of high compaction may exist. It seems likely that epigenetic modifications on the Xi, and not global compaction of the chromosome, are responsible for silencing and the increased incorporation of DNA intercalating dyes.

Despite occupying similar volumes, the shapes of the Xi and Xa are significantly different. When chromosome paints are used to delineate the Xa and Xi, the Xa is more elliptical, and the Xi is more spherical in interphase nuclei (Bischoff et al. 1993; Eils et al. 1996). In addition, the telomeres of the Xi are tenfold closer to each other than the telomeres of the Xa, confirming that the Xi shows different spatial organization from the Xa (Walker et al. 1991). A potential modulator of large-scale chromosome shape is SAF-A, a protein that binds both scaffold attachment regions and RNA. SAF-A is enriched on the Xi and might affect its overall shape and/or location within the nucleus by interacting with nuclear structural elements (Helbig and Fackelmayer 2003).

2.9
DNA Methylation

DNA methylation appears to play an important role in maintaining gene silencing on the Xi. Upstream sequences of genes on the X chromosome are hypermethylated on the Xi and hypomethylated on the Xa (Wolf et al. 1984; Pfeifer et al. 1990; Bartlett et al. 1991). Treatment with the DNA-demethylating agent 5-azadeoxycytidine results in reactivation of several X-linked genes (Mohandas et al. 1981).

ICF (immunodeficiency, centromeric instability and facial anomalies) syndrome results from a mutation in the DNA methyltransferase DNMT3b. Cells deficient for DNMT3b show hypomethylation of DNA and reactivation of genes on the Xi (Hansen et al. 2000). In addition, one class of repetitive elements, LINE-1 elements, which are normally hypermethylated on both the Xi and Xa, are hypomethylated exclusively on the Xi in DNMT3b mutant cells (Hansen 2003). In combination, these data indicate that DNMT3b contributes to DNA methylation and gene silencing on the Xi. The two other DNA methyltransferases, DNMT3a and DNMT1 (Bestor 2000), may also play a role in DNA methylation on the Xi.

In *Neurospora* and *Arabidopsis*, trimethylation of H3-K9 is required for DNA methylation (Tamaru and Selker 2001; Jackson et al. 2002). In *Neurospora*, HP1 is essential for DNA methylation, suggesting that HP1 binds methylated H3-K9 and recruits a DNA methyltransferase (Freitag et al. 2004). In mammalian cells, H3-K9 trimethylation is required for Dnmt3b-dependent DNA methylation at pericentric heterochromatin (Lehnertz et al. 2003). Therefore it is possible that H3-K9 methylation directs DNA methylation on the Xi as well. A mouse H3-K9 HMTase activity co-fractionates with Dnmt3a (Datta et al. 2003). Hdac1 also co-fractionates with Dnmt3a and the H3-K9 HMTase activity, indicating that histone acetylation, histone methylation and DNA methylation may be coordinately regulated in mammalian cells. The association of DNA methyltransferase and histone methyltransferase activities may be important for the spread of heterochromatin as it is possible that methyltransferases recruited to one nucleosome can modify adjacent nucleosomes. Although a functional interaction of methylation and deacetylation complexes has yet to be identified on the Xi, it seems likely that this observation will be extended to X-inactivation.

2.10
Late Replication Timing

Several types of heterochromatin, including silenced X chromatin, replicate late in S phase. Genes on the Xa replicate earlier than their counterparts on the Xi (Schmidt and Migeon 1990; Hansen et al. 1993, 1995; Torchia et al. 1994). Analysis of two replicons on the X chromosome showed that the same origins fire both on the Xi and the Xa (Cohen et al. 2003), suggesting that the same origins are differentially regulated on these chromosomes. This study raises an interesting question: What mechanisms are used to direct different behavior of the same origins on two homologous chromosomes within a single nucleus?

Studies in *Drosophila* may provide insight into the link between replication and silencing. Fly HP1 binds to components of the origin recognition complex (ORC) and flies mutant for an ORC protein show abnormalities in formation of heterochromatin (Pak et al. 1997). The three mammalian HP1 isoforms co-localize with heterochromatic regions, including the heterochromatin of the Xi (Chadwick and Willard 2003). HP1 is thought to nucleate the spread of heterochromatin by binding methylated H3-K9 and recruiting HMTases to methylate H3-K9 on neighboring nucleosomes. The interaction between HP1 and ORCs suggests two distinct models for the co-regulation of replication and silencing. Late-replicating origins on the Xi may recruit HP1, which mediates silencing. Alternatively, HP1 on the Xi could mediate a change in chromatin structure that affects both gene expression and replication timing.

Late replication timing and DNA methylation on the Xi show an intriguing relationship. Treatment with the DNA-demethylating agent 5-azadeoxycytidine can trigger early replication of the Xi and reactivation of X-linked loci (Hansen et al. 1996). In addition, cells deficient in DNMT3b show early replication of reactivated X-linked genes (Hansen et al. 2000). In DNMT3b mutant cells, a number of X-linked genes are unmethylated. A subset of these unmethylated genes replicate early and are expressed, suggesting that DNA methylation can influence replication timing.

2.11
Xist RNA

While most genes are silenced on the Xi and expressed from the Xa, the *XIST* gene (*XIST* in humans and *Xist* in mouse) shows the opposite expression pattern. *XIST* encodes a 17-Kb, spliced, polyadenylated, non-coding RNA that stably associates with the entire Xi, appearing to coat this chromosome (Brockdorff et al. 1992; Brown et al. 1992). *XIST* RNA's ability to coat the Xi depends on the tumor suppressor gene product BRCA1; in BRCA1 mutant cells, *XIST* RNA does not coat the Xi, although it is produced at normal levels (Ganesan et al. 2002). In cells that do not exhibit *XIST* RNA coating of the Xi, due to mutations in BRCA1 or deletion of the mouse *Xist* gene, macroH2A1.2 is not enriched on the Xi (Csankovszki et al. 1999; Beletskii et al. 2001; Ganesan et al. 2002). These results suggest that *XIST* RNA acts as a scaffold that coordinates at least two related activities, chromosome coating and regulation of chromatin structure.

2.12
Redundant Mechanisms Maintain Silencing

Thus far we have described a number of chromatin modifications specific to the Xi. Experiments were performed to address the importance of several of these modifications in stably maintaining the silent state of the Xi. Individually, deletion of *Xist*, DNA demethylation with 5-azadeoxycytidine or hyperacetylation of histones by treatment with the HDAC inhibitor TSA result in some reactivation of Xi-linked genes (Csankovszki et al. 2001). These three treatments in combination induce a significantly higher frequency of reactivation than any one alone. Thus, the chromatin modifications that characterize the Xi may work synergistically to maintain this chromosome in an inactive state, and redundant mechanisms may be employed for the extraordinarily stable maintenance of X chromosome silencing.

3
Chromatin at the *Xic*

Xist is the only gene expressed exclusively from the Xi, raising the question of how the *Xist* locus escapes chromosome-wide silencing. Early observations of metaphase chromosome spreads identified a sharp bend on the Xi at or near the *Xist* locus (Flejter et al. 1984), suggesting that there is unique chromatin structure at this site. Indeed, analysis of chromatin modifications at *Xist* and flanking sequences, which together comprise the *X-inactivation center* (*Xic*), indicates that the chromatin of the *Xic* on the Xi is different from that of the rest of the Xi and more similar to the Xa and autosomes.

3.1
Histone Modifications

As discussed earlier, H3–2 mK4 is underrepresented on much of the Xi (Table 1). However, allele-specific ChIP analysis for H3–2 mK4 shows that the expressed *Xist* gene and its promoter on the Xi show higher levels of H3–2 mK4 than adjacent silent genes, and that these levels are comparable to that seen on autosomal genes (Boggs et al. 2002; Goto et al. 2002). H3–2 mK4 is underrepresented on the silent *Xist* gene and its promoter region on the Xa (Goto et al. 2002). These results are consistent with the established correlation between enrichment for H3–2 mK4 and gene expression (Strahl et al. 1999). When a site 90 kb upstream of *Xist* was analyzed by the same method, both the Xi and Xa showed enrichment for H3–2 mK4, suggesting that on the Xi, this modification may delineate a chromatin boundary allowing for *Xist* expression from an otherwise transcriptionally silent X chromosome (Goto et al. 2002).

The interesting patterns of acetylation at or around *Xist* suggest that H3 and H4 acetylation also play a role in maintaining the transcriptional activity of this gene on the Xi. ChIP using an antibody directed against acetylated H4-K5, -K8, -K12 and -K18 showed that H4 acetylation is enriched at the *Xist* locus on the Xi, but not on the Xa (Goto et al. 2002). An antibody raised against acetylated H3-K9 and -K14 preferentially precipitated the *Xist* promoter on the Xi (Goto et al. 2002). This result demonstrates that the chromatin structure at the expressed *Xist* allele is typical of euchromatin. The unique pattern of acetylation at the *Xist* locus suggests that it may be involved in maintaining the active state of this gene. Experiments to delineate and investigate the boundaries of H3 and H4 acetylation may help to determine how Xist is oppositely regulated from the rest of the X chromosome.

3.2
DNA Methylation

DNA methylation at the promoters of X-linked genes is important to maintain their transcriptional repression on the Xi. DNA methylation of the *Xist* promoter, mediated by Dnmt1 and Dnmt3a, also contributes to silencing the *Xist* gene on the Xa (Panning and Jaenisch 1996; Chen et al. 2003). Male cells deficient in Dnmt1 exhibit hypomethylation of the *Xist* promoter region, ectopic expression of *Xist* RNA, and transcriptional silencing on the single X chromosome (Beard et al. 1995; Panning and Jaenisch 1996). The *Xist* promoter region is also hypomethylated and *Xist* is expressed in male mutant cells lacking Dnmt3a and Dnmt3b (Chen et al. 2003; Sado et al. 2004). Adding back Dnmt3a, but not Dnmt3b, to these double mutant cells restored wild-type levels of *Xist* promoter methylation (Chen et al. 2003). Together these results suggest that multiple methyltransferases prevent *Xist* expression on the Xa by maintaining methylation patterns at the *Xist* locus.

3.3
Replication Timing

Silenced genes on the Xi replicate later than their expressed homologues on the Xa, suggesting a correlation between gene expression and replication timing on the X chromosomes. It is unclear whether the *Xist* gene follows the general rule that expressed genes replicate early and silent genes replicate late. Fluorescence in situ hybridization (FISH) studies initially indicated that the expressed *XIST* gene on the Xi replicates early and the silent *XIST* gene on the Xa replicates late (Torchia et al. 1994; Torchia and Migeon 1995). However, additional studies using a physical assay for replication based on BrdU incorporation, in addition to FISH, showed that the expressed *XIST* gene on the Xi replicates late and the silent *XIST* gene on the Xa replicates early (Hansen et al. 1995; Gartler et al. 1999). As the ability of FISH to accurately measure replication timing depends on fixation conditions (Azuara et al. 2003), it is possible that slight differences in methodology produced these opposite conclusions.

4
Genes that Escape X-Inactivation

Some genes on the Xi are not subject to X-inactivation, and as a result are expressed from both the Xi and the Xa (Disteche et al. 2002). The chromatin state of the genes escaping X-inactivation more closely resembles that of expressed genes on the Xa and autosomes than that of neighboring silent genes on the Xi.

4.1
Histone Modifications

On the Xi, genes that escape X-inactivation lack many of the histone modifications that characterize the silent genes on this chromosome (Table 1). Escapees do not show enrichment for H3-2 mK9 or an underrepresentation of H3-2 mK4 (Boggs et al. 2002). Furthermore, the chromatin surrounding these genes is acetylated on histones H3 and H4 at levels comparable to that of autosomal genes (Jeppesen and Turner 1993). The absence of Xi-specific histone modifications on the subset of genes that escapes X-inactivation further indicates that these modifications may function in silencing most genes on the Xi.

4.2
DNA Methylation

CpG islands linked to genes that are subject to X-inactivation exhibit DNA methylation on their promoter regions. In contrast, CpG islands in genes that escape X-inactivation are unmethylated on the Xi (Goodfellow et al. 1988; Slim et al. 1993). In addition, these genes have fewer CpG islands within the sequence immediately 2 kb upstream of their open reading frames, suggesting that one reason these genes escape silencing is a lack of CpG dinucleotides to act as targets for DNA methylation (Ke and Collins 2003).

4.3
Replication Timing

Genes that escape X-inactivation tend to replicate earlier than genes that are subject to this process (Reddy et al. 1988). Even though both alleles of escapees replicate early, the allele on the Xa consistently replicates earlier than the allele on the Xi (Gilbert and Sharp 1999; Boggs et al. 2002). This result suggests that the two alleles are not equivalent despite the fact that both are transcriptionally active. This difference may reflect the heterochromatic nature of the Xi, or it may be indicative of a true epigenetic difference between the Xi and Xa alleles of genes that are expressed from both chromosomes.

4.4
Chromosome Organization

The genes that escape X-inactivation are found in clusters along the length of the X chromosome (Carrel et al. 1999). The clustering of these genes suggests that chromatin modifications are coordinately regulated over large regions of the X chromosome. Loci within the Xi may also be organized in three-dimensional space with respect to their transcriptional status. ANT2, an X-linked gene subject to silencing, lies in the interior of the sphere-like Xi. In contrast, ANT3 escapes silencing and is found at the periphery of the Xi territory. Both genes are transcribed on the Xa, and both are located on the surface of that chromosome (Dietzel et al. 1999). It is typical for expressed genes to lie on the surface of chromatin domains, but the reason for this correlation remains unclear (Lamond and Earnshaw 1998).

5
Developmental Regulation of X-Inactivation

Cells in early female embryos have two active X chromosomes, one of which becomes silenced in a developmentally regulated fashion. Analysis of the appearance of the chromatin modifications that characterize the Xi indicates that these modifications first occur during the transition from a transcriptionally active to a silent state. This correlation suggests that Xi chromatin modifications contribute to this transition. The first noticeable event is the spread of *Xist* RNA from its site of transcription to coat the X chromosome. This initial *cis*-spread correlates closely with chromosome-wide silencing (Panning et al. 1997; Sheardown et al. 1997; Wutz and Jaenisch 2000). *Xist* is necessary and sufficient for initiation of X chromosome silencing (Penny et al. 1996; Marahrens et al. 1997; Wutz and Jaenisch 2000). Female embryonic stem (ES) cells provide a valuable model system to study alterations in chromatin structure that occur during X-inactivation because this process is recapitulated when ES cells are induced to differentiate in vitro.

5.1
Three Stages of X-Inactivation

X-inactivation occurs in at least three stages, as characterized by the requirement for *Xist* (Fig. 2). The *cis*-spread of *Xist* RNA can be uncoupled from its developmental regulation in ES cells by expressing *Xist* from an inducible promoter. Normally *Xist* RNA-mediated silencing occurs 1–2 days after female ES cells begin differentiation (Panning et al. 1997; Sheardown et al. 1997; Wutz and Jaenisch 2000). However, differentiation is not required for *Xist*-mediated transcriptional inactivation, as silencing occurs in undifferentiated male ES cells when *Xist* expression is driven from an inducible promoter (Wutz and Jaenisch 2000). *Xist* RNA can coat the chromosome, but no longer causes silencing if expression is induced more than 36 h after the start of differentiation, suggesting that events that occur upon differentiation interfere with the ability of *Xist* RNA to mediate transcriptional silencing (Clemson et al. 1998; Tinker and Brown 1998; Wutz and Jaenisch 2000). Silencing is dependent on continued *Xist* expression for the first 2.5 days after differentiation. In contrast, silencing is not dependent on continued expression of *Xist* RNA in differentiated cells (Brown and Willard 1994; Csankovszki et al. 1999; Wutz and Jaenisch 2000). Taken in combination, these results indicate that silencing can be divided into three stages: initiation, establishment, and maintenance. In the initiation phase, *Xist* RNA can cause silencing de novo, and this silencing is *Xist*-dependent. During establishment, *Xist* expression can no longer trigger silencing, although silencing continues to be *Xist* RNA-dependent. During maintenance, the transcriptional state of the chromosome is stable in that silencing can neither be induced by *Xist* expression nor reversed by

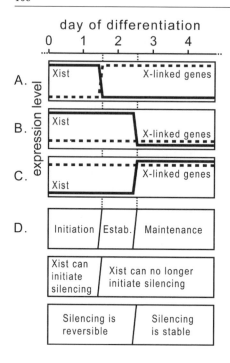

day of differentiation

Fig. 2. The X chromosome's transition from an active to an inactive state can be divided into three phases: initiation, establishment, and maintenance. Three experiments delineate the time frame of these phases (Wutz and Jaenisch 2000). **A** When Xist expression is induced early and then shut off, it causes transient silencing of linked genes. The initiation phase lasts until 1.5 days after differentiation, and is characterized by reversible, Xist-dependent silencing. **B** When Xist expression is induced for a longer period of time, stable silencing of linked genes results. Only when Xist is expressed continuously throughout the establishment phase can stable gene silencing be achieved. **C** When Xist is not expressed during the first 2.5 days of differentiation, linked genes become immune to Xist-mediated silencing. During the maintenance phase, Xist can no longer initiate silencing, and the expression status of linked genes is stable and Xist-independent. **D** The three phases are summarized. During initiation, but not the later phases, Xist can initiate silencing. During initiation and establishment, silencing is Xist-dependent and reversible; in maintenance, silencing is stable

loss of *Xist*. It seems likely that transition from one stage to the next is mediated by a precisely ordered series of chromatin modifications directed to the Xi by *Xist* RNA. In addition, these results suggest that the chromatin modifying activities present in a cell at the time *Xist* expression is upregulated determine whether silencing will occur. Thus, differentiated cells either lack appropriate chromatin modifying activities or are unable to recruit those activities to the Xi.

Ectopic *Xist* expression from an autosomal transgene in the differentiated HT-1080 human male fibrosarcoma cell line results in many of the same chromatin changes that are observed on the Xi in female cells (Hall et al. 2002). The autosome bearing an *XIST* transgene is coated by *XIST* RNA, hypoacetylated on histone H4, replicates late in S phase and shows chromosome-wide silencing. HT-1080 is the sole differentiated cell line in which activation of *XIST* expression has been reported to induce chromosome-wide silencing, indicating that cell lines differ in their ability to enact the large chromatin structural changes associated with inactivating an entire chromosome. It will be interesting to see if the chromatin modifying activities that normally direct initiation of X chromosome inactivation during differentiation are present in the already differentiated HT-1080 cell line.

5.2
Embryonic Stem Cells

Female ES cells undergo X chromosome silencing upon differentiation, facilitating temporal studies of alterations that occur during X-inactivation (Fig. 3). Changes in histone acetylation and methylation are the first chromatin modifications detected on the Xi in differentiating ES cells. H3–2 mK9 and H3–3 mK27 and deacetylation of H3-K9 are first detected on the Xi concomitant with or shortly after the initial *cis*-spread of *Xist* RNA (Keohane et al. 1996; Heard et al. 2001; Chaumeil et al. 2002; Mermoud et al. 2002; Plath et al. 2003; Silva et al. 2003). H4 hypoacetylation of the Xi occurs within the same time frame, but with slightly slower kinetics (Chaumeil et al. 2002). The decrease in acetylation and increase in methylation of H3-K9 on the Xi occur roughly simultaneously in differentiating ES cells (Heard et al. 2001; Chaumeil et al. 2002), suggesting that these modifications are mutually exclusive. As acetylated H3-K9 is a poor substrate for HMTases in vitro (Rea et al. 2000), it is possible that deacetylation must occur prior to methylation. It is unclear whether the deacetylation and methylation occur sequentially on the same

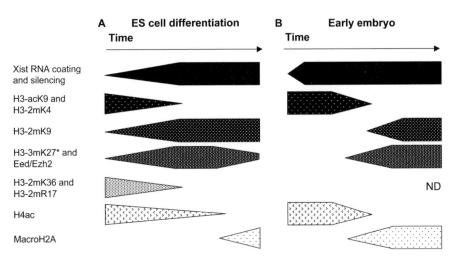

Fig. 3. Order of chromatin modifications appearance on and exclusion from the Xi during **A** random X-inactivation in differentiating ES cells in culture and **B** imprinted X-inactivation in the early embryo. *Bars* represent the overall trend, where the height of each region represents **A** the approximate percent of differentiating ES cells or **B** the approximate percent of cells in the early embryo that exhibit the modification. *Asterisk* H3–3 mK27 persists indefinitely in some differentiated cell types; *ND* not determined

histone or on different histones. H3-K4, H3-R17 and H3-K36 methylation disappear from the Xi with the same kinetics as H3-K9 deacetylation and H3-K9 methylation (Chaumeil et al. 2002), suggesting that H3 modifications are coordinately regulated during X-inactivation.

Enrichment of H3–3 mK27 is detected transiently on the Xi during ES cell differentiation (Plath et al. 2003). The Eed/Ezh2 HMTase complex is also transiently enriched on the Xi when X-inactivation is initiated in differentiating ES cells and in embryos (Plath et al. 2003; Silva et al. 2003). H3–3 mK27 and Eed/Ezh2 Xi-enrichment immediately follows *Xist* RNA coating. Cells expressing a mutant form of *Xist* RNA that coats but does not silence the X chromosome display Xi-enrichment of the Eed/Ezh2 complex and H3–3 mK27 (Plath et al. 2003), indicating that this modification is not sufficient for transcriptional silencing. As H3–3 mK27 Xi-enrichment can persist in some differentiated cell types (Gilbert et al. 2003), and since differentiated cells no longer show Eed-Ezh2 enrichment (Mak et al. 2002; Plath et al. 2003; Silva et al. 2003), it is possible that other HMTases may be required to maintain H3-K27 methylation on the Xi after it is first established by Eed/Ezh2.

Compared to the chromosome-wide enrichment of H3–2 mK9, recruitment of macroH2A1.2 to the Xi is a relatively late event in ES cell differentiation, indicating that this histone variant contributes to maintenance rather than initiation of X-inactivation. In differentiating ES cells, full *Xist* RNA coating is visible in a fraction of cells at day 1 and is visible in most cells by day 3 (Sheardown et al. 1997); macroH2A1.2 recruitment to the Xi begins in some cells at day 6 or 7 and is present in the majority of cells around day 9, indicating at least a 3-day lag between *Xist* coating and macroH2A recruitment (Mermoud et al. 1999; Rasmussen et al. 2001). The central region of *Xist* RNA is required for macroH2A recruitment to the Xi during ES cell differentiation (Wutz et al. 2002). Since the region of *Xist* RNA necessary for macroH2A recruitment is different from that necessary for transcriptional silencing, it seems likely that *Xist* RNA functions as a scaffold to direct multiple chromatin-modifying activities to the Xi in a developmentally regulated fashion.

5.3
Extraembryonic Cells

In the female mouse embryo and in ES cells, X chromosome inactivation is random in that either the maternally or the paternally inherited X chromosome (Xm and Xp, respectively) can be inactivated. However, in the extraembryonic, or placental, tissues of the mouse, X-inactivation is imprinted such that the Xp is always inactivated. Random and imprinted X-inactivation differ in the timing of appearance of histone modifications relative to *Xist* RNA coating and silencing (Fig. 3). During random X-inactivation in ES cells, histone modifications occur concomitantly or very shortly after the initial *cis*-spread of *Xist* RNA that triggers silencing. In contrast, there is a noticeable delay between *Xist*-mediated silencing and the acquisition of histone modifications during imprinted X-inactivation (Huynh and Lee 2003; Okamoto et al. 2004).

The Xi in extraembryonic cells shows the same chromatin modifications that are observed on the Xi in embryonic cells. However, the order in which these modifications appear on the Xi is different between random and imprinted X-inactivation (Fig. 3). In differentiating ES cells, hypoacetylation of H3-K9, hypomethylation of H3-K4 and enrichment of H3–3 mK27 and H3–2 mK9 on the Xi occur in the same time frame, and enrichment of macroH2A is a much later event. In contrast, during imprinted X-inactivation in early embryos, hypoacetylation of H4 and H3-K9 and hypomethylation of H3-K4 are observed first, followed by enrichment of H3–3 mK27 and macroH2A, and finally enrichment of H3–2 mK9 (Mak et al. 2004; Okamoto et al. 2004). Thus, H3–2 mK9 accumulation on the Xi appears to be coincident with H3–3 mK27 accumulation during random X-inactivation and is detected slightly later during imprinted X-inactivation. This apparent difference in timing of H3–2 mK9 accumulation may be due to the finer temporal resolution of the acquisition of histone modifications in embryos than in differentiating ES cells, as all the cells in embryos are synchronized for initiation of imprinted X-inactivation. The second difference between random and imprinted X-inactivation, the early appearance of macroH2A during imprinted X-inactivation, is too large to be explained by the difference in synchronization of initiation of X-inactivation during imprinted and random X-inactivation. The early appearance of macroH2A on the Xi during imprinted X-inactivation suggests that this variant histone may be involved in initiation of X-inactivation in extraembryonic cells.

5.4
Reactivation of the Xi

In the preimplantation embryo, which gives rise first to the placenta and subsequently to the embryo proper, all cells undergo imprinted X-inactivation. As a result, one would expect cells of both the embryonic and extraembryonic lineages to show imprinted X-inactivation (Fig. 4). As this is not the case, the Xp must undergo reactivation before random X-inactivation can occur. Indeed, reactivation of the Xp has been observed. In the subset of cells that will give rise to the embryo, *Xist* expression is downregulated and a number of associated chromatin modifications are reversed (Mak et al. 2004; Okamoto et al. 2004). As *Xist* levels drop, Eed/Ezh2 dissociates from the Xi, and following this, H3–3 mK27 is lost from this chromosome. A second example of reactivation during development occurs in the cells that give rise to gametes, also known as primordial germ cells (Nesterova et al. 2002). It will be interesting to determine whether these cells also establish and reverse chromatin modifications characteristic of the Xi.

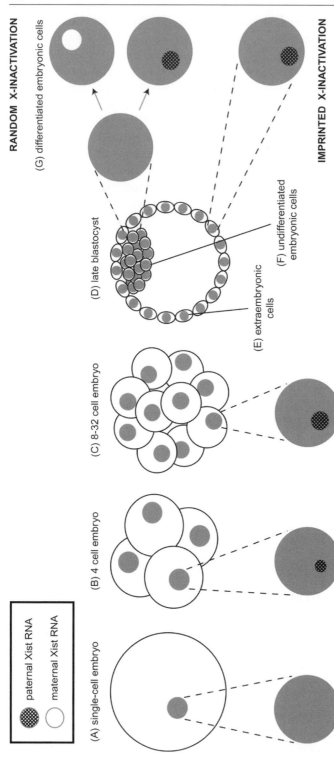

Fig. 4. X-inactivation in preimplantation mouse embryos. Cells of the early mouse embryo (**A–D**) undergo imprinted X-inactivation. Cells that give rise to the extraembryonic lineages maintain the inactive Xp (**E**). Reactivation of the Xp occurs in the cells that give rise to the embryo proper (**F**). These cells later undergo random X-inactivation upon differentiation (**G**)

6
Chromatin Features of the X Chromosomes Prior to X-Inactivation

Differential regulation of the *Xist* loci on the two X chromosomes is responsible for directing the different fates of two equivalent X chromosomes during differentiation. While the molecular mechanisms used to determine how *Xist* is silenced on one chromosome and upregulated on the other remain mysterious, it has been postulated that an early step in X-inactivation involves placing an epigenetic mark on one X chromosome to designate that chromosome as the future Xa (Plath et al. 2002).

6.1
Imprinted X-Inactivation

During imprinted X-inactivation the X chromosome inherited from the father is always silenced. This argues that an epigenetic mark(s) placed during gametogenesis designates the Xm as the future Xa and/or the Xp as the future Xi. Several lines of evidence suggest that the imprint lies on the Xm and is established during oogenesis (Huynh and Lee 2001; Plath et al. 2002). The imprint is likely to be DNA methylation, as almost all known imprinted loci are controlled by parent-of-origin-specific CpG methylation (Reik and Walter 2001).

No differentially methylated domain has yet been shown to be required for establishing and maintaining imprinted X-inactivation in extraembryonic tissues. Two CpG-rich regions have been postulated to contain the imprint. First, DNA methylation of the *Xist* promoter on the Xm could ensure that *Xist* remains silent (Ariel et al. 1995; Zuccotti and Monk 1995, 1996). It remains unclear whether the *Xist* promoter is differentially methylated in the male and female germlines, as different techniques to analyze DNA methylation at the *Xist* promoter yielded conflicting results (Ariel et al. 1995; Zuccotti and Monk 1995; McDonald et al. 1998).

A second CpG-rich region, the DXPas34 locus, resides 15 kb downstream of *Xist* (Debrand et al. 1999). This sequence is implicated in the regulation of imprinted X-inactivation because it encompasses the major transcription initiation site for *Tsix* (Lee et al. 1999; Mise et al. 1999). *Tsix* encodes a transcript antisense to *Xist* and is expressed exclusively from the Xm in cells undergoing imprinted X-inactivation (Lee 2000; Sado et al. 2001). *Tsix* blocks the ability of *Xist* RNA to coat and silence the Xm during imprinted X-inactivation (Lee 2000; Sado et al. 2001). Deletion of DXPas34 on the Xm decreases *Tsix* transcript levels and results in improper silencing of the Xm in extraembryonic tissues (Lee 2000; Sado et al. 2001). Therefore, either DXPas34 is the site where the imprint is placed or it functions downstream to read out the imprint. While DXPas34 is hypermethylated on the Xa in differentiated cells, there are no differences in CpG methylation at DXPas34 between oocytes and spermatocytes, suggesting that DXPas34 does not contain a CpG methylation imprint (Prissette et al. 2001). DXPas34 contains multiple binding sites for the CCCTC binding factor (CTCF), a transcriptional regulator that preferentially binds unmethylated DNA (Bell and Felsenfeld 2000; Hark et al. 2000; Kanduri et al. 2000). CTCF binds these sites in vitro (Chao et al. 2002). Parent-of-origin differences in methylation of CTCF binding sites can direct imprinted gene expression (Bell and Felsenfeld 2000; Hark et al. 2000; Szabo et al. 2000), suggesting that this mechanism may also be employed to regulate imprinted X-inactivation. However, CpG methylation does not affect the ability of CTCF to bind to DXPas34; rather, methylation of all cytosines blocks this interaction in vitro (Chao et al. 2002), suggesting that if the imprint resides at DXPas34, it is likely to be non-CpG methylation.

Although the location of the imprint remains unknown, the timing of its readout is well characterized. RT-PCR for X-linked genes in female embryos manipulated such that they contain two maternally or two paternally inherited X chromosomes indicates that there is silencing of genes near the *Xic* on the paternally inherited X as early as the eight-cell stage (Latham and Rambhatla 1995). *Xist* RNA partially coats the Xp in eight-cell embryos (Sheardown et al. 1997), suggesting that localized spread of *Xist* RNA mediates local silencing of X-linked genes. Indeed, two groups demonstrated that restricted *Xist* RNA coating and silencing on the Xp begins at the two-cell (Huynh and Lee 2003) or four-cell stage (Okamoto et al. 2004). These results raise the possibility that during spermatogenesis the Xp acquires an epigenetic mark that predisposes it to silencing in the early embryo. However, given the evidence for an imprint that prevents inactivation of the Xm, the alternative possibility is that expression of Xist and X-chromosome silencing is the default fate. The Xp is subject to this silencing and an imprint on the Xm prevents its inactivation. The imprint may repress *Xist* expression on the Xm and/or upregulate *Tsix*, which is expressed exclusively from the Xm in early embryos (Lee 2000; Sado et al. 2001).

6.2
Random X-Inactivation

Female ES cells do not carry an imprint that determines which X chromosome will be active and which will be silenced. However, it seems likely that some type of epigenetic mark is used to distinguish the future active X from the future inactive X in embryonic cells which are about to initiate X-inactivation. While no differences between the two Xa's in female ES cells have been reported, both the X chromosomes share a number of unusual chromatin features that may be important for X-inactivation to occur upon differentiation. A hotspot of H3–2 mK9 can be detected on the single X chromosome in male ES cells and both X chromosomes in female ES cells. It is approximately

100 kb in length and centered 55 kb upstream of the *Xist* promoter (Heard et al. 2001). When differentiation is induced, H3-K9 methylation appears to spread from this site, closely following *Xist* RNA coating of Xi (Heard et al. 2001), suggesting that this hotspot may serve as a nucleation center for the spread of H3-2 mK9 during differentiation. These data suggest that this hotspot is an unusual regulatory element that directs alterations in chromatin structure in *cis*. Several key mechanistic questions about the hotspot remain: How is it established? Does *Xist* RNA play a role in its establishment? And how is the spread of H3-2 mK9 blocked on the active X chromosome?

There is an intriguing possibility that ES cells can sense whether they have one or two X chromosomes and that this information is used to direct the pattern of chromatin modifications on the X chromosomes. ChIP experiments suggest that each X chromosome in female ES cells is distinguished from the single X chromosome in male ES cells by an increase in H3-K4 methylation and a decrease in H3-K9 methylation (O'Neill et al. 2003). In addition, each X chromosome in female cells displays an increase in acetylation on all four core histones relative to the single X chromosome in males (O'Neill et al. 2003). These differences in histone methylation and acetylation may act to designate the cell as female so that X-inactivation can occur upon differentiation.

A recent study indicates that H3-2 mK4 is a modification that precedes monoallelic expression of X-linked genes. H3-2 mK4 is enriched at the promoters, but not the transcribed region, of X-linked genes in male and female ES cells (Rougeulle et al. 2003). In contrast, most biallelically expressed genes exhibit a more uniform distribution of H3-2 mK4 at the promoter and in the coding sequences. Interestingly, several imprinted genes, which, like X-linked genes, show monoallelic expression in somatic cells, are also enriched for H3-2 mK4 at their promoters in ES cells. In addition, *Smcx*, an X-linked gene that escapes X-inactivation, does not show promoter-specific H3-K4 methylation in ES cells. These results indicate that the enrichment of H3-2 mK4 exclusively at the promoter in ES cells may mark genes for monoallelic expression upon differentiation. H3-K4 methylation is generally associated with transcriptional activity in differentiated cells (Strahl et al. 1999), suggesting that this modification may have different roles before and after differentiation of pluripotent stem cells.

The *Xist* gene also shows monoallelic expression in differentiated cells, but does not exhibit promoter-specific enrichment of H3-2 mK4 in ES cells. Instead, *Xist* is enriched for H3-2 mK4 in both its promoter and transcribed sequences (Rougeulle et al. 2003; Morey et al. 2004). Therefore the *Xist* gene is unusual among the genes that are monoallelically expressed in differentiated cells in that its pattern of H3-2 mK4 enrichment in ES cells is more similar to that exhibited by biallelically expressed genes. In ES cells the *Xist* locus is transcribed in both the sense and antisense orientations, generating *Xist* and *Tsix* transcripts (Lee and Lu 1999; Mise et al. 1999). These two non-coding RNAs are simultaneously expressed at low levels from both Xa's. Upon differentiation, *Xist* RNA coats and silences the Xi, and *Tsix* expression is extinguished on this chromosome (Lee and Lu 1999). *Tsix* and *Xist* continue to be expressed from the Xa in a brief window after X-inactivation is initiated, and *Tsix* is thought to block *Xist*-mediated silencing of the Xa during this period (Huynh and Lee 2001; Lee and Lu 1999). *Tsix* is implicated in regulating the H3–2 mK4 distribution at the *Xist* locus in undifferentiated ES cells, as the *Xist* gene body enrichment of H3–2 mK4 in ES cells requires the *Tsix* promoter region (Morey et al. 2004). Therefore, this non-coding RNA may regulate chromatin structure at the *Xist* locus in ES cells. In yeast, elevated levels of H3–2 mK4 result in part from the association of the Set1 H3 HMTase with elongating RNA polymerase (Krogan et al. 2003; Ng et al. 2003), raising the possibility that the *Tsix* promoter may direct assembly of a transcription complex containing a mammalian H3-K4 HMTase.

7
Conclusion

The Xi differs from the Xa and autosomes in differentiated cells, as it is characterized by a unique combination of epigenetic features. The non-coding RNAs *Xist* and *Tsix* are crucial in establishing this difference. *Xist* is believed to act as a scaffold to coordinately recruit multiple chromatin-modifying activities to the Xi, including histone methyltransferases, histone deacetylases and DNA methyltransferases. These activities are recruited during development in a temporally regulated manner that appears to be tissue-specific. The marked presence or absence of specific chromatin modifications on the Xi suggests that these modifications are involved in establishing and/or maintaining a transcriptionally silent state on the Xi. Studies of these modifications show that they act in combination, underlining the importance of multiple redundant mechanisms to regulate the X-inactivation process. Identification of the enzymatic activities that mediate the changes in histone methylation and acetylation that occur during X chromosome silencing will be crucial to understanding how these activities are targeted to the Xi. Elucidating the mechanisms by which these marks are established and how they act to mediate transcriptional silencing are the next major challenges in the study of how chromatin structure regulates gene expression.

Acknowledgements. We thank Geeta Narlikar, Hiten Madhani, Mary Kate Alexander, Angela Anderson, Cecile de la Cruz, Susanna Mlynarczyk-Evans and Dmitri Nusinow for critical reading of this manuscript and helpful suggestions. B.P. is a Pew Scholar and is funded by grants from the National Institute of Health, Howard Hughes Medical Institute and the Sandler Family Foundation. H.R.C. is supported by a National Science Foundation graduate fellowship.

References

Angelov D, Molla A, Perche PY, Hans F, Cote J, Khochbin S, Bouvet P, Dimitrov S (2003) The histone variant macroH2A interferes with transcription factor binding and SWI/SNF nucleosome remodeling. Mol Cell 11:1033–1041

Ariel M, Robinson E, McCarrey JR, Cedar H (1995) Gamete-specific methylation correlates with imprinting of the murine Xist gene. Nat Genet 9:312–315

Azuara V, Brown KE, Williams RR, Webb N, Dillon N, Festenstein R, Buckle V, Merkenschlager M, Fisher AG (2003) Heritable gene silencing in lymphocytes delays chromatid resolution without affecting the timing of DNA replication. Nat Cell Biol 5:668–674

Bannister AJ, Zegerman P, Partridge JF, Miska EA, Thomas JO, Allshire RC, Kouzarides T (2001) Selective recognition of methylated lysine 9 on histone H3 by the HP1 chromo domain. Nature 410:120–124

Bannister AJ, Schneider R, Kouzarides T (2002) Histone methylation: dynamic or static? Cell 109:801–806

Barr M, Carr D (1961) Correlations between sex chromatin and sex chromosomes. Acta Cytol 6:34–45

Bartlett MH, Adra CN, Park J, Chapman VM, McBurney MW (1991) DNA methylation of two X chromosome genes in female somatic and embryonal carcinoma cells. Somat Cell Mol Genet 17:35–47

Beard C, Li E, Jaenisch R (1995) Loss of methylation activates Xist in somatic but not in embryonic cells. Genes Dev 9:2325–2334

Beletskii A, Hong YK, Pehrson J, Egholm M, Strauss WM (2001) PNA interference mapping demonstrates functional domains in the noncoding RNA Xist. Proc Natl Acad Sci USA 98:9215–9220

Bell AC, Felsenfeld G (2000) Methylation of a CTCF-dependent boundary controls imprinted expression of the Igf2 gene. Nature 405:482–485

Belyaev ND, Keohane AM, Turner BM (1996) Histone H4 acetylation and replication timing in Chinese hamster chromosomes. Exp Cell Res 225:277–285

Bestor TH (2000) The DNA methyltransferases of mammals. Hum Mol Genet 9:2395–2402

Bischoff A, Albers J, Kharboush I, Stelzer E, Cremer T, Cremer C (1993) Differences of size and shape of active and inactive X-chromosome domains in human amniotic fluid cell nuclei. Microsci Res Tech 25:68–77

Boggs BA, Connors B, Sobel RE, Chinault AC, Allis CD (1996) Reduced levels of histone H3 acetylation on the inactive X chromosome in human females. Chromosoma 105:303–309

Boggs BA, Cheung P, Heard E, Spector DL, Chinault AC, Allis CD (2002) Differentially methylated forms of histone H3 show unique association patterns with inactive human X chromosomes. Nat Genet 30:73–76

Brockdorff N, Ashworth A, Kay GF, McCabe VM, Norris DP, Cooper PJ, Swift S, Rastan S (1992) The product of the mouse Xist gene is a 15 kb inactive X-specific transcript containing no conserved ORF and located in the nucleus. Cell 71:515–526

Brown CJ, Willard HF (1994) The human X-inactivation centre is not required for maintenance of X-chromosome inactivation. Nature 368:154–156

Brown CJ, Hendrich BD, Rupert JL, Lafreniere RG, Xing Y, Lawrence J, Willard HF (1992) The human XIST gene: analysis of a 17 kb inactive X-specific RNA that contains conserved repeats and is highly localized within the nucleus. Cell 71:527–542

Cao R, Wang L, Wang H, Xia L, Erdjument-Bromage H, Tempst P, Jones RS, Zhang Y (2002) Role of histone H3 lysine 27 methylation in Polycomb-group silencing. Science 298:1039–1043

Carrel L, Cottle AA, Goglin KC, Willard HF (1999) A first-generation X-inactivation profile of the human X chromosome. Proc Natl Acad Sci USA 96:14440–14444

Chadwick BP, Willard HF (2001a) Histone H2A variants and the inactive X chromosome: identification of a second macroH2A variant. Hum Mol Genet 10:1101–1113

Chadwick BP, Willard HF (2001b) A novel chromatin protein, distantly related to histone H2A, is largely excluded from the inactive X chromosome. J Cell Biol 152:375–384

Chadwick BP, Willard HF (2003) Chromatin of the Barr body: histone and non-histone proteins associated with or excluded from the inactive X chromosome. Hum Mol Genet 12:2167–2178

Chadwick BP, Valley CM, Willard HF (2001) Histone variant macroH2A contains two distinct macrochromatin domains capable of directing macroH2A to the inactive X chromosome. Nucleic Acids Res 29:2699–2705

Changolkar LN, Pehrson JR (2002). Reconstitution of nucleosomes with histone macroH2A1.2. Biochemistry 41:179–184

Chao W, Huynh KD, Spencer RJ, Davidow LS, Lee JT (2002) CTCF, a candidate trans-acting factor for X-inactivation choice. Science 295:345–347

Chaumeil J, Okamoto I, Guggiari M, Heard E (2002) Integrated kinetics of X chromosome inactivation in differentiating embryonic stem cells. Cytogenet Genome Res 99:75–84

Chen C, Yang TP (2001) Nucleosomes are translationally positioned on the active allele and rotationally positioned on the inactive allele of the HPRT promoter. Mol Cell Biol 21:7682–7695

Chen T, Ueda Y, Dodge JE, Wang Z, Li E (2003) Establishment and maintenance of genomic methylation patterns in mouse embryonic stem cells by Dnmt3a and Dnmt3b. Mol Cell Biol 23:5594–5605

Clemson CM, Chow JC, Brown CJ, Lawrence JB (1998) Stabilization and localization of Xist RNA are controlled by separate mechanisms and are not sufficient for X inactivation. J Cell Biol 142:13–23

Clemson CM, McNeil JA, Willard HF, Lawrence JB (1996) XIST RNA paints the inactive X chromosome at interphase: evidence for a novel RNA involved in nuclear/chromosome structure. J Cell Biol 132:259–275

Cohen SM, Brylawski BP, Cordeiro-Stone M, Kaufman DG (2003) Same origins of DNA replication function on the active and inactive human X chromosomes. J Cell Biochem 88:923–931

Costanzi C, Pehrson JR (1998) Histone macroH2A1 is concentrated in the inactive X chromosome of female mammals. Nature 393:599–601

Costanzi C, Pehrson JR (2001) MACROH2A2, a new member of the MARCOH2A core histone family. J Biol Chem 276:21776–21784

Csankovszki G, Nagy A, Jaenisch R (2001) Synergism of Xist RNA, DNA methylation, and histone hypoacetylation in maintaining X chromosome inactivation. J Cell Biol 153:773–784

Csankovszki G, Panning B, Bates B, Pehrson JR, Jaenisch R (1999) Conditional deletion of Xist disrupts histone macroH2A localization but not maintenance of X inactivation (letter). Nat Genet 22:323–324

Czermin B, Melfi R, McCabe D, Seitz V, Imhof A, Pirrotta V (2002) Drosophila enhancer of Zeste/ESC complexes have a histone H3 methyltransferase activity that marks chromosomal Polycomb sites. Cell 111:185–196

Datta J, Ghoshal K, Sharma SM, Tajima S, Jacob ST (2003) Biochemical fractionation reveals association of DNA methyltransferase (Dnmt) 3b with Dnmt1 and that of Dnmt 3a with a histone H3 methyltransferase and Hdac1. J Cell Biochem 88:855–864

Debrand E, Chureau C, Arnaud D, Avner P, Heard E (1999) Functional analysis of the DXPas34 locus, a 3′ regulator of Xist expression. Mol Cell Biol 19:8513–8525

Denisenko O, Shnyreva M, Suzuki H, Bomsztyk K (1998) Point mutations in the WD40 domain of Eed block its interaction with Ezh2. Mol Cell Biol 18:5634–5642

Dietzel S, Schiebel K, Little G, Edelmann P, Rappold GA, Eils R, Cremer C, Cremer T (1999) The 3D positioning of ANT2 and ANT3 genes within female X chromosome territories correlates with gene activity. Exp Cell Res 252:363–375

Disteche CM, Filippova GN, Tsuchiya KD (2002) Escape from X inactivation. Cytogenet Genome Res 99:36–43

Eils R, Dietzel S, Bertin E, Schrock E, Speicher MR, Ried T, Robert-Nicoud M, Cremer C, Cremer T (1996) Three-dimensional reconstruction of painted human interphase chromosomes: active and inactive X chromosome territories have similar volumes but differ in shape and surface structure. J Cell Biol 135:1427–1440

Erhardt S, Su IH, Schneider R, Barton S, Bannister AJ, Perez-Burgos L, Jenuwein T, Kouzarides T, Tarakhovsky A, Surani MA (2003) Consequences of the depletion of zygotic and embryonic enhancer of zeste 2 during preimplantation mouse development. Development 130:4235–4248

Fischle W, Wang Y, Jacobs SA, Kim Y, Allis CD, Khorasanizadeh S (2003) Molecular basis for the discrimination of repressive methyl-lysine marks in histone H3 by Polycomb and HP1 chromodomains. Genes Dev 17:1870–1881

Flejter WL, van Dyke DL, Weiss L (1984) Bends in human mitotic metaphase chromosomes, including a bend marking the X-inactivation center. Am J Hum Genet 36:218–226

Francis NJ, Kingston RE (2001) Mechanisms of transcriptional memory. Nat Rev Mol Cell Biol 2:409–421

Freitag M, Hickey PC, Khlafallah TK, Read ND, Selker EU (2004) HP1 is essential for DNA methylation in *Neurospora*. Mol Cell 13:427–434

Ganesan S, Silver DP, Greenberg RA, Avni D, Drapkin R, Miron A, Mok SC, Randrianarison V, Brodie S, Salstrom J, Rasmussen TP, Klimke A, Marrese C, Marahrens Y, Deng CX, Feunteun J, Livingston DM (2002) BRCA1 supports XIST RNA concentration on the inactive X chromosome. Cell 111:393–405

Gartler SM, Goldstein L, Tyler-Freer SE, Hansen RS (1999) The timing of XIST replication: dominance of the domain. Hum Mol Genet 8:1085–1089

Gilbert N, Boyle S, Sutherland H, de Las Heras J, Allan J, Jenuwein T, Bickmore WA (2003) Formation of facultative heterochromatin in the absence of HP1. EMBO J 22:5540–5550

Gilbert SL, Sharp PA (1999) Promoter-specific hypoacetylation of X-inactivated genes. Proc Natl Acad Sci USA 96:13825–13830

Goodfellow PJ, Mondello C, Darling SM, Pym B, Little P, Goodfellow PN (1988) Absence of methylation of a CpG-rich region at the 5′ end of the MIC2 gene on the active X, the inactive X, and the Y chromosome. Proc Natl Acad Sci USA 85:5605–5609

Goto Y, Gomez M, Brockdorff N, Feil R (2002) Differential patterns of histone methylation and acetylation distinguish active and repressed alleles at X-linked genes. Cytogenet Genome Res 99:66–74

Hall LL, Byron M, Sakai K, Carrel L, Willard HF, Lawrence JB (2002) An ectopic human XIST gene can induce chromosome inactivation in postdifferentiation human HT-1080 cells. Proc Natl Acad Sci USA 99:8677–8682

Hansen RS (2003) X inactivation-specific methylation of LINE-1 elements by DNMT3B: implications for the Lyon repeat hypothesis. Hum Mol Genet 12:2559–2567

Hansen RS, Canfield TK, Lamb MM, Gartler SM, Laird CD (1993) Association of fragile X syndrome with delayed replication of the FMR1 gene. Cell 73:1403–1409

Hansen RS, Canfield TK, Gartler SM (1995) Reverse replication timing for the XIST gene in human fibroblasts. Hum Mol Genet 4:813–820

Hansen RS, Canfield TK, Fjeld AD, Gartler SM (1996) Role of late replication timing in the silencing of X-linked genes. Hum Mol Genet 5:1345–1353

Hansen RS, Stoger R, Wijmenga C, Stanek AM, Canfield TK, Luo P, Matarazzo MR, D'Esposito M, Feil R, Gimelli G, Weemaes CM, Laird CD, Gartler SM (2000) Escape from gene silencing in ICF syndrome: evidence for advanced replication time as a major determinant. Hum Mol Genet 9:2575–2587

Hark AT, Schoenherr CJ, Katz DJ, Ingram RS, Levorse JM, Tilghman SM (2000) CTCF mediates methylation-sensitive enhancer-blocking activity at the H19/Igf2 locus. Nature 405:486–489

Heard E, Rougeulle C, Arnaud D, Avner P, Allis CD, Spector DL (2001) Methylation of histone H3 at Lys-9 is an early mark on the X chromosome during X inactivation. Cell 107:727–738

Helbig R, Fackelmayer FO (2003) Scaffold attachment factor A (SAF-A) is concentrated in inactive X chromosome territories through its RGG domain. Chromosoma 112:173–182

Huynh KD, Lee JT (2001) Imprinted X inactivation in eutherians: a model of gametic execution and zygotic relaxation. Curr Opin Cell Biol 13:690–697

Huynh KD, Lee JT (2003) Inheritance of a pre-inactivated paternal X chromosome in early mouse embryos. Nature 426:857–862

Jackson JP, Lindroth AM, Cao X, Jacobsen SE (2002) Control of CpNpG DNA methylation by the KRYPTONITE histone H3 methyltransferase. Nature 416:556–560

Jenuwein T, Allis CD (2001) Translating the histone code. Science 293:1074–1080

Jeppesen P, Turner BM (1993) The inactive X chromosome in female mammals is distinguished by a lack of histone H4 acetylation, a cytogenetic marker for gene expression. Cell 74:281–289

Kanduri C, Pant V, Loukinov D, Pugacheva E, Qi CF, Wolffe A, Ohlsson R, Lobanenkov VV (2000) Functional association of CTCF with the insulator upstream of the H19 gene is parent of origin-specific and methylation-sensitive. Curr Biol 10:853–856

Ke X, Collins A (2003) CpG islands in human X-inactivation. Ann Hum Genet 67:242–249

Keohane AM, O'Neill LP, Belyaev ND, Lavender JS, Turner BM (1996) X-inactivation and histone H4 acetylation in embryonic stem cells. Dev Biol 180:618–630

Krogan NJ, Kim M, Tong A, Golshani A, Cagney G, Canadien V, Richards DP, Beattie BK, Emili A, Boone C, Shilatifard A, Buratowski S, Greenblatt J (2003) Methylation of histone H3 by Set2 in Saccharomyces cerevisiae is linked to transcriptional elongation by RNA polymerase II. Mol Cell Biol 23:4207–4218

Kuzmichev A, Nishioka K, Erdjument-Bromage H, Tempst P, Reinberg D (2002) Histone methyltransferase activity associated with a human multiprotein complex containing the Enhancer of Zeste protein. Genes Dev 16:2893–2905

Lachner M, O'Carroll D, Rea S, Mechtler K, Jenuwein T (2001) Methylation of histone H3 lysine 9 creates a binding site for HP1 proteins. Nature 410:116–120

Lachner M, O'Sullivan RJ, Jenuwein T (2003) An epigenetic road map for histone lysine methylation. J Cell Sci 116:2117–2124

Lamond AI, Earnshaw WC (1998) Structure and function in the nucleus. Science 280:547–553

Landry J, Sutton A, Hesman T, Min J, Xu RM, Johnston M, Sternglanz R (2003) Set2-catalyzed methylation of histone H3 represses basal expression of GAL4 in Saccharomyces cerevisiae. Mol Cell Biol 23:5972–5978

Latham KE, Rambhatla L (1995) Expression of X-linked genes in androgenetic, gynogenetic, and normal mouse preimplantation embryos. Dev Genet 17:212–222

Leach TJ, Mazzeo M, Chotkowski HL, Madigan JP, Wotring MG, Glaser RL (2000) Histone H2A.Z is widely but nonrandomly distributed in chromosomes of Drosophila melanogaster. J Biol Chem 275:23267–23272

Lee JT (2000) Disruption of imprinted X inactivation by parent-of-origin effects at Tsix. Cell 103:17–27

Lee JT, Lu N (1999) Targeted mutagenesis of Tsix leads to nonrandom X inactivation. Cell 99:47–57

Lee JT, Davidow LS, Warshawsky D (1999) Tsix, a gene antisense to Xist at the X-inactivation centre. Nat Genet 21:400–404

Lehnertz B, Ueda Y, Derijck AA, Braunschweig U, Perez-Burgos L, Kubicek S, Chen T, Li E, Jenuwein T, Peters AH (2003) Suv39h-mediated histone H3 lysine 9 methylation directs DNA methylation to major satellite repeats at pericentric heterochromatin. Curr Biol 13:1192–1200

Li B, Howe L, Anderson S, Yates JR 3rd, Workman JL (2003) The Set2 histone methyltransferase functions through the phosphorylated carboxyl-terminal domain of RNA polymerase II. J Biol Chem 278:8897–8903

Litt MD, Hansen RS, Hornstra IK, Gartler SM, Yang TP (1997) 5-Azadeoxycytidine-induced chromatin remodeling of the inactive X-linked HPRT gene promoter occurs prior to transcription factor binding and gene reactivation. J Biol Chem 272:14921–14926

Lyon MF (1961) Gene action in the X-chromosome of the mouse (*Mus musculus* L). Nature 190:372–373

Maison C, Bailly D, Peters AH, Quivy JP, Roche D, Taddei A, Lachner M, Jenuwein T, Almouzni G (2002) Higher-order structure in pericentric heterochromatin involves a distinct pattern of histone modification and an RNA component. Nat Genet 30:329–334

Mak W, Baxter J, Silva J, Newall AE, Otte AP, Brockdorff N (2002) Mitotically stable association of polycomb group proteins eed and enx1 with the inactive X chromosome in trophoblast stem cells. Curr Biol 12:1016–1020

Mak W, Nesterova TB, de Napoles M, Appanah R, Yamanaka S, Otte AP, Brockdorff N (2004) Reactivation of the paternal X chromosome in early mouse embryos. Science 303:666–669

Marahrens Y, Panning B, Dausman J, Strauss W, Jaenisch R (1997) Xist-deficient mice are defective in dosage compensation but not spermatogenesis. Genes Dev 11:156–166

McDonald LE, Paterson CA, Kay GF (1998) Bisulfite genomic sequencing-derived methylation profile of the *Xist* gene throughout early mouse development. Genomics 54:379–386

Meneghini MD, Wu M, Madhani HD (2003) Conserved histone variant H2A.Z protects euchromatin from the ectopic spread of silent heterochromatin. Cell 112:725–736

Mermoud JE, Costanzi C, Pehrson JR, Brockdorff N (1999) Histone macroH2A1.2 relocates to the inactive X chromosome after initiation and propagation of X-inactivation. J Cell Biol 147:1399–1408

Mermoud JE, Popova B, Peters AH, Jenuwein T, Brockdorff N (2002) Histone h3 lysine 9 methylation occurs rapidly at the onset of random X chromosome inactivation. Curr Biol 12:247–251

Mise N, Goto Y, Nakajima N, Takagi N (1999) Molecular cloning of antisense transcripts of the mouse *Xist* gene. Biochem Biophys Res Commun 258:537–541

Mohandas T, Sparkes RS, Shapiro LJ (1981) Reactivation of an inactive human X chromosome: evidence for X inactivation by DNA methylation. Science 211:393–396

Morey C, Navarro P, Debrand E, Avner P, Rougeulle C, Clerc P (2004) The region 3′ to Xist mediates X chromosome counting and H3 Lys-4 dimethylation within the *Xist* gene. EMBO J 23:594–604

Muller J, Hart CM, Francis NJ, Vargas ML, Sengupta A, Wild B, Miller EL, O'Connor MB, Kingston RE, Simon JA (2002) Histone methyltransferase activity of a *Drosophila* Polycomb group repressor complex. Cell 111:197–208

Nesterova TB, Mermoud JE, Brockdorff N, Hilton K, McLaren A, Surani MA, Pehrson J (2002) Xist expression and macroH2A1.2 localisation in mouse primordial and pluripotent embryonic germ cells. Differentiation 69:216–225

Ng HH, Robert F, Young RA, Struhl K (2003) Targeted recruitment of Set1 histone methylase by elongating Pol II provides a localized mark and memory of recent transcriptional activity. Mol Cell 11:709–719

O'Neill LP, Randall TE, Lavender J, Spotswood HT, Lee JT, Turner BM (2003) X-linked genes in female embryonic stem cells carry an epigenetic mark prior to the onset of X inactivation. Hum Mol Genet 12:1783–1790

Okamoto I, Otte AP, Allis CD, Reinberg D, Heard E (2004) Epigenetic dynamics of imprinted X inactivation during early mouse development. Science 303:644–649

Pak DT, Pflumm M, Chesnokov I, Huang DW, Kellum R, Marr J, Romanowski P, Botchan MR (1997) Association of the origin recognition complex with heterochromatin and HP1 in higher eukaryotes. Cell 91:311–323

Panning B, Jaenisch R (1996) DNA hypomethylation can activate Xist expression and silence X-linked genes. Genes Dev 10:1991–2002

Panning B, Dausman J, Jaenisch R (1997) X chromosome inactivation is mediated by Xist RNA stabilization. Cell 90:907–916

Pehrson JR, Costanzi C, Dharia C (1997) Developmental and tissue expression patterns of histone macroH2A1 subtypes. J Cell Biochem 65:107–113

Pehrson JR, Fried VA (1992) MacroH2A, a core histone containing a large nonhistone region. Science 257:1398–1400

Pehrson JR, Fuji RN (1998) Evolutionary conservation of histone macroH2A subtypes and domains. Nucleic Acids Res 26:2837–2842

Penny GD, Kay GF, Sheardown SA, Rastan S, Brockdorff N (1996) Requirement for Xist in X chromosome inactivation. Nature 379:131–137

Perche PY, Vourc'h C, Konecny L, Souchier C, Robert-Nicoud M, Dimitrov S, Khochbin S (2000) Higher concentrations of histone macroH2A in the Barr body are correlated with higher nucleosome density. Curr Biol 10:1531–1534

Peters AH, Mermoud JE, O'Carroll D, Pagani M, Schweizer D, Brockdorff N, Jenuwein T (2002) Histone H3 lysine 9 methylation is an epigenetic imprint of facultative heterochromatin. Nat Genet 30:77–80

Peters AH, Kubicek S, Mechtler K, O'Sullivan RJ, Derijck AA, Perez-Burgos L, Kohlmaier A, Opravil S, Tachibana M, Shinkai Y, Martens JH, Jenuwein T (2003) Partitioning and plasticity of repressive histone methylation states in mammalian chromatin. Mol Cell 12:1577–1589

Pfeifer GP, Tanguay RL, Steigerwald SD, Riggs AD (1990) In vivo footprint and methylation analysis by PCR-aided genomic sequencing: comparison of active and inactive X chromosomal DNA at the CpG island and promoter of human PGK-1. Genes Dev 4:1277–1287

Plath K, Mlynarczyk-Evans S, Nusinow DA, Panning B (2002) Xist RNA and the mechanism of X chromosome inactivation. Annu Rev Genet 36:233–278

Plath K, Fang J, Mlynarczyk-Evans SK, Cao R, Worringer KA, Wang H, de la Cruz CC, Otte AP, Panning B, Zhang Y (2003) Role of histone H3 lysine 27 methylation in X inactivation. Science 300:131–135

Prissette M, El-Maarri O, Arnaud D, Walter J, Avner P (2001) Methylation profiles of DXPas34 during the onset of X-inactivation. Hum Mol Genet 10:31–38

Rangasamy D, Berven L, Ridgway P, Tremethick DJ (2003) Pericentric heterochromatin becomes enriched with H2A.Z during early mammalian development. EMBO J 22:1599–1607

Rasmussen TP, Huang T, Mastrangelo MA, Loring J, Panning B, Jaenisch R (1999) Messenger RNAs encoding mouse histone macroH2A1 isoforms are expressed at similar levels in male and female cells and result from alternative splicing. Nucleic Acids Res 27:3685–3689

Rasmussen TP, Wutz AP, Pehrson JR, Jaenisch RR (2001) Expression of Xist RNA is sufficient to initiate macrochromatin body formation. Chromosoma 110:411–420

Rea S, Eisenhaber F, O'Carroll D, Strahl BD, Sun ZW, Schmid M, Opravil S, Mechtler K, Ponting CP, Allis CD, Jenuwein T (2000) Regulation of chromatin structure by site-specific histone H3 methyltransferases. Nature 406:593–599

Reddy KS, Savage JR, Papworth DG (1988) Replication kinetics of X chromosomes in fibroblasts and lymphocytes. Hum Genet 79:44–48

Reik W, Walter J (2001) Genomic imprinting: parental influence on the genome. Nat Rev Genet 2:21–32

Rice JC, Briggs SD, Ueberheide B, Barber CM, Shabanowitz J, Hunt DF, Shinkai Y, Allis CD (2003) Histone methyltransferases direct different degrees of methylation to define distinct chromatin domains. Mol Cell 12:1591–1598

Richards EJ, Elgin SC (2002) Epigenetic codes for heterochromatin formation and silencing: rounding up the usual suspects. Cell 108:489–500

Rougeulle C, Navarro P, Avner P (2003) Promoter-restricted H3 Lys 4 di-methylation is an epigenetic mark for monoallelic expression. Hum Mol Genet 12:3343–3348

Sado T, Wang Z, Sasaki H, Li E (2001) Regulation of imprinted X-chromosome inactivation in mice by Tsix. Development 128:1275–1286

Sado T, Okano M, Li E, Sasaki H (2004) De novo DNA methylation is dispensable for the initiation and propagation of X chromosome inactivation. Development 131:975–982

Schaft D, Roguev A, Kotovic KM, Shevchenko A, Sarov M, Neugebauer KM, Stewart AF (2003) The histone 3 lysine 36 methyltransferase, SET2, is involved in transcriptional elongation. Nucleic Acids Res 31:2475–2482

Schmidt M, Migeon BR (1990) Asynchronous replication of homologous loci on human active and inactive X chromosomes. Proc Natl Acad Sci USA 87:3685–3689

Shao Z, Raible F, Mollaaghababa R, Guyon JR, Wu CT, Bender W, Kingston RE (1999) Stabilization of chromatin structure by PRC1, a Polycomb complex. Cell 98:37–46

Sheardown SA, Duthie SM, Johnston CM, Newall AE, Formstone EJ, Arkell RM, Nesterova TB, Alghisi GC, Rastan S, Brockdorff N (1997) Stabilization of Xist RNA mediates initiation of X chromosome inactivation. Cell 91:99–107

Silva J, Mak W, Zvetkova I, Appanah R, Nesterova TB, Webster Z, Peters AH, Jenuwein T, Otte AP, Brockdorff N (2003) Establishment of histone h3 methylation on the inactive X chromosome requires transient recruitment of Eed-Enx1 polycomb group complexes. Dev Cell 4:481–495

Simon JA, Tamkun JW (2002) Programming off and on states in chromatin: mechanisms of Polycomb and trithorax group complexes. Curr Opin Genet Dev 12:210–218

Slim R, Levilliers J, Ludecke HJ, Claussen U, Nguyen VC, Gough NM, Horsthemke B, Petit C (1993) A human pseudoautosomal gene encodes the ANT3 ADP/ATP translocase and escapes X-inactivation. Genomics 16:26–33

Strahl BD, Allis CD (2000) The language of covalent histone modifications. Nature 403:41–45

Strahl BD, Ohba R, Cook RG, Allis CD (1999) Methylation of histone H3 at lysine 4 is highly conserved and correlates with transcriptionally active nuclei in Tetrahymena. Proc Natl Acad Sci USA 96:14967–14972

Strahl BD, Grant PA, Briggs SD, Sun ZW, Bone JR, Caldwell JA, Mollah S, Cook RG, Shabanowitz J, Hunt DF, Allis CD (2002) Set2 is a nucleosomal histone H3-selective methyltransferase that mediates transcriptional repression. Mol Cell Biol 22:1298–1306

Szabo P, Tang SH, Rentsendorj A, Pfeifer GP, Mann JR (2000) Maternal-specific footprints at putative CTCF sites in the H19 imprinting control region give evidence for insulator function. Curr Biol 10:607–610

Tamaru H, Selker EU (2001) A histone H3 methyltransferase controls DNA methylation in Neurospora crassa. Nature 414:277–283

Tinker AV, Brown CJ (1998) Induction of XIST expression from the human active X chromosome in mouse/human somatic cell hybrids by DNA demethylation. Nucleic Acids Res 26:2935–2940

Torchia BS, Migeon BR (1995) The XIST locus replicates late on the active X, and earlier on the inactive X based on FISH DNA replication analysis of somatic cell hybrids. Somat Cell Mol Genet 21:327–333

Torchia BS, Call LM, Migeon BR (1994) DNA replication analysis of FMR1, XIST, and factor 8C loci by FISH shows nontranscribed X-linked genes replicate late. Am J Hum Genet 55:96–104

Turner BM (2000) Histone acetylation and an epigenetic code. Bioessays 22:836–845

Van der Vlag J, Otte AP (1999). Transcriptional repression mediated by the human poly-comb-group protein EED involves histone deacetylation. Nat Genet 23:474–478

Walker CL, Cargile CB, Floy KM, Delannoy M, Migeon BR (1991) The Barr body is a looped X chromosome formed by telomere association. Proc Natl Acad Sci USA 88:6191–6195

Wang H, An W, Cao R, Xia L, Erdjument-Bromage H, Chatton B, Tempst P, Roeder RG, Zhang Y (2003) mAM facilitates conversion by ESET of dimethyl to trimethyl lysine 9 of histone H3 to cause transcriptional repression. Mol Cell 12:475–487

Wang J, Mager J, Chen Y, Schneider E, Cross JC, Nagy A, Magnuson T (2001) Imprinted X inactivation maintained by a mouse Polycomb group gene. Nat Genet 28:371–375

Wolf SF, Jolly DJ, Lunnen KD, Friedmann T, Migeon BR (1984) Methylation of the hypoxan-thine phosphoribosyltransferase locus on the human X chromosome: implications for X-chromosome inactivation. Proc Natl Acad Sci USA 81:2806–2810

Wutz A, Jaenisch R (2000) A shift from reversible to irreversible X inactivation is triggered during ES cell differentiation. Mol Cell 5:695–705

Wutz A, Rasmussen TP, Jaenisch R (2002) Chromosomal silencing and localization are mediated by different domains of Xist RNA. Nat Genet 30:167–174

Xiao T, Hall H, Kizer KO, Shibata Y, Hall MC, Borchers CH, Strahl BD (2003) Phosphoryla-tion of RNA polymerase II CTD regulates H3 methylation in yeast. Genes Dev 17:654–663

Yang L, Mei Q, Zielinska-Kwiatkowska A, Matsui Y, Blackburn ML, Benedetti D, Krumm AA, Taborsky GJ Jr, Chansky HA (2003) An ERG (ets-related gene)-associated histone methyltransferase interacts with histone deacetylases 1/2 and transcription co-repres-sors mSin3A/B. Biochem J 369:651–657

Zuccotti M, Monk M (1995) Methylation of the mouse Xist gene in sperm and eggs corre-lates with imprinted Xist expression and paternal X-inactivation. Nat Genet 9:316–320

Zuccotti M, Monk M (1996) The mouse Xist gene: a model for studying the gametic imprint-ing phenomenon. Acta Genet Med Gemellol (Roma) 45:199–204

Chromatin Mechanisms in *Drosophila* Dosage Compensation

Mikko Taipale, Asifa Akhtar

Abstract   Dosage compensation ensures that males and females equalize the expression of the X-linked genes and therefore provides an exquisite model system to study chromosome-wide transcription regulation. In *Drosophila*, this is achieved by hyper-transcription of the genes on the male X chromosome. This process requires an RNA/protein-containing dosage compensation complex. Here, we discuss the current status of the known *Drosophila* complex members as well as the recent views on targeting, assembly and spreading mechanisms.

1
Introduction

Males and females differ by number or type of sex chromosomes in many animal species. Since aneuploidy is lethal to organisms in most circumstances, animals have evolved ways to ensure equal expression of genes on the sex chromosomes. These processes are collectively referred to as dosage compensation. Currently, there are three model organisms in which dosage compensation mechanisms have been studied at a molecular level. These studies have shown that dosage compensation has evolved independently several times. Despite the separate evolutionary origins, the mechanisms show striking similarities in several aspects.

In mammals, one X chromosome in females (XX) is inactivated during early embryogenesis, in order to balance gene dosage. A counting mechanism that is still poorly understood ensures that the single male X chromosome stays active and that female cells have only one active X chromosome (Avner and Heard 2001). The nematode *Caenorhabditis elegans* has evolved a repressive mechanism for dosage compensation. It is achieved by 50 % downregula-

M. Taipale, A. Akhtar
European Molecular Biology Laboratory, Gene Expression Programme, Meyerhofstrasse 1, 69117 Heidelberg, Germany, e-mail: asifa.akhtar@embl.de

Progress in Molecular and Subcellular Biology
P. Jeanteur (Ed.)
Epigenetics and Chromatin
© Springer-Verlag Berlin Heidelberg 2005

tion of transcription from the two hermaphrodite X chromosomes compared to the single X chromosome of the male worm (XO) (Meyer 2000). In contrast, in male *Drosophila*, the single X chromosome is transcriptionally upregulated two-fold to ensure similar gene expression dose with females, who carry two X chromosomes.

Mammals, fruit flies and nematodes represent only a small fraction of all animal clades, and very little is known about dosage compensation in other species. In some cases, there is clearly no need for dosage compensation. Many species have homomorphic sex chromosomes (e.g. the housefly *Musca domestica*), or their sex is determined by environmental factors (e.g. many fishes, reptiles and some insects), and therefore these species lack dosage compensation (Charlesworth 1996). Even though female birds have a degenerated W chromosome, it was thought that they do not compensate for sex-linked gene expression. This view, which was based on results from a single gene (Baverstock et al. 1982), was overturned by a recent observation that many Z-linked genes are expressed in equal levels in male (ZZ) and female (ZW) chick embryos (McQueen et al. 2001). The molecular mechanism by which dosage compensation occurs is still unknown, except that it is unlikely to happen by inactivation of one Z chromosome in males (Ellegren 2002; Kuroiwa et al. 2002). Some studies suggest that butterflies with ZW/ZZ sex determination system do not have dosage compensation (Suzuki et al. 1999), but until more extensive analyses have been carried out, these results should be interpreted with caution.

2
The MSL Complex

Dosage compensation is the prototypic example of gene regulation by epigenetic mechanisms. Since the X chromosomes of females and males have no sequence differences, the gene expression machinery is posed with several problems: how to impose a mechanism that functions sex-specifically; how to distinguish the X chromosome from the autosomes; and how to establish and maintain correct level of gene expression compared to the other sex, throughout development. It is evident from comparative studies on dosage compensation that compensated genes are regulated en bloc by specialized protein complexes, rather than genes being individually dosage-compensated by a variety of mechanisms and trans-acting regulators.

Elegant genetic and biochemical experiments have provided a framework for understanding dosage compensation in *Drosophila*. To date, six proteins (MSL-1, MSL-2, MSL-3, MOF, MLE, JIL-1) and two non-coding RNAs (roX1, roX2) have been associated with dosage compensation. Instead of a biochemical cascade, these proteins form an approximately 2-MDa multiprotein complex (Copps et al. 1998), commonly referred to as the dosage compensation complex (DCC), the MSL complex or the compensasome. Protein components, with the exception of JIL-1, are often referred to as male-specific lethals (MSLs).

MSL proteins bind specifically to hundreds of sites along the male X chromosome. This is beautifully illustrated by immunofluorescence studies on giant polytene chromosomes of *Drosophila* larvae (Fig. 1A). In addition to protein components, the X chromosome is characterized by acetylation of lysine K16 of histone H4 (H4K16Ac) (Turner et al. 1992), and enrichment of phosphorylated serine S10 of histone H3 (H3S10P) (Wang et al. 2001).

Drosophila dosage compensation genes were initially found in genetic screens, aimed at finding mutations conferring male-specific lethality. To date, all dosage compensation genes (except JIL-1) have proven to be dispensable in females, but this does not exclude the possibility that at least some of them have a function in females. Because of redundancy with other chromatin remodelling factors, the phenotype in females could be suppressed. Most MSLs have been found in several independent genetic screens, suggesting that it is unlikely that there are more dosage compensation genes with a male-specific lethal phenotype in the *Drosophila* genome (Baker et al. 1994). The MSL complex could, however, have other components that show no male-specific phenotypes. These factors could be regulated, for example, by sex-specific recruitment or post-translational modifications.

It is important to note that not all genes on the X chromosome are dosage-compensated by the MSL complex. For example, larval serum protein *LSP1α* is not dosage-compensated, and consequently females have higher amounts of *LSP1α* mRNA (Ghosh et al. 1989). In addition, there is evidence that some genes are dosage-compensated independently of the MSL complex. These genes are most likely compensated directly by Sex-lethal, the master sex-

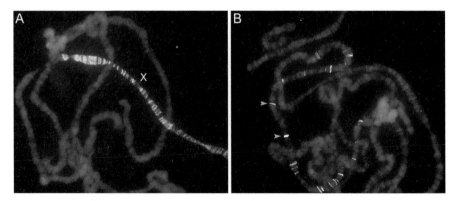

Fig. 1. A The MSL complex is found localized on hundreds of sites along the male X chromosome in wild-type flies. **B** Chromatin entry sites can be visualized as approximately 35 distinct bands on the X chromosome in *msl-3* mutant background. *Arrowheads* indicate two of the entry sites. The polytene chromosomes were stained with α-MSL-1 antibody (X chromosome-specific signal) and Hoechst 33342 (to visualize DNA)

determining gene in *Drosophila* (Baker et al. 1994; Kelley et al. 1995; Cline and Meyer 1996). Here, we shall focus solely on dosage compensation by MSL complex in *Drosophila*.

2.1
MSL-1

MSL-1 protein is 1,039 amino acids (aa) long; however, the amino acid sequence offers few clues as to its biochemical function (Palmer et al. 1993). Recently, detailed evolutionary and sequence analysis has shown that it contains two conserved domains. There is a putative leucine zipper in the N-terminus, and the C-terminus contains a novel PEHE domain with an unknown function (Marin 2003). In addition, MSL-1 has an acidic region and an S/T/P-rich region in its N-terminus (Palmer et al. 1993; Scott et al. 2000). The role of these regions is also unknown.

In yeast two-hybrid assays, MSL-1 residues 85–186 are required for interaction with the RING finger of MSL-2 (Copps et al. 1998). Residues 760–1,039 mediate MSL-1 interaction with MSL-3 in vivo and in vitro. The C-terminal region of MSL-1 has also been shown to interact with MOF in vivo (Scott et al. 2000). MSL-1 can also be co-immunoprecipitated with MSL-2, MSL-3 and MLE from SL2 cells (Kelley et al. 1995; Copps et al. 1998). Overexpression of C- or N-terminal fragments of MSL-1 leads to delocalization of the MSL complex from the X chromosome and male-specific lethality. In particular, the FΔ84 construct with an 84-aa N-terminal deletion, which still interacts with MSL-2 but fails to localize the X chromosome, caused male-specific lethality. This could be enhanced by deletion of one copy of MSL-2 or suppressed by overexpression of MSL-2. These results suggest that the N-terminus of MSL-1 is important for its targeting to the X chromosome in vivo (Scott et al. 2000).

The MSL complex initially nucleates on approximately 35 'high affinity' or 'chromatin entry' sites, and then spreads to the surrounding chromatin in *cis*, coating the male X chromosome (Fig. 1B; Lyman et al. 1997; Kelley et al. 1999). Nucleation and spreading processes can be genetically separated. Based on genetic evidence, MSL-1 forms the core dosage compensation complex together with MSL-2. The MSL complex is unable to nucleate on the X chromosome without MSL-1 protein (Lyman et al. 1997). Chang and Kuroda (1998) have suggested that the role of MSL-1 is to tether the MSL complex to chromatin, while MSL-2 regulates sex specificity of dosage compensation and targeting to the X chromosome. The *msl-1* gene is transcribed in both sexes, but its translation is partly inhibited by the Sex-lethal protein in females. MSL-1 is also regulated at the level of protein stability. It is unstable in MSL-2 mutants, and if MSL-2 is ectopically expressed in females, MSL-1 protein is stabilized (Chang and Kuroda 1998).

2.2
MSL-2

msl-2 locus encodes an acidic, 773-aa protein with a RING finger, a proline-rich region and a cysteine-rich region known as the CXC motif, found in many chromatin-associated proteins like Enhancer of zeste (Bashaw and Baker 1995; Kelley et al. 1995; Zhou et al. 1995; Marin 2003). Several results indicate that RING finger is essential for MSL-2 function. First, MSL-2 interacts directly with MSL-1 with its RING finger, and mutations disrupting RING finger integrity abolish this interaction (Copps et al. 1998). Second, constructs with point mutations in the MSL-2 RING finger fail to rescue *msl-2* flies (Lyman et al. 1997). Third, two original mutant alleles of *msl-2* contain a mutation in the RING finger region (Zhou et al. 1995).

In contrast to the RING finger, the CXC motif does not seem to be essential for MSL-2 function. Transgenic constructs carrying point mutations in the conserved cysteine residues of the motif can still rescue *msl-2* mutant males (Lyman et al. 1997). Thus, the function of CXC motif in dosage compensation is unclear.

MSL-2 is the link between dosage compensation and sex determination. Translation of MSL-2 mRNA is under the control of the master switch gene *Sex-lethal* (*Sxl*), the primary regulator of female sexual fate in *Drosophila* (Bashaw and Baker 1997; Kelley et al. 1997). In females, Sex-lethal binds to the 3′ and 5′ untranslated regions of MSL-2 transcript and represses its translation by inhibiting stable association of the 40S ribosomal subunit (Gebauer et al. 2003; Grskovic et al. 2003). Sxl is not expressed in males, and consequently, msl-2 is translated and it can induce assembly of the MSL complex on the X chromosome.

MSL-2 is the primary determinant of the dosage compensation complex assembly. In the absence of MSL-2, other MSL proteins are not associated with the X chromosome. Furthermore, ectopic expression of MSL-2 in females causes assembly of the MSL complex on the X chromosomes, developmental delays and lethality, most probably as a result of overexpression of genes on the X chromosome (Kelley et al. 1995). Lethality is suppressed in *msl-1* heterozygous mutant background, illustrating the importance of correct stoichiometry for proper MSL complex function (Kelley et al. 1995).

2.3
MSL-3

MSL-3 is a 512-aa protein that contains an N-terminal chromodomain and a C-terminal domain that is a diverged chromodomain or a leucine zipper (Marin and Baker 2000; Bertram and Pereira-Smith 2001). These two domains reveal little about its function; hence the role of MSL-3 in dosage compensation has been elusive. However, recent data suggest that it may have a function in spreading of the MSL complex.

Previously, it was believed that RNA helicase MLE is the link between the MSL complex and roX RNAs. Therefore, it was a surprising finding that MSL-3 binds RNA via its chromodomains in vitro (Akhtar et al. 2000). Recently, MLS-3 has also been shown to associate with roX2 RNA in Schneider (SL2) cells (Buscaino et al. 2003). This association has a functional role, since RNase treatment of cultured cells leads to delocalization of MSL3 from the X chromosome. Interestingly, MSL-3 protein can be acetylated by the histone acetyltransferase MOF both in vitro and in vivo. Acetylation of lysine K116 in MSL3 results in specific loss of interaction between MSL-3 and roX2 RNA in vitro. Furthermore, inhibition of histone deacetylase activity in SL2 cells decreases the amount of roX2 RNA in MSL-3 immunoprecipitates, and leads to delocalization of MSL-3 from the X chromosome. To complete the story, Buscaino et al. (2003) showed that histone deacetylase RPD3 can be co-immunoprecipitated with MSL-3 under low-stringency conditions, and that RPD3 complex can deacetylate MSL-3 in vitro (Buscaino et al. 2003). Taken together, the results suggest that there is a constant acetylation–deacetylation cycle of MSL-3 and this, in turn, could regulate the spreading of the complex along the X chromosome. Consistent with this, the MSL complex is unable to spread to flanking chromatin from entry sites in homozygous *msl3* mutant male flies.

2.4
MOF

MOF (males absent on the first) is a 827-aa histone acetyltransferase that belongs to the MYST (<u>MO</u>Z/<u>Y</u>BF2/<u>S</u>AS2/<u>T</u>ip60) family. This group of acetyltransferases are characterized by a C_2HC-type zinc finger embedded in their catalytic domain. Intact zinc finger is essential for enzymatic activity (Akhtar and Becker 2001). MOF also contains a chromodomain close to the HAT domain (Hilfiker et al. 1997).

Compared to most acetyltransferases, MOF is a very specific enzyme. Recombinant MOF or partially purified MSL complex acetylates only lysine 16 of the histone H4 tail in nucleosomes (Akhtar and Becker 2000; Smith et al. 2000). A glycine to glutamic acid substitution that causes male-specific lethality in flies also renders the recombinant enzyme and the MSL complex inactive. MOF activates transcription in both in vitro transcription assays and in vivo in a heterologous GAL4 activation assay (Akhtar and Becker 2000). This requires an enzymatically active HAT domain, suggesting a causal link between H4K16 acetylation and transcriptional activation (Akhtar and Becker 2000). However, these studies did not address whether acetylation influences transcription initiation or elongation, so the exact mechanism remains elusive.

MOF has only low affinity for DNA, but it can interact with both RNA and nucleosomes. RNA interaction requires an intact chromodomain (Akhtar et al. 2000), whereas the C_2HC zinc finger regulates nucleosome binding (Akhtar and Becker 2001). MOF protein associates with roX2 in vivo, since they can be co-immunoprecipitated from SL2 cell extracts. Similar to MSL-3, MOF association with the X chromosome is sensitive to RNase treatment (Akhtar et al. 2000).

Many chromatin-associated proteins contain chromodomains, and recently some of these domains have been shown to bind specifically methylated lysines of histone H3 or DNA (Jacobs and Khorasanizadeh 2002; Fischle et al. 2003). MOF chromodomain resembles that of MSL3, and both lack methyl-lysine-interacting residues and conserved hydrophobic residues (Jacobs and Khorasanizadeh 2002). Thus, they may define a novel chromodomain subfamily that binds RNA.

Besides transcriptional activation and regulation of MSL3, MOF plays a role in spreading and targeting of the MSL complex. In the absence of enzymatically active MOF, intact MSL complex is formed, but it is unable to spread beyond entry sites (Gu et al. 2000).

2.5
MLE

MLE (maleless) is the other MSL complex member possessing enzymatic activity. *mle* locus encodes a 1,293-aa protein that contains two dsRNA-binding domains and a helicase/NTPase domain of the DExH subfamily type. Furthermore, the C-terminal portion MLE contains nine glycine-rich heptad repeats (Kuroda et al. 1991). Full-length MLE has helicase activity, and it can resolve RNA:RNA, RNA:DNA and DNA:DNA hybrids with 3′ ssDNA overhangs in vitro. It can also utilize all common NTPs in helicase assays with similar K_m values, illustrating its broad specificity for substrates (Lee et al. 1997).

The function of MLE in dosage compensation is unclear. Transgenic fly lines carrying mutant versions of MLE helicase domain have revealed some requirements for its function. The mutant form of MLE (MLE[GET]) that abolishes its ATPase activity but retains ssDNA and ssRNA binding ability cannot rescue *mle* mutant flies. Richter et al. (1996) mutated the same GKT motif, but their construct (MLE[GNT]) could restore male viability to approximately 50 %. Point mutation in the conserved DExD box of MLE severely reduced the rescue ability of the transgene. MLE[DQID] mutant retains ATPase binding ability but abrogates ATPase and helicase activity. No effect was seen when the conserved SAT motif was changed to AAA, even though the mutation is postulated to abolish RNA helicase activity.

When MLE[GET] construct was expressed in *mle* null mutant background, the MSL complex could nucleate on entry sites, but it was unable to spread to the surrounding chromatin (Lee et al. 1997; Gu et al. 2000). Helicase activity is thus not strictly required for correct localization. In addition, roX RNAs are not stable in the absence of MLE helicase activity, which suggests that MLE is required for proper complex maintenance but not core complex assembly (Gu et al. 2000). Failure to spread could also be indirect, caused by the lack of roX RNAs.

Maternally supplied MLE stabilizes roX1 transcripts in early embryos (Meller 2003), but this is not essential for dosage compensation, as zygotic MLE is sufficient for male viability. roX2 transcripts associate with MLE in vivo (Meller et al. 2000), and similar to MOF and MSL3, MLE association with the X chromosome is sensitive to RNase treatment. Surprisingly, the MLE C-terminal portion (amino acids 941–1,293) containing only glycine-rich repeats can associate with chromatin in an RNase-sensitive manner (Richter et al. 1996).

Taken together, these results illustrate the interdependence between roX1, roX2 and MLE. First, MLE is required for roX1 and roX2 stability (Gu et al. 2000; Kageyama et al. 2001). Second, an RNA component (either roX RNAs or a yet unidentified RNA) maintains MLE association with the X chromosome (Richter et al. 1996).

MLE is the only MSL complex protein with male-specific lethal phenotype that has been shown to have an additional function not related to dosage compensation. *nap* (no action potential) is an allele of the *mle* locus that shows no male-specific lethality (Kernan et al. 1991). Instead, flies are paralyzed in non-permissive (high) temperatures (Wu et al. 1978). Since this phenotype is very similar to *Drosophila para* mutants, Reenan et al. (2000) reasoned that MLE could be involved in processing of para transcript. Indeed, para mRNA is aberrantly spliced in *mle^nap* mutant flies, such that only 17 % of transcripts are correctly spliced. para mRNA is predicted to form extensive secondary structures, which suggests that MLE is involved in resolving these structures, allowing correct splicing. Strikingly, *mle* alleles with male-specific lethal phenotype have no effect on para mRNA splicing (Reenan et al. 2000). These results could have several different explanations. First, *mle^nap* is a gain-of-function allele and wild-type MLE has no function in para splicing. Second, it could be a hypomorphic *mle* allele epistatic to other RNA helicases. In complete absence of mle, other helicases could resolve para secondary structures, but *mle^nap* still retains some function and prevents the access of other helicases to para mRNA. The third explanation is that nap and male-specific lethal phenotypes map to different domains of MLE protein. However, this is unlikely, as *mle^nap* mutation maps in the next amino acid after the GKT motif that was shown to be important for MSL complex spreading (Kernan et al. 1991; Lee et al. 1997).

Biochemical data also support the idea that MLE may have other functions in addition to dosage compensation. First, co-immunoprecipitation experiments have revealed that MLE is only loosely associated with the MSL complex (Copps et al. 1998; Buscaino et al. 2003). Second, the bulk of MLE exists as a monomer in SL2 nuclear extracts, while only a small proportion of it co-fractionates with the 2-MDa MSL complex (Copps et al. 1998).

2.6
JIL-1

JIL-1 was cloned as a protein that is recognized by a monoclonal antibody mAb2A (Jin et al. 1999). The 1,207-aa protein contains two tandemly arranged serine/threonine kinase domains, and it can phosphorylate histone H3 in vitro (Jin et al. 1999). JIL-1 is required for maintenance of chromatin structure in flies. Null mutation of *jil-1* is lethal. Mutant flies have strongly decreased H3S10 phosphorylation and defects in chromatin structure. Interestingly, weak alleles of JIL-1 show distortion of the sex ratio, implicating that male flies are more vulnerable to partial loss of the protein (Wang et al. 2001). In addition to chromosome morphology phenotypes, jil-1 mutant flies have posterior-to-anterior homeotic transformations (Zhang et al. 2003a). Trithorax group (*trxG*) members *brahma* and *trithorax* enhance the homeotic phenotype, suggesting that JIL-1 participates in regulation of the BX-C locus with *trxG* genes (Zhang et al. 2003a). Recently, JIL-1 was shown to interact with a splice variant from the *lola* locus, but it is unknown whether this interaction is significant for dosage compensation (Zhang et al. 2003b).

JIL-1 is enriched approximately two-fold on the male X chromosome (Jin et al. 1999), and this localization coincides with H3S10 phosphorylation and phosphoacetylated (S10P/K14Ac) histone H3 (Wang et al. 2001). Epitope-tagged JIL-1 can be co-immunoprecipitated with MSL-1, MSL-2 and MSL-3 from *Drosophila* SL2 cells. Furthermore, in vitro pull-down experiments have shown that JIL-1 interaction with MSL-1 and MSL-3 is mediated by its two kinase domains (Jin et al. 2000).

In contrast to other MSL complex members, MSL-2 localization is not disrupted in JIL-1 mutants, suggesting that JIL-1 is not required for targeting, assembly or spreading of the complex (Wang et al. 2001). However, in *jil-1* mutants, the X chromosome morphology is more severely affected than that of autosomes (Wang et al. 2001). Taken together, these results indicate that it is a bona fide member of the MSL complex, but as there is no direct evidence linking it to hypertranscription of the X chromosome, further studies need to address this suggestive link.

2.7
roX1 and *roX2*

Amrein and Axel (1997) cloned *roX1* and *roX2* while isolating transcripts specifically expressed in adult males. Independently, Meller et al. (1997) cloned *roX1* from an enhancer trap screen. Both genes lack significant open reading frames, and therefore they were postulated to function as non-coding RNAs (Amrein and Axel 1997; Meller et al. 1997). roX1 has one intron, and the spliced transcript is 3.7 kb long (Kageyama et al. 2001). RoX2 has two introns and undergoes alternative splicing, the major isoform being approximately

600 nucleotides (Park et al. 2003). It is not known whether roX RNAs are capped and polyadenylated or whether they are transcribed by RNA polymerase II (Stuckenholz et al. 2003).

roX1 and roX2 are redundant in function, even though they differ in size and bear very little sequence similarity. They have a dual role in dosage compensation. First, roX RNAs colocalize with MSL proteins on the male X chromosome and physically associate with the MSL complex (Franke and Baker 1999; Meller et al. 2000). Second, roX1 and rox2 genomic loci are chromatin entry sites for the MSL complex (Kelley et al. 1999). This function is independent of transcription (Kageyama et al. 2001; Park et al. 2003).

2.7.1
roX Genes as Non-Coding RNAs

Deletion of one roX gene has no phenotype, but lack of both transcripts causes male-specific lethality with very few male escapers. MSL complex is not properly targeted to the X chromosome in roX1/roX2 double mutant males (Franke and Baker 1999; Meller and Rattner 2002). Unexpectedly, simultaneous overexpression of MSL1 and MSL2 in roX1 roX2 mutant background can induce the assembly of MSL complex on the X chromosome and increase viability of roX1 roX2 mutant males (Oh et al. 2003). Non-coding RNAs are therefore necessary for dosage compensation in a wild-type condition, but overexpression studies show that they are not the only components mediating targeting of the complex.

Dissection of roX function has been complicated by the fact that RNA forms extensive secondary structures. Furthermore, computational secondary structure prediction for large RNAs is still unreliable. Stuckenholz et al. (2003) used a genetic approach to find functional domains in roX1. They generated sequential deletions of roX1, and tested the rescue ability of the constructs in roX1 roX2 double mutant background. Even though roX1 appears to tolerate short deletions (spanning 10% of transcript length), Stuckenholz and colleagues (2003) could identify a stem-loop structure that at least partially accounts for the function of roX1. As a part of the study, they also showed that the frequency of roX1 roX2 escapers depends on the genetic background, and more precisely on the proximal part of the X chromosome. These results imply that there are other unknown factors modifying the function of roX RNAs.

2.7.2
roX Loci as Chromatin Entry Sites

Initial observation of roX genes as chromatin entry sites came from transgenic studies. Inserting roX genes on autosomes results in ectopic recruitment of MSL complex on the site of the transgene (Kelley et al. 1999). This was supported by the fact that endogenous *roX1* and *roX2* loci map cytologically to previously mapped entry sites (Lyman et al. 1997). The role of *roX* genes as entry sites is separate from their function as non-coding RNAs in MSL complex. First, sequences needed for MSL complex attraction are dispensable for *roX* function as non-coding RNAs. Second, transcription is not required for ectopic recruitment of MSLs (Kageyama et al. 2001; Park et al. 2003). Furthermore, entry site sequence in the *roX2* locus does not overlap with *roX2* primary transcript (Park et al. 2003).

To date, *roX1* and *roX2* loci are the only entry sites that have been cloned and characterized. In both cases, a 200- to 300-bp DNA fragment can recruit MSL complex to the autosomes (Kageyama et al. 2001; Park et al. 2003). Entry sites have several distinct genetic and biochemical properties. In the case of *roX1* and *roX2*, these sequences are sensitive to DNase I treatment in males but not in females. This implicates that the chromatin structure at entry sites is different, perhaps more accessible, from the surrounding regions. Chromatin immunoprecipitation experiments from cultured SL2 cells have shown that *roX1* entry site is enriched in MSL proteins, compared to autosomal control genes (Kageyama et al. 2001). The *roX2* locus is enriched in H4K16 acetylation and MSL1, as assessed by chromatin immunoprecipitation from embryo extracts (Smith et al. 2001).

By carefully comparing *roX1* and *roX2* entry site sequences, Park et al. (2003) could find a putative consensus sequence. Mutation of the consensus sequence had a modest but reproducible effect on MSL complex recruitment. Using computational analysis, the authors could find other consensus sequences on the X chromosome, but they failed to recruit the MSL complex. Thus, until more entry sites have been cloned, sequence requirements for MSL complex recruitment remain a mystery.

3
Targeting, Assembly and Spreading of the MSL Complex

3.1
Targeting and Assembly

The initial step in dosage compensation is the targeting of the chromatin-modifying MSL complex to the X chromosome. As already discussed, this is achieved by *trans*-acting factors, namely *roX* RNAs and MSL-1/MSL-2, and *cis*-acting DNA sequences, the chromatin entry sites. Chromatin entry sites play a central role in correct targeting of the complex. Despite being very short stretches of DNA, they contain all the necessary information to attract the MSL complex. This is reminiscent of insulator elements, short DNA sequences that are able to prevent the spreading of chromatin states (West et al. 2002). Whether chromatin entry sites work at the level of DNA sequence or chromatin structure remains to be determined.

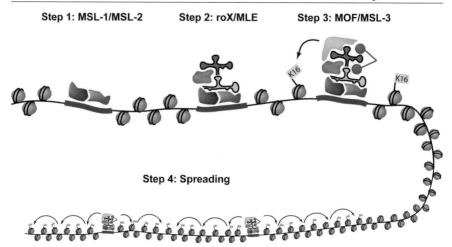

Fig. 2. Assembly and spreading of the MSL complex. MSL-1 and MSL-2 form the core complex that nucleates on entry sites (*step 1*). Current evidence suggests that *roX1* and *roX2* together with MLE are the next ones to enter the complex (*step 2*). MOF and MSL-3 are integrated into the complex in the end (*step 3*), which leads to acetylation of adjacent nucleosomes and spreading of the complex to flanking chromatin (*step 4*)

MSL-1 and MSL-2 are clearly the core components of the MSL complex (Fig. 2). Both are required for the nucleation of the complex, while other MSLs only affect the subsequent step, spreading in *cis*. Consistently, depletion of MSL-2 in SL2 cells disrupts the whole complex, whereas depletion of MSL-3, MLE or MOF has no visible effect on MSL-1 or MSL-2 localization (Buscaino et al. 2003).

The subsequent events in the assembly of the MSL complex are still not completely understood. MLE is required for MOF localization on the X chromosome, but only as a structural component. Enzymatically inactive but full-length MLE can still recruit MOF to the X chromosome, but in *mle^{y205}* mutants that carry a truncated version of MLE, MOF is not present on the X (Gu et al. 1998). Also MSL-3 is largely absent in *mle* mutant background (Gorman et al. 1995). MOF, on the other hand, is required for the correct localization of MLE and MSL-3, and similar to MLE, enzymatic activity does not appear to be necessary. Gu et al. (1998) have suggested that MLE enters the complex before MOF and MSL-3, because in *mof^2* mutant females ectopically expressing MSL-2, still some MLE but no MSL-3 can be seen localized to the X chromosome. In SL2 cells, however, depletion of MSL-3 or MOF by RNA interference leads to dissociation of MLE from the complex (Buscaino et al. 2003). This apparent contradiction could reflect the different nature of SL2 cells and larval polytene chromosomes, or it may simply be a demonstration of strict interdependence of different factors.

MSL-3, in turn, appears to enter the complex before MOF. MSL-3 depletion by RNAi dissociates MOF from the X chromosomes, whereas MOF depletion has no effect on MSL-3 localization (Buscaino et al. 2003).

3.2
Spreading

Subsequent to targeting and assembly, the MSL complex spreads to the chromatin, strictly *in cis* (Fig. 2). Kuroda and colleagues have started to address the mechanism of spreading in elegant genetic experiments. They have illustrated the importance of the balance between MSL proteins and roX RNAs in regulating the extent of spreading (Park et al. 2002; Oh et al. 2003).

Spreading to adjacent chromatin domains rarely occurs from autosomal roX transgenes (Kelley et al. 1999). If the dose of *roX* RNAs is reduced by deleting endogenous *roX* loci on the X chromosome, the MSL complex spreads more efficiently from the transgenes to the flanking chromatin. The extent of spreading from the roX transgene inversely correlates with the number of transcribed roX genes in the genome (Park et al. 2002). This suggests that individual roX RNAs compete for MSL binding and spreading. Furthermore, spreading of the MSL complex is associated with increased stability of transgenic roX RNAs. Given the requirement of MSL-1 and MSL-2 for roX stability, it is not surprising that simultaneous overexpression of MSL-1 and MSL-2 can also enhance spreading from the transgenes (Park et al. 2002). If they are coexpressed in wild-type flies, MSL proteins concentrate on the vicinity of *roX* loci, while distal regions have very low MSL levels, showing that modulating MSL/*roX* balance can disrupt normal distribution of dosage compensation proteins (Oh et al. 2003).

It is important to note that spreading requires *roX* transcription, but the recruitment of the MSL complex does not require RNA. Consequently, promoterless *roX* transgenes are also unable to compete for MSL binding (Park et al. 2002). This again illustrates separation of the entry site and the spreading functions of *roX* genes.

It is tempting to speculate that roX RNA stability is a determinant of spreading *in cis*. When the concentration of MSL proteins is low, roX RNA can diffuse away from its site of transcription, but it is quickly degraded in the nucleoplasm. If MSL proteins bind nascent roX transcripts, RNA is stabilized and it is located in the immediate vicinity of the X chromosome, allowing spreading *in cis*. Interestingly, Xist, a non-coding RNA that paints the inactive X chromosome of female mammals, is also unstable when not correctly tethered to the X chromosome (Wutz et al. 2002). Perhaps rapid degradation of RNA can ensure its packaging into a stabilizing protein complex primarily in the immediate vicinity of the locus. However, the model outlined here cannot account for spreading on its own, since roX transgenes can also function *in trans. roX* loci may also be distinct from other entry sites in their function.

Cloning and characterization of other entry sites will tell us whether all entry sites are created equal, and shed light on the molecular mechanism of spreading.

4
Cracking the Code X

4.1
Establishing the Code

Covalent histone modifications and their role in chromatin structure have been extensively studied in the last decade. Allis and Jenuwein have put forward an attractive hypothesis as to how different states of chromatin are established and maintained. The 'histone code' hypothesis predicts that covalent modifications of histone tails act combinatorially and sequentially, and that protein modules specifically recognize these modifications and translate them to functional states by interacting with other factors (Jenuwein and Allis 2001). There are also alternative hypotheses of histone modifications and chromatin structure, which are not necessarily mutually exclusive (Schreiber and Bernstein 2002). However, many predictions of the histone code hypothesis have been shown to be correct (Turner 2002).

In the case of *Drosophila* dosage compensation, the most striking modification is acetylation of lysine K16 of histone H4, catalyzed by the histone acetyltransferase MOF. MOF acetylates H4K16 in vitro, and mutation in the *mof* gene abolishes K16 acetylation in vivo, demonstrating the direct requirement of wild-type *mof* function for H4K16 acetylation (Hilfiker et al. 1997; Akhtar and Becker 2000; Smith et al. 2000). H4K16Ac coats the hyperactive male X chromosome and, together with recently observed phosphorylation of H3S10, defines it as a distinct chromatin structure (Turner et al. 1992; Wang et al. 2001).

The histone H4 tail contains four closely spaced lysines (K5, K8, K12, K16) that are subject to acetylation by various histone acetyltransferases. Histone acetylation is generally regarded as an activating modification (Eberharter and Becker 2002). Chromatin immunoprecipitation (ChIP) experiments have shown that during gene activation, acetylation of different N-terminal lysines is regulated in a very specific manner, and that each modification has a specific function (Agalioti et al. 2002). In *Drosophila* polytene chromosomes, K5Ac, K8Ac and K12Ac localize in discrete bands, but unlike K16Ac, they are not enriched on the hyperactive X chromosome (Turner et al. 1992).

The role of histone phosphorylation in dosage compensation is poorly understood. Histone phosphorylation has been implicated in gene activation, mitotic chromosome condensation and apoptosis (Iizuka and Smith 2003). H3S10 phosphorylation, specifically, serves both as an activating signal in transcription and as a condensation signal during mitosis. Phosphorylation is not required for MSL complex assembly or spreading, since MSLs are correctly localized in *jil-1* mutant background (Wang et al. 2001). Therefore, its function is more likely in regulation of chromatin structure of the X chromosome. It should be noted, however, that it is not currently possible to distinguish between the roles of H4K16 acetylation and H3S10 phosphorylation in transcriptional regulation of the X chromosome.

4.2
Reading the Code

Enrichment of H4K16 acetylation is the hallmark of active X chromatin, but is it only because the X chromosome is hyperactive? That is, does acetylation only reflect the hyperactive state, or is it the cause for specific upregulation? The latter seems to fit experimental results better. First, transcriptionally active regions of autosomes are not enriched in acetylated H4K16, as assessed by immunofluorescence on polytene chromosomes (Turner et al. 1992). Second, H4K16-specific histone acetyltransferase MOF can activate transcription both in vivo and in vitro (Akhtar and Becker 2000). However, some contradicting data make a straightforward connection unlikely. For example, overexpression of MOF redistributes the MSL complex to the autosomes, suggesting that acetylation plays a role in targeting and spreading of the complex (Gu et al. 2000). Moreover, ectopic activation of transcription on X chromosome or on autosomes with the yeast transactivator Gal4 system can in some cases lead to recruitment of the MSL complex on the site of transcription, implicating that the MSL complex targets active regions (Sass et al. 2003).

It is clear that MOF sets the mark on the hyperactive X chromosome, but what reads it? Based on the histone code hypothesis, the most obvious candidate would be a bromodomain-containing protein. Bromodomains are found in many chromatin-associated proteins, and they can specifically bind acetylated lysines (Zeng and Zhou 2002), but none of the known MSL complex members contains bromodomains. However, it cannot be excluded that acetylated H4K16 is specifically recognized by a protein without a bromodomain.

Another possibility is that acetylation of lysine K16 does not create a binding site for another module, but, in contrast, inhibits binding of a protein. Recent results on chromatin remodeling factor ISWI point in this direction. ISWI is the catalytic subunit of three chromatin remodeling complexes, NURF, CHRAC and ACF (Langst and Becker 2001). In polytene chromosomes ISWI preferentially localizes to RNA polymerase II-poor regions, suggesting that its main function in vivo is repressive (Deuring et al. 2000). *ISWI²* mutant males have grossly abnormal X chromosomes, but this phenotype can be suppressed by a mutation in *mle* (Corona et al. 2002). Corona et al. (2002) have shown that acetylation of H4K12 or H4K16 inhibits both ISWI interaction with nucleosomes and its ATPase activity. These data indicate that ISWI and dosage compensation complex may have opposing effects on chromatin structure of the X chromosome.

One function of H4K16 acetylation could be derepression by means of exclusion of ISWI-containing repressive chromatin remodeling complexes. However, if the MSL complex is tethered to autosomes with an ectopic entry site, it can open chromatin structure and activate transcription (Henry et al. 2001; Kelley and Kuroda 2003). It is therefore unlikely that H4K16 has only a derepressive role.

5
Molecular Mechanism of Dosage Compensation

In his insightful paper, over 40 years before the first MSL gene was cloned (Muller 1948), Herman Muller phrased the general problem of dosage compensation thus: "the compensation mechanism must be concerned with the equalization of exceedingly minute differences." Still today, perhaps the most persistent enigma of dosage compensation is how to achieve two-fold upregulation of a large number of genes on the X chromosome. Whatever the mechanism, it must adapt to a vast linear scale. Abundance of transcripts in a single cell varies by over five orders of magnitude, and the dosage compensation machinery must accurately regulate most of these genes (Velculescu et al. 1999). Furthermore, the machinery should respond to normal regulatory signals during development, allowing dynamic changes in gene expression but still maintaining dosage compensation.

An aspect of *Caenorhabditis elegans* dosage compensation provides an interesting case study in modulation of transcript levels. Not only do *C. elegans* dosage compensation proteins downregulate X chromosomal gene expression in hermaphrodites by 50%, they also repress transcription of a single autosomal gene, *her-1*, 20-fold (Meyer 2000). Remarkably, the same complex is associated with both chromosome-wide (X chromosome) and gene-specific (*her-1*) repression, with distinct outcomes (Chu et al. 2002). The two functions can be genetically separated, and biochemically the complex composition differs at least in respect to one dosage compensation protein (Yonker and Meyer 2003). It is currently not known precisely how this difference in repression levels is brought about, but elucidation of the molecular mechanisms will help us to understand how a specific level of gene expression can be achieved.

5.1
Initiation Versus Elongation

It is not known at which stage of transcription the MSL complex exerts its function. It could, for example, either increase the transcription initiation rate or enhance polymerase elongation. To address this question, Smith et al. (2001) performed chromatin immunoprecipitation experiments with antibodies against acetylated H4K16. Even though their study included only a few genes, the general trend was that acetylation levels were significantly increased on coding regions of genes. However, promoters had only modest enrichment of acetylation. Within the limitations of the study, the results of Smith et al. (2000) suggest that the MSL complex targets transcription elongation rather than initiation (Smith et al. 2001).

Nucleosomes can inhibit transcription in vitro, but in vivo transcription through chromatin occurs very rapidly. Elongation factors associate with RNA polymerase II and help it overcome this obstacle (Svejstrup 2002). Histone acetylation has been known to enhance transcription elongation (Orphanides and Reinberg 2000). Furthermore, HAT activity of yeast Elongator complex is required for its function in vivo (Winkler et al. 2002), illustrating the connection between histone acetylation and the elongation process. It is important to note that a direct connection between elongation and H4K16 acetylation in dosage compensation has not been made, and therefore chromatin immunoprecipitation results are still only suggestive.

It is obvious that dosage compensation is linked to basal gene expression machinery of the cell. In very simple terms, dosage compensation proteins may provide a signal for basal machinery to modulate gene expression. This need not be a direct link; interaction could also occur at a general level of chromatin structure. Genetic studies on *Drosophila* dosage compensation have not been able to address this question. Therefore, biochemical approaches would greatly increase our understanding of the mechanism of action of dosage compensation. The endogenous MSL complex has not been purified to homogeneity. Consequently, the enzymatic properties of the complex are not completely known, and potential accessory factors have not been characterized.

The possibility remains that the MSL complex contains other proteins that do not have a male-specific function. Rather, they could link the complex to essential factors that are present in every cell in both sexes. Genetic approaches are unlikely to reveal these interactions, since they would not have striking sex-specific phenotypes. A reoccurring theme in chromatin remodeling complexes is that they often share subunits with other complexes (Lusser and Kadonaga 2003). Thus far, MSL complex subunits have not been associated with other complexes, and it remains to be seen whether the MSL complex truly is exceptional in this sense.

5.2
The Inverse Effect Hypothesis

It is widely accepted that the function of the MSL complex is to upregulate X-chromosomal genes. However, some data suggest that this is too simplified a model of dosage compensation. For example, dosage compensation occurs not only in males (1X:2A) but also in metamales (1X:3A), metafemales (3X:2A) and triploid intersexes (2X:3A). This phenomenon cannot be explained purely in terms of MSL complex upregulating the X chromosome. Birchler and colleagues have suggested that the MSL complex actually sequesters activating factors, most likely MOF, away from the autosomes, in order to prevent male-specific lethality from inverse dosage effects (Birchler et al. 2003).

Deletion of large segments of chromosomes (segmental aneuploidy) not only often influences the genes in the region, but also causes genome-wide effects in gene expression. The most common effect is upregulation of unlinked genes, i.e. those not located in the region of aneuploidy. This has been explained as being due to inverse dosage effect. That is, deleted regions often contain negative regulators (for example, repressors), which leads to upregulation of unlinked genes (Birchler et al. 2001).

Since the single X chromosome in males effectively creates an aneuploid state, it would be expected to cause an inverse dosage effect on autosomes. Birchler and colleagues argue that the MSL complex sequesters MOF and H4K16 acetylation to the X chromosome to counteract these deleterious effects. Furthermore, the MSL complex would modify the X chromosome and/or MOF activity such that genes on the X could no longer respond to high acetylation levels (Birchler et al. 2003).

Birchler and coworkers addressed this issue by comparing gene expression patterns in flies mutant for MSLs. They observed that in *mle* mutant males, many autosomal transgenes are upregulated, but transgenes on the X remain dosage compensated (Hiebert and Birchler 1994). The same trend was seen in *mof* mutants (Bhadra et al. 1999). Furthermore, ectopic expression of MSL2 in females did not lead to increased expression of the X chromosomal transgenes they tested (Bhadra et al. 1999). Bhadra et al. (1999) also examined the level of MOF protein and H4K16 acetylation in female and male polytene chromosomes. MOF protein is present at equal levels in males and females, but it is enriched on the X chromosome only in males. In females, autosomal MOF and H4K16 acetylation signals are increased, suggesting that MOF is redistributed to autosomes if it is not recruited to the MSL complex (Bhadra et al. 1999).

It is difficult to distinguish between the two models of dosage compensation. Current data are still consistent with both models. It has been argued, however, that the ability of ectopic MSL complex on the autosomes to modify local chromatin structure and upregulate adjacent genes is inconsistent with the inverse effect hypothesis (Henry et al. 2001; Park et al. 2002; Kelley and Kuroda 2003).

Chiang and Kurnit (2003) used a sensitive quantitative RT-PCR method to examine the effects of MSL mutations on the expression of several genes. The general trend was that in *mle* or *mof* mutant males, expression of both autosomal and X-chromosomal genes was decreased. However, the amount of change varied significantly between individual genes. In addition, some genes on the X showed no difference, whereas one autosomal gene was significantly upregulated in *mle* and *mof* mutants (Chiang and Kurnit 2003). The inherent problem in quantitative analysis is the possibility of indirect effects. Chiang and Kurnit (2003) analyzed gene expression in early and late larvae, but dosage compensation already occurs in early embryos (Rastelli et al. 1995; Franke et al. 1996). Since X chromosome contains dosage-sensitive regulators, these will highly likely have secondary effects on gene expression at a genome-wide scale.

6
The Origin and Evolution of the MSL Complex

Many features of metazoan development are conserved across phyla (Carroll et al. 2001). However, dosage compensation appears to have evolved independently several times. Sex determination mechanisms evolve very rapidly (Charlesworth 1996; Marin and Baker 1998), and since dosage compensation is intimately linked to sex determination, it is not surprising that nature has found different solutions to the same problem. Comparative studies have shown that *Drosophila* dosage compensation system, where the MSL complex decorates the X chromosome, is at least 50–60 million years old (Bone and Kuroda 1996; Marin et al. 1996). Consistently, also Sex-lethal function is estimated to be at least 60 million years old (Marin and Baker 1998). In the fungus gnat *Sciara ocellaris*, which belongs to the Nematocera suborder of Diptera, dosage compensation is also achieved by hypertranscription of the X chromosome in males, but *Sciara* homologues of MSL proteins appear to play no role in this process (da Cunha et al. 1994; Ruiz et al. 2000). *Anopheles gambiae*, another nematoceran insect, is separated from *Drosophila* by 250 million years (Gaunt and Miles 2002). This suggests that *Drosophila* dosage compensation system evolved at least 50, but less than 250, million years ago.

Studies on the three model organisms (*C. elegans, D. melanogaster, M. musculus*) have shown that animals have co-opted different chromatin remodeling complexes to function in dosage compensation. Mammalian polycomb proteins Eed and Ezh2/Enx1 coat the inactive X chromosome transiently during the initiation phase of inactivation (Plath et al. 2003; Silva et al. 2003). Eed/Ezh2 complex has been shown to have H3K27-specific histone methyltransferase activity (Cao et al. 2002; Kuzmichev et al. 2002). Similar to H4K16 acetylation in flies, methylated H3K27 is a hallmark of inactive X chromatin. The orthologous ESC/E(z) complex in *Drosophila* is required for maintenance of homeotic gene repression (Czermin et al. 2002; Muller et al. 2002).

C. elegans has adopted an unrelated way to mediate repression of X chromosomes. *C. elegans* dosage compensation complex is very similar to *Xenopus laevis* 13S condensin complex (Meyer 2000). Condensins are involved in

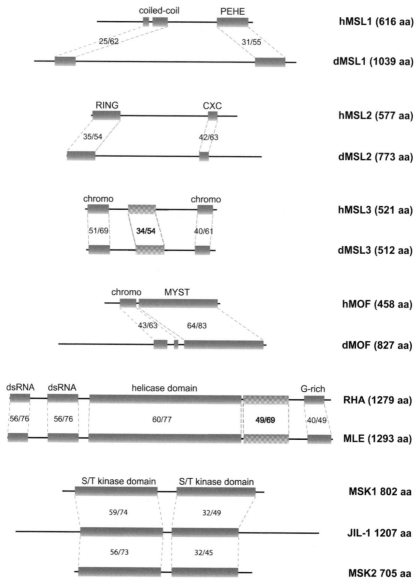

Fig. 3. Mammalian orthologs of MSL complex proteins, drawn in scale. Conserved domains are indicated by *boxes*. Amino acid identity/similarity percentage is denoted in the *middle*. MSL3 and MLE contain conserved regions that bear no homology to known protein domains (*checkered boxes*). Note that dMOF chromodomain and hMSL1 coiled-coil domain contain insertions that are absent in their orthologs

sister chromatid cohesion and separation during mitosis and meiosis. Accordingly, some *C. elegans* dosage compensation proteins have a dual function; they also play an essential role in mitosis and meiosis (Hagstrom and Meyer 2003).

Thus, it is not surprising that also the MSL complex contains evolutionary conserved proteins. All MSL proteins have orthologs in mammals (Fig. 3) suggesting an ancestral role for these proteins in chromatin regulation. The yeast ortholog of MOF, Sas2p, counteracts the silencing function of the histone deacetylase Sir2p and regulates telomeric heterochromatin boundaries (Kimura et al. 2002; Suka et al. 2002). Human genome contains a single MOF ortholog, hMOF, that has been shown to acetylate histones H2A, H3 and H4 in vitro (Neal et al. 2000). The *C. elegans* ortholog of MOF, K03D10.3, is still uncharacterized. In genome-wide RNA interference screen, depletion of K03D10.3 had no visible phenotype (Fraser et al. 2000). The mammalian ortholog of MLE, RNA helicase A, is involved in cAMP-dependent transcriptional activation (Nakajima et al. 1997), and is essential for mouse development (Lee et al. 1998). Currently, the functions of the MSL-1, MSL-2 and MSL-3 orthologs in humans are unknown (Prakash et al. 1999; Marin 2003).

It is not known whether there is a homologous MSL complex in mammalian cells, but given the evolutionary conservation of all proteins, it would be very surprising if *Drosophila* complex had been assembled from scratch. Characterizing the putative mammalian MSL complex would shed light on evolution of dosage compensation mechanisms.

Acknowledgements. We are grateful to Jop Kind for performing the polytene squashes used in Fig. 1. We also thank colleagues in the laboratory for their support.

References

Agalioti T, Chen G, Thanos D (2002) Deciphering the transcriptional histone acetylation code for a human gene. Cell 111:381–392

Akhtar A, Becker PB (2000) Activation of transcription through histone H4 acetylation by MOF, an acetyltransferase essential for dosage compensation in *Drosophila*. Mol Cell 5:367–375

Akhtar A, Becker PB (2001) The histone H4 acetyltransferase MOF uses a C2HC zinc finger for substrate recognition. EMBO Rep 2:113–118

Akhtar A, Zink D, Becker PB (2000) Chromodomains are protein-RNA interaction modules. Nature 407:405–409

Amrein H, Axel R (1997) Genes expressed in neurons of adult male *Drosophila*. Cell 88:459–469

Avner P, Heard E (2001) X-chromosome inactivation: counting, choice and initiation. Nat Rev Genet 2:59–67

Baker BS, Gorman M, Marin I (1994) Dosage compensation in *Drosophila*. Annu Rev Genet 28:491–521

Bashaw GJ, Baker BS (1995) The msl-2 dosage compensation gene of *Drosophila* encodes a putative DNA-binding protein whose expression is sex specifically regulated by Sex-lethal. Development 121:3245–3258

Bashaw GJ, Baker BS (1997) The regulation of the *Drosophila* msl-2 gene reveals a function for Sex-lethal in translational control. Cell 89:789–798

Baverstock PR, Adams M, Polkinghorne RW, Gelder M (1982) A sex-linked enzyme in birds – Z-chromosome conservation but no dosage compensation. Nature 296:763–766

Bertram MJ, Pereira-Smith OM (2001) Conservation of the MORF4 related gene family: identification of a new chromo domain subfamily and novel protein motif. Gene 266:111–121

Bhadra U, Pal-Bhadra M, Birchler JA (1999) Role of the male specific lethal (msl) genes in modifying the effects of sex chromosomal dosage in *Drosophila*. Genetics 152:249–268

Birchler JA, Bhadra U, Bhadra MP, Auger DL (2001) Dosage-dependent gene regulation in multicellular eukaryotes: implications for dosage compensation, aneuploid syndromes, and quantitative traits. Dev Biol 234:275–288

Birchler JA, Pal-Bhadra M, Bhadra U (2003) Dosage dependent gene regulation and the compensation of the X chromosome in *Drosophila* males. Genetica 117:179–190

Bone JR, Kuroda MI (1996) Dosage compensation regulatory proteins and the evolution of sex chromosomes in *Drosophila*. Genetics 144:705–713

Buscaino A, Kocher T, Kind JH, Holz H, Taipale M, Wagner K, Wilm M, Akhtar A (2003) MOF-regulated acetylation of MSL-3 in the *Drosophila* dosage compensation complex. Mol Cell 11:1265–1277

Cao R, Wang L, Wang H, Xia L, Erdjument-Bromage H, Tempst P, Jones RS, Zhang Y (2002) Role of histone H3 lysine 27 methylation in Polycomb-group silencing. Science 298:1039–1043

Carroll SB, Grenier JK, Weatherbee SD (2001) From DNA to diversity: molecular genetics and the evolution of animal design. Blackwell, Boston

Chang KA, Kuroda MI (1998) Modulation of MSL1 abundance in female *Drosophila* contributes to the sex specificity of dosage compensation. Genetics 150:699–709

Charlesworth B (1996) The evolution of chromosomal sex determination and dosage compensation. Curr Biol 6:149–162

Chiang PW, Kurnit DM (2003) Study of dosage compensation in *Drosophila*. Genetics 165:1167–1181

Chu DS, Dawes HE, Lieb JD, Chan RC, Kuo AF, Meyer BJ (2002) A molecular link between gene-specific and chromosome-wide transcriptional repression. Genes Dev 16:796–805

Cline TW, Meyer BJ (1996) Vive la difference: males vs females in flies vs worms. Annu Rev Genet 30:637–702

Copps K, Richman R, Lyman LM, Chang KA, Rampersad-Ammons J, Kuroda MI (1998) Complex formation by the *Drosophila* MSL proteins: role of the MSL2 RING finger in protein complex assembly. EMBO J 17:5409–5417

Corona DF, Clapier CR, Becker PB, Tamkun JW (2002) Modulation of ISWI function by site-specific histone acetylation. EMBO Rep 3:242–247

Czermin B, Melfi R, McCabe D, Seitz V, Imhof A, Pirrotta V (2002) *Drosophila* enhancer of Zeste/ESC complexes have a histone H3 methyltransferase activity that marks chromosomal Polycomb sites. Cell 111:185–196

Da Cunha PR, Granadino B, Perondini AL, Sanchez L (1994) Dosage compensation in sciarids is achieved by hypertranscription of the single X chromosome in males. Genetics 138:787–790

Deuring R, Fanti L, Armstrong JA, Sarte M, Papoulas O, Prestel M, Daubresse G, Verardo M, Moseley SL, Berloco M, Tsukiyama T, Wu C, Pimpinelli S, Tamkun JW (2000) The ISWI chromatin-remodeling protein is required for gene expression and the maintenance of higher order chromatin structure in vivo. Mol Cell 5:355–365

Eberharter A, Becker PB (2002) Histone acetylation: a switch between repressive and permissive chromatin. Second in review series on chromatin dynamics. EMBO Rep 3:224–229

Ellegren H (2002) Dosage compensation: do birds do it as well? Trends Genet 18:25–28

Fischle W, Wang Y, Jacobs SA, Kim Y, Allis CD, Khorasanizadeh S (2003) Molecular basis for the discrimination of repressive methyl-lysine marks in histone H3 by Polycomb and HP1 chromodomains. Genes Dev 17:1870–1881

Franke A, Baker BS (1999) The rox1 and rox2 RNAs are essential components of the compensasome, which mediates dosage compensation in *Drosophila*. Mol Cell 4:117–122

Franke A, Dernburg A, Bashaw GJ, Baker BS (1996) Evidence that MSL-mediated dosage compensation in *Drosophila* begins at blastoderm. Development 122:2751–2760

Fraser AG, Kamath RS, Zipperlen P, Martinez-Campos M, Sohrmann M, Ahringer J (2000) Functional genomic analysis of *C. elegans* chromosome I by systematic RNA interference. Nature 408:325–330

Gaunt MW, Miles MA (2002) An insect molecular clock dates the origin of the insects and accords with palaeontological and biogeographic landmarks. Mol Biol Evol 19:748–761

Gebauer F, Grskovic M, Hentze MW (2003) *Drosophila* sex-lethal inhibits the stable association of the 40S ribosomal subunit with msl-2 mRNA. Mol Cell 11:1397–1404

Ghosh S, Chatterjee RN, Bunick D, Manning JE, Lucchesi JC (1989) The LSP1-alpha gene of *Drosophila melanogaster* exhibits dosage compensation when it is relocated to a different site on the X chromosome. EMBO J 8:1191–1196

Gorman M, Franke A, Baker BS (1995) Molecular characterization of the male-specific lethal-3 gene and investigations of the regulation of dosage compensation in *Drosophila*. Development 121:463–475

Grskovic M, Hentze MW, Gebauer F (2003) A co-repressor assembly nucleated by Sex-lethal in the 3'UTR mediates translational control of *Drosophila* msl-2 mRNA. EMBO J 22:5571–5581

Gu W, Szauter P, Lucchesi JC (1998) Targeting of MOF, a putative histone acetyl transferase, to the X chromosome of *Drosophila melanogaster*. Dev Genet 22:56–64

Gu W, Wei X, Pannuti A, Lucchesi JC (2000) Targeting the chromatin-remodeling MSL complex of *Drosophila* to its sites of action on the X chromosome requires both acetyl transferase and ATPase activities. EMBO J 19:5202–5211

Hagstrom KA, Meyer BJ (2003) Condensin and cohesin: more than chromosome compactor and glue. Nat Rev Genet 4:520–534

Henry RA, Tews B, Li X, Scott MJ (2001) Recruitment of the male-specific lethal (MSL) dosage compensation complex to an autosomally integrated roX chromatin entry site correlates with an increased expression of an adjacent reporter gene in male *Drosophila*. J Biol Chem 276:31953–31958

Hiebert JC, Birchler JA (1994) Effects of the maleless mutation on X and autosomal gene expression in *Drosophila melanogaster*. Genetics 136:913–926

Hilfiker A, Hilfiker-Kleiner D, Pannuti A, Lucchesi JC (1997) mof, a putative acetyl transferase gene related to the Tip60 and MOZ human genes and to the SAS genes of yeast, is required for dosage compensation in *Drosophila*. EMBO J 16:2054–2060

Iizuka M, Smith MM (2003) Functional consequences of histone modifications. Curr Opin Genet Dev 13:154–160

Jacobs SA, Khorasanizadeh S (2002) Structure of HP1 chromodomain bound to a lysine 9-methylated histone H3 tail. Science 295:2080–2083

Jenuwein T, Allis CD (2001) Translating the histone code. Science 293:1074–1080

Jin Y, Wang Y, Walker DL, Dong H, Conley C, Johansen J, Johansen KM (1999) JIL-1: a novel chromosomal tandem kinase implicated in transcriptional regulation in *Drosophila*. Mol Cell 4:129–135

Jin Y, Wang Y, Johansen J, Johansen KM (2000) JIL-1, a chromosomal kinase implicated in regulation of chromatin structure, associates with the male specific lethal (MSL) dosage compensation complex. J Cell Biol 149:1005–1010

Kageyama Y, Mengus G, Gilfillan G, Kennedy HG, Stuckenholz C, Kelley RL, Becker PB, Kuroda MI (2001) Association and spreading of the *Drosophila* dosage compensation complex from a discrete roX1 chromatin entry site. EMBO J 20:2236–2245

Kelley RL, Kuroda MI (2003) The *Drosophila* roX1 RNA gene can overcome silent chromatin by recruiting the male-specific lethal dosage compensation complex. Genetics 164:565–574

Kelley RL, Solovyeva I, Lyman LM, Richman R, Solovyev V, Kuroda MI (1995) Expression of msl-2 causes assembly of dosage compensation regulators on the X chromosomes and female lethality in *Drosophila*. Cell 81:867–877

Kelley RL, Wang J, Bell L, Kuroda MI (1997) Sex lethal controls dosage compensation in *Drosophila* by a non-splicing mechanism. Nature 387:195–199

Kelley RL, Meller VH, Gordadze PR, Roman G, Davis RL, Kuroda MI (1999) Epigenetic spreading of the *Drosophila* dosage compensation complex from roX RNA genes into flanking chromatin. Cell 98:513–522

Kernan MJ, Kuroda MI, Kreber R, Baker BS, Ganetzky B (1991) napts, a mutation affecting sodium channel activity in *Drosophila*, is an allele of mle, a regulator of X chromosome transcription. Cell 66:949–959

Kimura A, Umehara T, Horikoshi M (2002) Chromosomal gradient of histone acetylation established by Sas2p and Sir2p functions as a shield against gene silencing. Nat Genet 32:370–377

Kuroda MI, Kernan MJ, Kreber R, Ganetzky B, Baker BS (1991) The maleless protein associates with the X chromosome to regulate dosage compensation in *Drosophila*. Cell 66:935–947

Kuroiwa A, Yokomine T, Sasaki H, Tsudzuki M, Tanaka K, Namikawa T, Matsuda Y (2002) Biallelic expression of Z-linked genes in male chickens. Cytogenet Genome Res 99:310–314

Kuzmichev A, Nishioka K, Erdjument-Bromage H, Tempst P, Reinberg D (2002) Histone methyltransferase activity associated with a human multiprotein complex containing the Enhancer of Zeste protein. Genes Dev 16:2893–2905

Langst G, Becker PB (2001) Nucleosome mobilization and positioning by ISWI-containing chromatin-remodeling factors. J Cell Sci 114:2561–2568

Lee CG, Chang KA, Kuroda MI, Hurwitz J (1997) The NTPase/helicase activities of *Drosophila* maleless, an essential factor in dosage compensation. EMBO J 16:2671–2681

Lee CG, da Costa Soares V, Newberger C, Manova K, Lacy E, Hurwitz J (1998) RNA helicase A is essential for normal gastrulation. Proc Natl Acad Sci USA 95:13709–13713

Lusser A, Kadonaga JT (2003) Chromatin remodeling by ATP-dependent molecular machines. Bioessays 25:1192–1200

Lyman LM, Copps K, Rastelli L, Kelley RL, Kuroda MI (1997) *Drosophila* male-specific lethal-2 protein: structure/function analysis and dependence on MSL-1 for chromosome association. Genetics 147:1743–1753

Marin I (2003) Evolution of chromatin-remodeling complexes: comparative genomics reveals the ancient origin of "novel" compensasome genes. J Mol Evol 56:527–539

Marin I, Baker BS (1998) The evolutionary dynamics of sex determination. Science 281:1990–1994

Marin I, Baker BS (2000) Origin and evolution of the regulatory gene male-specific lethal-3. Mol Biol Evol 17:1240–1250

Marin I, Franke A, Bashaw GJ, Baker BS (1996) The dosage compensation system of *Drosophila* is co-opted by newly evolved X chromosomes. Nature 383:160–163

McQueen HA, McBride D, Miele G, Bird AP, Clinton M (2001) Dosage compensation in birds. Curr Biol 11:253–257

Meller VH (2003) Initiation of dosage compensation in *Drosophila* embryos depends on expression of the roX RNAs. Mech Dev 120:759–767

Meller VH, Rattner BP (2002) The roX genes encode redundant male-specific lethal transcripts required for targeting of the MSL complex. EMBO J 21:1084–1091

Meller VH, Wu KH, Roman G, Kuroda MI, Davis RL (1997) roX1 RNA paints the X chromosome of male *Drosophila* and is regulated by the dosage compensation system. Cell 88:445–457

Meller VH, Gordadze PR, Park Y, Chu X, Stuckenholz C, Kelley RL, Kuroda MI (2000) Ordered assembly of roX RNAs into MSL complexes on the dosage-compensated X chromosome in *Drosophila*. Curr Biol 10:136–143

Meyer BJ (2000) Sex in the worm: counting and compensating X-chromosome dose. Trends Genet 16:247–253

Muller HJ (1948) Evidence of the precision of genetic adaptation. Harvey Lect 43:165–229

Muller J, Hart CM, Francis NJ, Vargas ML, Sengupta A, Wild B, Miller EL, O'Connor MB, Kingston RE, Simon JA (2002) Histone methyltransferase activity of a *Drosophila* Polycomb group repressor complex. Cell 111:197–208

Nakajima T, Uchida C, Anderson SF, Lee CG, Hurwitz J, Parvin JD, Montminy M (1997) RNA helicase A mediates association of CBP with RNA polymerase II. Cell 90:1107–1112

Neal KC, Pannuti A, Smith ER, Lucchesi JC (2000) A new human member of the MYST family of histone acetyl transferases with high sequence similarity to *Drosophila* MOF. Biochim Biophys Acta 1490:170–174

Oh H, Park Y, Kuroda MI (2003) Local spreading of MSL complexes from roX genes on the *Drosophila* X chromosome. Genes Dev 17:1334–1339

Orphanides G, Reinberg D (2000) RNA polymerase II elongation through chromatin. Nature 407:471–475

Palmer MJ, Mergner VA, Richman R, Manning JE, Kuroda MI, Lucchesi JC (1993) The male-specific lethal-one (msl-1) gene of *Drosophila melanogaster* encodes a novel protein that associates with the X chromosome in males. Genetics 134:545–557

Park Y, Kelley RL, Oh H, Kuroda MI, Meller VH (2002) Extent of chromatin spreading determined by roX RNA recruitment of MSL proteins. Science 298:1620–1623

Park Y, Mengus G, Bai X, Kageyama Y, Meller VH, Becker PB, Kuroda MI (2003) Sequence-specific targeting of *Drosophila* roX genes by the MSL dosage compensation complex. Mol Cell 11:977–986

Plath K, Fang J, Mlynarczyk-Evans SK, Cao R, Worringer KA, Wang H, de la Cruz CC, Otte AP, Panning B, Zhang Y (2003) Role of histone H3 lysine 27 methylation in X inactivation. Science 300:131–135

Prakash SK, van den Veyver IB, Franco B, Volta M, Ballabio A, Zoghbi HY (1999) Characterization of a novel chromo domain gene in xp22.3 with homology to *Drosophila* msl-3. Genomics 59:77–84

Rastelli L, Richman R, Kuroda MI (1995) The dosage compensation regulators MLE, MSL-1 and MSL-2 are interdependent since early embryogenesis in *Drosophila*. Mech Dev 53:223–233

Reenan RA, Hanrahan CJ, Barry G (2000) The mle(napts) RNA helicase mutation in *Drosophila* results in a splicing catastrophe of the para Na⁺ channel transcript in a region of RNA editing. Neuron 25:139–149

Richter L, Bone JR, Kuroda MI (1996) RNA-dependent association of the *Drosophila* male-less protein with the male X chromosome. Genes Cells 1:325–336

Ruiz MF, Esteban MR, Donoro C, Goday C, Sanchez L (2000) Evolution of dosage compensation in Diptera: the gene maleless implements dosage compensation in *Drosophila* (Brachycera suborder) but its homolog in *Sciara* (Nematocera suborder) appears to play no role in dosage compensation. Genetics 156:1853–1865

Sass GL, Pannuti A, Lucchesi JC (2003) Male-specific lethal complex of *Drosophila* targets activated regions of the X chromosome for chromatin remodeling. Proc Natl Acad Sci USA 100:8287–8291

Schreiber SL, Bernstein BE (2002) Signaling network model of chromatin. Cell 111:771–778

Scott MJ, Pan LL, Cleland SB, Knox AL, Heinrich J (2000) MSL1 plays a central role in assembly of the MSL complex, essential for dosage compensation in *Drosophila*. EMBO J 19:144–155

Silva J, Mak W, Zvetkova I, Appanah R, Nesterova TB, Webster Z, Peters AH, Jenuwein T, Otte AP, Brockdorff N (2003) Establishment of histone h3 methylation on the inactive X chromosome requires transient recruitment of Eed-Enx1 polycomb group complexes. Dev Cell 4:481–495

Smith ER, Pannuti A, Gu W, Steurnagel A, Cook RG, Allis CD, Lucchesi JC (2000) The *Drosophila* MSL complex acetylates histone H4 at lysine 16, a chromatin modification linked to dosage compensation. Mol Cell Biol 20:312–318

Smith ER, Allis CD, Lucchesi JC (2001) Linking global histone acetylation to the transcription enhancement of X-chromosomal genes in *Drosophila* males. J Biol Chem 276:31483–31486

Stuckenholz C, Meller VH, Kuroda MI (2003) Functional redundancy within roX1, a noncoding RNA involved in dosage compensation in *Drosophila melanogaster*. Genetics 164:1003–1014

Suka N, Luo K, Grunstein M (2002) Sir2p and Sas2p opposingly regulate acetylation of yeast histone H4 lysine16 and spreading of heterochromatin. Nat Genet 32:378–383

Suzuki MG, Shimada T, Kobayashi M (1999) Bm kettin, homologue of the *Drosophila* kettin gene, is located on the Z chromosome in *Bombyx mori* and is not dosage compensated. Heredity 82(2):170–179

Svejstrup JQ (2002) Chromatin elongation factors. Curr Opin Genet Dev 12:156–161

Turner BM (2002) Cellular memory and the histone code. Cell 111:285–291

Turner BM, Birley AJ, Lavender J (1992) Histone H4 isoforms acetylated at specific lysine residues define individual chromosomes and chromatin domains in *Drosophila* polytene nuclei. Cell 69:375–384

Velculescu VE, Madden SL, Zhang L, Lash AE, Yu J, Rago C, Lal A, Wang CJ, Beaudry GA, Ciriello KM, Cook BP, Dufault MR, Ferguson AT, Gao Y, He TC, Hermeking H, Hiraldo SK, Hwang PM, Lopez MA, Luderer HF, Mathews B, Petroziello JM, Polyak K, Zawel L, Zhang W, Zhang X, Zhou W, Haluska FG, Jen J, Sukumar S, Landes GM, Riggins GJ, Vogelstein B, Kinzler KW (1999) Analysis of human transcriptomes. Nat Genet 23:387–388

Wang Y, Zhang W, Jin Y, Johansen J, Johansen KM (2001) The JIL-1 tandem kinase mediates histone H3 phosphorylation and is required for maintenance of chromatin structure in *Drosophila*. Cell 105:433–443

West AG, Gaszner M, Felsenfeld G (2002) Insulators: many functions, many mechanisms. Genes Dev 16:271–288

Winkler GS, Kristjuhan A, Erdjument-Bromage H, Tempst P, Svejstrup JQ (2002) Elongator is a histone H3 and H4 acetyltransferase important for normal histone acetylation levels in vivo. Proc Natl Acad Sci USA 99:3517–3522

Wu CF, Ganetzky B, Jan LY, Jan YN (1978) A *Drosophila* mutant with a temperature-sensitive block in nerve conduction. Proc Natl Acad Sci USA 75:4047–4051

Wutz A, Rasmussen TP, Jaenisch R (2002) Chromosomal silencing and localization are mediated by different domains of Xist RNA. Nat Genet 30:167–174

Yonker SA, Meyer BJ (2003) Recruitment of *C. elegans* dosage compensation proteins for gene-specific versus chromosome-wide repression. Development 130:6519–6532

Zeng L, Zhou MM (2002) Bromodomain: an acetyl-lysine binding domain. FEBS Lett 513:124–128

Zhang W, Jin Y, Ji Y, Girton J, Johansen J, Johansen KM (2003a) Genetic and phenotypic analysis of alleles of the *Drosophila* chromosomal JIL-1 kinase reveals a functional requirement at multiple developmental stages. Genetics 165:1341–1354

Zhang W, Wang Y, Long J, Girton J, Johansen J, Johansen KM (2003b) A developmentally regulated splice variant from the complex lola locus encoding multiple different zinc finger domain proteins interacts with the chromosomal kinase JIL-1. J Biol Chem 278:11696–11704

Zhou S, Yang Y, Scott MJ, Pannuti A, Fehr KC, Eisen A, Koonin EV, Fouts DL, Wrightsman R, Manning JE, Lucchesi JC (1995) Male-specific lethal 2, a dosage compensation gene of *Drosophila*, undergoes sex-specific regulation and encodes a protein with a RING finger and a metallothionein-like cysteine cluster. EMBO J 14:2884–2895

DNA Methylation in Epigenetic Control of Gene Expression

Aharon Razin, Boris Kantor

Abstract   Over three decades ago DNA methylation had been suggested to play a role in the regulation of gene expression. This chapter reviews the development of this field of research over the last three decades, from the time when this idea was proposed up until now when the molecular mechanisms involved in the effect of DNA methylation on gene expression are becoming common knowledge. The dynamic changes that the DNA methylation pattern undergoes during gametogenesis and embryo development have now been revealed. The three-way connection between DNA methylation, chromatin structure and gene expression has been recently clarified and the interrelationships between DNA methylation and histone modification are currently under investigation. DNA methylation is implicated in developmental processes such as X-chromosome inactivation, genomic imprinting and disease, including tumor development. This chapter discusses all these issues in depth.

1 Introduction

The involvement of DNA methylation in cell differentiation and gene function was suggested more than three decades ago by Scarano (1971) and later by Holliday and Pugh (1975) and Riggs (1975). This suggestion was substantiated by some early data and formulated in a working hypothesis by Razin and Riggs in 1980. A large body of experimental data has been accumulated over the past two decades clearly indicating that epigenetic control of gene expression in mammals involves DNA methylation and that this control of expression is associated with gene-specific methylation patterns (Yeivin and Razin 1993). DNA methylation happens to be a perfect clonally inherited epigenetic feature of the genome. De novo methylation by the de novo methyltrans-

A. Razin, B. Kantor
The Hebrew University Medical School, P.O. Box 12272, Jerusalem, 91120, Israel,
e-mail: razina@md.huji.ac.il

Progress in Molecular and Subcellular Biology
P. Jeanteur (Ed.)
Epigenetics and Chromatin
© Springer-Verlag Berlin Heidelberg 2005

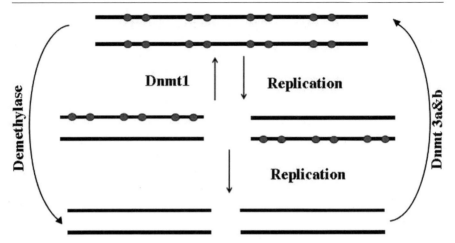

Fig. 1. Metabolism of DNA methylation. *Dnmt1* The DNA methyltransferase-maintenance enzyme; *Dnmt3a&b* de novo methyltransferases; *filled circles* methylated CpG sites

ferases Dnmt3a and Dnmt3b and demethylation by a yet uncharacterized demethylase establish the methylation patterns which are then maintained by a maintenance methyltransferase (Dnmt1) (Fig. 1; Razin and Kafri 1994).

Methylated cytosine residues in mammalian DNA (5 metCyt) are almost always found in the small palindrome CpG thus symmetrically positioned on the two DNA strands (Sinsheimer 1955). Methyltransferase activity, which is present in all cells, acts at the replication fork to restore the fully methylated state of the DNA (Gruenbaum et al. 1982, 1983). In the absence of this maintenance methylation, the DNA may lose its methylation by a so-called passive demethylation mechanism. After two rounds of replication without methylation, 50 % of the DNA molecules will be unmethylated on both strands and the other 50 % will be hemimethylated. However, in most systems studied so far, the demethylation process in the cell involves an active mechanism in which specific methylated sites undergo active demethylation, not necessarily associated with replication (Razin and Kafri 1994). On the other hand, fully unmethylated DNA can undergo de novo methylation by one of several de novo methyltransferases which are present primarily in embryonic cells (Okano et al. 1999; Fig. 1).

The mammalian genome is characterized by a bimodal pattern of DNA methylation. Seventy to 80 % of all CpG dinucleotides in the genome are methylated (Ehrlich et al. 1982). The remaining CpGs that are constantly unmethylated are clustered in CpG-rich islands. These unmethylated CpG islands are usually found in promoter regions extending to the first exon of housekeeping genes (Bird 1986). Unmethylated CpG islands are found in somatic cells in the gametes and in embryonic cells throughout embryogenesis and gametogenesis (Kafri et al. 1992). The unmethylated state of CpG islands is rigorously maintained by a mechanism that involves efficient island demethylation activity (Frank et al. 1991; Brandeis et al. 1994). In contrast to housekeeping genes, tissue-specific genes lack CpG islands and are generally methylated in nonexpressing cells, but are unmethylated in their cell type of expression (Yeivin and Razin 1993). The bimodal methylation pattern in the mammalian genome is faithfully maintained by the DNA methyltransferase 1 (Dnmt1) (Bestor and Ingram 1983). This maintenance methyltransferase methylates hemimethylated DNA during replication (Gruenbaum et al. 1983), and therefore propagates the methylation pattern for many generations of actively dividing cells (Razin and Riggs 1980). Gene-specific methylation patterns which are observed in adult tissues reflect the result of dynamic changes in methylation that are known to take place during embryogenesis (Razin and Kafri 1994).

2
Changes in Gene-Specific Methylation Patterns During Early Embryo Development

The gene-specific methylation patterns in oocyte and sperm, which are acquired during differentiation of the germ cells (Kafri et al. 1992), contribute to a combined methylation pattern in the zygote which is erased by an active demethylation mechanism during the first two or three cleavages (Kafri et al. 1993). This undermethylated state of the genome that results from this demethylation process persists through the blastula stage. This had been shown for both embryo genomic DNA (Monk et al. 1987) and specific gene sequences (Kafri et al. 1992). Global de novo methylation takes place post implantation, leaving CpG islands unmethylated. Primordial germ cells (PGCs) emerge from the unmethylated epiblast escaping the global de novo methylation that takes place following implantation. Germ cells that are first seen at day 7.5 post coitum (p.c.) in the base of the allantois (McCarrey 1993) are unmethylated (Monk et al. 1987). The germ cells remain undermethylated until after cells populate the gonads at 11.5–12.5 days p.c. De novo methylation and gene-specific demethylation take place during gametogenesis after differentiation to male and female gonads (Kafri et al. 1992; Fig. 2). This de novo methylation that is followed by gene-specific demethylations is much like what happens in the embryo proper during gastrulation and further development (Kafri et al. 1992).

In the gastrula, a process of gene-specific demethylation starts, concomitant with cell differentiation (Benvenisty et al. 1985; Shemer et al. 1991). This process continues well into adult life and ends in the fully differentiated cell with the final gene-specific methylation patterns that are observed in adult tissue (Yeivin and Razin 1993).

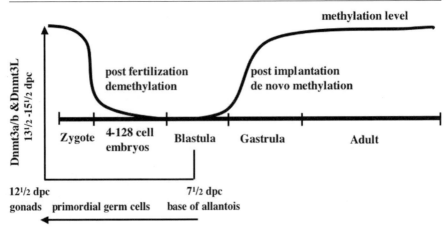

Fig. 2. DNA methylation changes during gametogenesis and early embryogenesis

3
Effect of Methylation on Gene Expression

DNA methylation has been implicated as playing a role in the multi-level hierarchy of control mechanisms that govern gene expression in mammals (Razin and Cedar 1984). Three major lines of evidence led to this conclusion: (1) genes tend to be undermethylated in the tissue of expression and stably methylated in all other tissues (Yeivin and Razin 1993); (2) genes that are inactive in vivo can be activated by treatment with the potent demethylating agent, 5-azacytidine (Jones 1984); and (3) in vitro methylated genes introduced into fibroblasts in culture remain methylated and suppressed (Stein et al. 1982; Yisraeli et al. 1988). The inverse correlation between DNA methylation and gene activity that was demonstrated for a large number of genes suggested that DNA methylation acts to repress gene activity. The transcription suppression by CpG methylation can be accomplished by two basic mechanisms: (1) *direct* interference with the binding of transcription factors and (2) *binding of multiprotein repressory complexes* resulting in the formation of an inactive chromatin structure.

3.1
Direct Transcription Inhibition

Early experiments established that local cytosine methylation of particular sequences could directly interfere with binding of a transcription factor (Tate and Bird 1993). In an in vivo footprinting experiment, methyl groups interfered with the binding of liver-specific factors to the Tat gene (Becker et al.

1987). Methylation at the downstream region of the late E2A promoter of adenovirus type 2 prevents protein binding (Hermann et al. 1989), and methylation of a site in the promoter region of the human proenkephalin gene inhibits expression and binding of the transcription factor AP2 (Comb and Goodman 1990). Cytosine methylation also prevents binding of a HeLa cell transcription factor required for optimal expression of the adenovirus major late promoter (MLP) (Watt and Molloy 1988). Maybe the most prominent example is that of a chromatin boundary element binding protein (CTCF) that binds only to its unmethylated site and can block the interaction between the enhancer and its promoter when placed between the two elements (Bell et al. 1999). The effects demonstrated with the examples described above suggest that methylation at the 5′ end of the gene is frequently sufficient to suppress gene activity (Keshet et al. 1985). However, it should be noted that a direct effect of DNA methylation on binding of specific factors has not always been observed. For example, the transcription factor, Sp1, binds and facilitates transcription even when Sp1 sites are fully methylated (Holler et al. 1988). Another example of a transcription factor that binds efficiently to its recognition site, even in its methylated state, is the chloramphenicol acetyltransferase transcription factor (CTF) (Ben-Hattar et al. 1989). In spite of the lack of effect of methylation on the binding of CTF in vitro, methylation of the site reduced

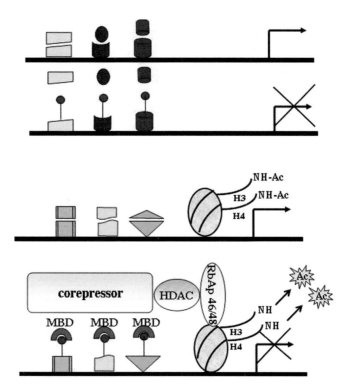

Fig. 3. a,b (*Legend see page 156*)

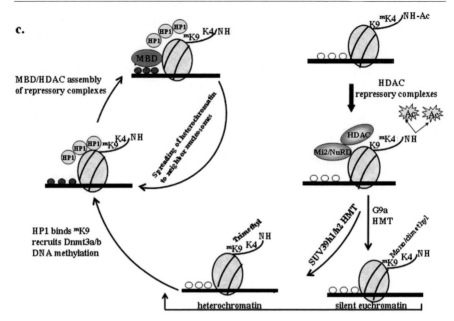

Fig. 3. Effect of methylation on gene expression. **a** Direct gene repression. Methylated transcription factor binding sites (*blue lollipops*) prevent binding of transcription factors to the corresponding sites, thus inhibiting transcription initiation (*horizontal arrow*). **b** Methyl binding domain proteins (*MBDs*) bind to methylated CpG sites (*blue lollipops*), recruit corepressory proteins that bind to histone deacetylases (*HDACs*), which in turn are bridged to nucleosomal histone (*NH*) H3 and H4 by RbAp46/48, resulting in deacetylation of acetylated lysine residues, causing repression of transcription initiation (*horizontal arrow*). **c** Epigenetic modifications of nucleosomal histones and DNA form silent euchromatin or heterochromatin depending on the histone methyltransferase (*HMT*) that methylated lysine K9. *$^mK4$* Methylated lysine K4; *Ac* acetyl; *NH* amino tail of histone H3; *K9* lysine K9; *open circles* unmethylated CpG sites; *closed blue circles* methylated CpG sites; *HDAC* histone deacetylase; *MBD* methyl binding domain protein

the activity of the gene in vivo. The discrepancy between the in vitro and in vivo results suggests that although methylation of *cis* elements may affect promoter activity, other mechanisms may be involved in promoter repression as well (Fig. 3a).

3.2
Indirect Transcription Inhibition

For a long time studies on DNA methylation and gene expression did not address the question of how chromatin structure might affect gene function. This is in spite of the fact that nucleosomal DNA had already been known to be richer in 5-methylcytosine than internucleosomal DNA (Razin and Cedar

1977; Ball et al. 1983) and that DNaseI sensitivity is a property of transcriptionally active regions of chromatin and correlated with undermethylation of transcribed genes (Sweet et al. 1982). Almost a decade later microinjection and transfection experiments using in vitro methylated gene sequences revealed that DNA methylation results in the formation of inactive chromatin (Keshet et al. 1986), and that the silencing effect exerted by CpG methylation is observed only after the methylated DNA acquired its appropriate chromatin structure (Buschhausen et al. 1987). Moreover, whereas the repressed state of the gene, which is exerted by DNA methylation alone, can be alleviated by a strong activator such as GAL4-VP16, the activator cannot overcome repression once chromatin is assembled on the methylated template (Kass et al. 1997).

These observations clearly indicate that silencing of a gene by methylation frequently involves the generation of a condensed chromatin structure that may, among other things, limit promoter accessibility to the transcription machinery. Although it has become increasingly clear that DNA methylation and chromatin structure correlate, these two epigenetic marks have only recently been connected mechanistically (Razin 1998). The discovery of the two methyl binding proteins, MeCP1 (Meehan et al. 1989) and MeCP2 (Lewis et al. 1992), helped us to realize that DNA methylation is connected with chromatin structure and gene expression. MeCP1 was later shown to be a multiprotein repressory complex (Feng and Zhang 2001). However, MeCP2, being a purifiable protein, had been cloned and characterized (Meehan et al. 1992; Nan et al. 1993). MeCP2 that is known to bind to methylated CpG-rich heterochromatin (Nan et al. 1996) contains a methyl binding domain (MBD) and a transcriptional repressory domain (TRD) (Nan et al. 1997). MeCP2 binds to the methylated DNA by its MBD and recruits the corepressor Sin3A through its TRD (Jones et al. 1998; Nan et al. 1998). MeCP2 anchors to the DNA a multiprotein repressory complex that causes histone deacetylation and chromatin remodeling by two histone deacetylase activities, HDAC1 and HDAC2. Transcriptional inactivation caused by this deacetylation could be alleviated by the HDAC inhibitor trichostatin A (TSA) (Eden et al. 1998; Jones et al. 1998; Nan et al. 1998). These observations finally solved the long suspected three-way connection between DNA methylation, condensed chromatin structure and gene silencing (Razin 1998). A search of EST databases with the methyl-binding domain (MBD) sequence of MeCP2 revealed four additional MBD proteins, MBD1–MBD4 (Hendrich and Bird 1998).

Recent studies in mammalian systems, where DNA methylation clearly plays a role in gene silencing, have shown that methyl binding proteins when bound to methylated CpG residues recruit corepressory proteins that in turn interact with two histone deacetylases HDAC1 and HDAC2. These histone deacetylases act on specific acetylated lysine residues within histones H3 and H4. This process is mediated by the RbAp46/48 proteins that serve as a bridge between the repressory complexes and the nucleosomal histones (Razin 1998; Kantor and Razin 2001). Histone deacetylation appears to be required for stable gene silencing (Eden et al. 1998; Jones et al. 1998; Nan et al. 1998) perhaps by shaping an inheritable 'closed' chromatin structure (Kantor et al. 2003; Fig. 3b).

It is now clear that MeCP2 is not the sole multiprotein repressory complex involved in transcriptional repression. MeCP1, which was discovered over a decade ago (Meehan et al. 1989), turns out to be a histone deacetylase multiprotein complex composed of ten components that include two MBD proteins, MBD2 and MBD3 (Ng et al. 1999; Feng and Zhang 2001). MeCP1 was found to share most of its components with a third multiprotein repressory complex Mi2/NuRD. This histone deacetylase-chromatin remodeling complex contains MBD3 (Zhang et al. 1999) which had been shown to play a distinctive role in mouse development (Hendrich et al. 2001).

The most studied multiprotein histone deacetylase repressory complexes to date are Mi2/NuRD, MeCP1 and MeCP2/Sin3A. The MBD proteins associated with these complexes are MBD3, MBD2 and MeCP2, respectively (Kantor and Razin 2001). Yet, many other histone deacetylase repressory complexes may exist.

Interestingly, DNA methyltransferases were also found to be components of histone deacetylase repressory complexes. The de novo DNA methyltransferase Dnmt3a binds deacetylases and is recruited by the sequence-specific repressory DNA binding protein RP58 to silence transcription (Fuks et al. 2001). The repressory activity of this complex is independent of Dnmt3a methyltransferase activity. The human maintenance DNA methyltransferase (DNMT1) forms a complex with Rb, E2F1 and HDAC1 and represses transcription from E2F-responsive promoters (Robertson et al. 2000). In parallel, DNMT1 binds HDAC2 and the corepressory DMAP1 to form a complex at replication foci (Rountree et al. 2000). The interaction between DNMT1 and histone deacetylase suggests that histone deacetylases or histone acetylation patterns can play a role in targeting DNA methylation, as well. Alternatively, DNMT1 may directly target deacetylation to regions that should become methylated. In any event, apparently methylation and deacetylation could act in concert to potentiate the repressed state. Furthermore, methyltransferases possess two activities, DNA methylation and transcriptional repression.

Although MBD1 has been shown to cause methylation-mediated transcription silencing in euchromatin (Fujita et al. 1999; Ng et al. 2000), it is not known to participate in any of the known histone deacetylase multiprotein repressory complexes. Nevertheless, its repressory effect can be alleviated by TSA, suggesting that MBD1 is part of a yet unknown histone deacetylase repressory complex. For a comprehensive discussion of all methylation-associated repressory complexes, see the review by Kantor and Razin (2001).

It has recently become clear that the flow of epigenetic information may be bidirectional. DNA methylation affects histone modification which in turn can affect DNA methylation. The interaction between these covalent modifications of chromatin may shed light on the yet unsolved mechanisms concerning the establishment of heterochromatin, its spreading along large domains of the genome and its stable inheritance. The DNA methylation–heterochromatin interrelationship including the three epigenetic marks, DNA methylation, histone methylation and acetylation, is described in Fig. 3 c.

The DNA in an active chromatin domain is unmethylated and histone H3 is acetylated, methylated on lysine at position K4 but unmethylated at lysine k9. Binding of a histone deacetylase repressory complex to the promoter region results in histone deacetylation and K4 demethylation. Two different histone methyl transferases methylate K9 depending on the position of the gene. The G9a HMT methylates K9 to the monomethyl and dimethyl level if the gene is in euchromatin, and the other methylase, Suv39h1 and h2, methylates K9 to the trimethyl level if positioned in heterochromatin. The heterochromatic protein HP1 binds to the dimethyl and trimethyl K9 and recruits the de novo methyltransferases Dnmt3a and b. This methylated structure recruits more HP1 to neighboring nucleosomes, thereby leading to a continuous process of spreading of the heterochromatic structure.

4
DNA Methylation and Genomic Imprinting

DNA methylation was implicated in several developmental processes such as X-chromosome inactivation and genomic imprinting. DNA methylation in X-inactivation is discussed in Chapter 4 (this Vol.). Genomic imprinting is perhaps the best studied developmental process in which DNA methylation plays a pivotal role. The fact that the parental genomes contribute unequally to the development of a mammalian fetus was first demonstrated by nuclear transfer studies (McGrath and Solter 1984; Surani et al. 1984). While androgenetic embryos that contain two paternal copies of the genome show very poor embryonic development with normally developed extraembryonic tissues, parthenogenetic embryos, which contain two maternal genomes, show normal embryo development and underdeveloped extraembryonic tissues. These observations suggested that both sets of parental genomes are required for proper development. This conclusion gained further support by genetic experiments in which mice containing uniparental duplications of subchromosomal regions showed phenotypes that were either embryonic lethal or developmentally retarded (Cattanach and Kirk 1985).

This phenomenon suggested that expression of a subset of genes depends on their parental origin and that such differential expression of genes must be controlled by epigenetic modifications that take place during gametogenesis when the parental alleles are in separate compartments. The epigenetic mark could then serve as a signal to discriminate between the two parental alleles post fertilization and help to maintain this discrimination during embryonic development and adult life. The mark should then be erased and created anew during gametogenesis according to the gender of the offspring.

What epigenetic modification could best fulfill these requirements? DNA methylation is a good candidate since it can be erased by demethylation, established through de novo methylation and propagated by the maintenance methyltransferase (see Fig. 1). In fact, differentially methylated regions (DMRs) were found in all imprinted genes (Razin and Cedar 1994). Interestingly, DMRs are usually located in CpG islands. This is in contrast to the general rule that characterizes CpG islands being invariably unmethylated. The monoallelic methylation of DMRs is established in the gametes by Dnmt3a, Dnmt3b and Dnmt3L. Dnmt3L which in itself lacks methyltransferase activity is believed to recruit Dnmt3a and Dnmt3b to methylate the maternal allele in imprinted genes (Bourc'his et al. 2001; Hata et al. 2002). The differential methylation that is established in the gametes or early in embryo development is maintained throughout development (Fig. 4). Therefore methylation of the DMRs must escape the genome-wide demethylation that takes place in the early embryo and also the global de novo methylation that takes place post implantation (Kafri et al. 1992; see Fig. 2). However, the imprint must be erased by an as yet unknown demethylase that should work at the time period when the primordial germ cells migrate to the gonads (Fig. 4).

In fact, several imprinted genes (Igf2r, H19, Snrpn and Xist) among others have been shown to obey these rules (Brandeis et al. 1993; Stoger et al. 1993; Ariel et al. 1995; Tremblay et al. 1995; Shemer et al. 1997). DNA methylation may therefore play a dual role in the imprinting process. The differential methylation may mark the parental alleles, allowing the transcriptional machinery of the cells to distinguish between the parental alleles. In addition, DNA methylation can directly affect promoter activity of the imprinted genes.

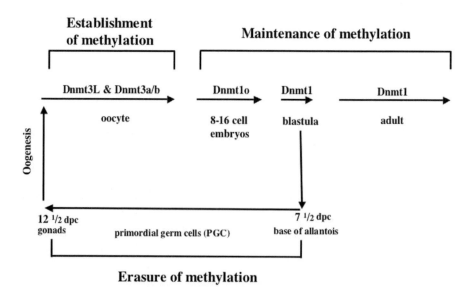

Fig. 4. DNA methylation changes of imprinted genes during gametogenesis and embryo development

Alternatively, DNA methylation can indirectly affect monoallelic expression of a gene by silencing a promoter of an antisense gene, thus allowing production of antisense RNA only from the allele on which the imprinted gene is repressed (Rougeulle et al. 1998; Lyle et al. 2000).

Loss of function of DMRs, or defects in their methylation, frequently cause loss of imprinting associated with neurobehavioral disorders. Such epigenetic defects within a 2-Mb domain on human chromosome 15q11–q13 cause two different syndromes. Prader-Willi syndrome is caused by loss of function of a large number of paternally expressed genes (Buiting et al. 1995), while silencing of the maternally expressed genes within the domain causes Angelman's syndrome (Reis et al. 1994). The imprinting of this entire domain is regulated by an imprinting center that constitutes a DMR within the 5′ region of the imprinted SNRPN gene. Individuals with deletions of this region on the paternal allele have Prader-Willi while another sequence located 35 kb upstream of SNRPN confers methylation of the SNRPN DMR on the maternal allele, thereby inactivating the paternally expressed genes on the maternal allele. When this upstream region is deleted on the maternal allele, the SNRPN DMR does not become methylated (Perk et al. 2002). Consequently, the entire domain on the maternal allele remains unmethylated and all paternally expressed genes on the maternal allele are activated while maternally expressed genes are silenced, thus causing Angelman's syndrome. A model was proposed suggesting that the upstream sequence together with the SNRPN DMR constitute a complex imprinting box responsible for both the establishment and maintenance of the imprinting state at PWS/AS domain on both alleles (Shemer et al. 2000).

Altered allelic methylation and expression patterns of the imprinted gene IGF2 have been found in Beckwith-Wiedemann syndrome (BWS) patients. BWS is a pre- and post-natal growth syndrome associated with predisposition for childhood tumors. Translocation breakpoints in a number of BWS patients map to the imprinted gene KCNQ1 which is located in the center of the 800-kb BWS region on human chromosome 11p15.5. The translocations in BWS are associated with loss of imprinting of IGF2 but not H19 (Brown et al. 1996). It appears that this impairment in imprinting involves the differentially methylated intronic CpG island in KCNQ1. In a small number of BWS patients, hypomethylation of the KCNQ1 CpG island correlated with biallelic expression of IGF2 (Smilinich et al. 1999; Paulsen et al. 2000). Deletion of this CpG island on the paternal chromosome 11 leads to silencing of KCNQ1 antisense transcript and activation of KCNQ1, p57^{KIP2} and SMS4 which are located downstream on the normally repressed paternal allele. It is therefore possible that this CpG island is at least part of an imprinting center on human chromosome 11p15.5 and its orthologous region on mouse chromosome 7.

5
DNA Methylation and Disease

Many examples exist where DNA methylation in non-imprinted genes goes awry, causing neurodevelopmental disorders, such as ATR-X, ICF, Rett and Fragile-X syndromes, or the imprinting disorders Prader-Willi, Angelman's and Beckwith-Wiedemann syndromes described above. This clearly suggests that control of gene expression that is associated with DNA methylation is particularly important in brain development and function. ATR-X (α-thalassemia, mental retardation, X-linked) patients are known to have methylation defects that result from high methyltransferase activity in neurons (Goto et al. 1994). Such high activity of methyltransferase in mice results in delayed ischemic brain damage (Endres et al. 2000) associated with hypo- or hypermethylation of repetitive sequences (Gibbons et al. 2000).

ICF (immunodeficiency, centromeric instability and facial anomaly) syndrome is linked to mutations in the de novo methyltransferase gene DNMT3B (mapped to chromosome 20q) affecting its carboxy terminal catalytic domain (Hansen et al. 1999), resulting in hypomethylation of the normally heavy methylated repetitive sequences (Kondo et al. 2000) and single copy sequences on the inactive X-chromosomes (Miniou et al. 1994; Bourc'his et al. 1999). Mice knocked out in Dnmt3b show similar demethylation and could therefore serve as an experimental ICF model (Okano et al. 1999). How DNMT3B deficiency affects brain development remains to be elucidated.

Another syndrome that is manifested in mental retardation and is associated with methylation-dependent gene silencing is the X-linked Rett syndrome that results from mutations in the MeCP2 gene (Amir et al. 1999). Mutations that cause the disease disrupt the integrity of the methyl binding domain (MBD) or the transcription repressory domain (TRD) of MeCP2, whose function in gene repression has been discussed above. How MeCP2 mutations lead to developmental defects in the brain is currently a matter under investigation.

The most common form of inherited mental retardation after Down syndrome is Fragile-X syndrome. The X-linked gene that is associated with the disease, Fragile-X mental retardation 1 (FMR1), contains highly polymorphic CGG repeats with an average length of 29 repeats in normal individuals and 200–600 repeats in Fragile-X patients. In addition, the CpG island at the 5' end of the gene in patients is abnormally methylated and histone deacetylated, causing silencing of the gene (Oberle et al. 1991; Coffee et al. 1999). The reasons for this de novo methylation and the mechanisms driving this de novo methylation are, as yet, unclear.

6
Concluding Remarks

DNA methylation may have evolved as a luxury device for setting up central biological processes. Central to all processes that involve DNA methylation is the control of gene expression. The high complexity of the mammalian genome required a multilevel hierarchy of mechanisms that control gene expression. One of these levels of regulation involves DNA methylation. DNA methylation is a flexible epigenetic feature of the genome that can be established, maintained and erased. Being flexible, this feature had been successfully employed to serve the dynamic changes the cells undergo during gametogenesis and development of the embryo. One striking example of how DNA methylation functions in development is its being an epigenetic mark in discrimination between the alleles in phenomena such as X-chromosome inactivation and genomic imprinting. The importance of DNA methylation in the well-being of the cell is clearly reflected in the large number of genetic diseases, including cancer, that occur when DNA methylation goes awry. Being conserved in evolution, DNA methylation must have had an evolutionary advantage that outweighs the price paid by the cell in the form of genetic disorders. The recent discoveries in the field of epigenetic modifications of nucleosomal histones revealed the role played by DNA methylation in shaping chromatin structure, thereby affecting formation of silent domains in the genome.

References

Amir RE, Van den Veyver IB, Wan M, Tran CQ, Francke U, Zoghbi HY (1999) Rett syndrome is caused by mutations in X-linked MECP2, encoding methyl-CpG-binding protein 2. Nature Genet 23:185–188

Ariel M, Robinson E, McCarrey JR, Cedar H (1995) Gamete-specific methylation imprints on the Xist gene. Nature Genet 9:312–315

Ball DJ, Gross DS, Garrard WT (1983) 5-methylcytosine is localized in nucleosomes that contain histone H1. Proc Natl Acad Sci USA 80:5490–5494

Becker PB, Ruppert S, Schultz G (1987) Genomic footprinting reveals cell type-specific DNA binding of ubiquitous factors. Cell 51:435–443

Bell AC, West AG, Felsenfeld G (1999) The protein CTCF is required for the enhancer blocking activity of vertebrate insulators. Cell 98:387–396

Ben-Hattar J, Beard P, Jiricny J (1989) Cytosine methylation in CTF and Sp1 recognition sites of an HSV tk promoter: effects on transcription in vivo and on factor binding in vitro. Nucleic Acids Res 17:10179–10190

Benvenisty N, Mencher D, Meyuchas O, Razin A, Reshef L (1985) Sequential changes in DNA methylation patterns of the rat phosphoenolpyruvate carboxykinase gene during development. Proc Natl Acad Sci USA 82:267–271

Bestor TH, Ingram VM (1983) Two DNA methyltransferases from murine erythroleukemia cells: purification, sequence specificity, and mode of interaction with DNA. Proc Natl Acad Sci USA 80:5559–5563

Bird AP (1986) CpG-rich islands and the function of DNA methylation. Nature 321:209–213

Bourc'his D, Miniou P, Jeanpierre M, Molina Gomes D, Dupont J, De Saint-Basile G, Maraschio P, Tiepolo L, Viegas-Pequignot E (1999) Abnormal methylation does not prevent X inactivation in ICF patients. Cytogenet Cell Genet 84:245–252

Bourc'his D, Xu GL, Lin CS, Bollman B, Bestor TH (2001) Dnmt3L and the establishment of maternal genomic imprints. Science 294:2536–2539

Brandeis M, Kafri T, Ariel M, Chaillet JR, McCarrey J, Razin A, Cedar H (1993) The ontogeny of allele-specific methylation associated with imprinted genes in the mouse. EMBO J 12:3669–3677

Brandeis M, Frank D, Keshet I, Siegfried Z, Mendelsohn M, Nemes A, Temper V, Razin A, Cedar H (1994) Sp1 elements protect a CpG island from de novo methylation. Nature 371:435–438

Brown KW, Villar AJ, Bickmore W, Clayton-Smith J, Catchpoole D, Maher ER, Reik W (1996) Imprinting mutation in the Beckwith-Wiedemann syndrome leads to biallelic IGF2 expression through an H19-independent pathway. Hum Mol Genet 5:2027–2032

Buiting K, Saitoh S, Gross S, Dittrich B, Schwartz S, Nicholls DR, Horsthemke B (1995) Inherited microdeletions in the Angelman and Prader-Willi syndromes define an imprinting centre on human chromosome 15. Nature Genet 9:395–400

Buschhausen G, Wittig B, Graessmann M, Graessmann A (1987) Chromatin structure is required to block transcription of the methylated herpes simplex virus thymidine kinase gene. Proc Natl Acad Sci USA 84:1177–1181

Cattanach BM, Kirk M (1985) Differential activity of maternally and paternally derived chromosome regions in mice. Nature 315:496–498

Coffee B, Zhang F, Warren ST, Reines D (1999) Acetylated histones are associated with FMR1 in normal but not fragile X-syndrome cells. Nature Genet 22:98–101

Comb M, Goodman HM (1990) CpG methylation inhibits proenkephalin gene expression and binding of the transcription factor AP-2. Nucleic Acids Res 18:3975–3982

Eden S, Hashimshony T, Keshet I, Thorne AW, Cedar H (1998) DNA methylation models histone acetylation. Nature 394:842–843

Ehrlich M, Gama-Sosa MA, Huang LH, Midgett RM, Kuo KC, McCune RA, Gehrke C (1982) Amount and distribution of 5-methylcytosine in human DNA from different types of tissues of cells. Nucleic Acids Res 10:2709–2721

Endres M, Meisel A, Biniszkiewicz D, Namura S, Prass K, Ruscher K, Lipski A, Jaenisch R, Moskowitz MA, Dirnagl U (2000) DNA methyltransferase contributes to delayed ischemic brain injury. J Neurosci 20:3175–3181

Feng Q, Zhang Y (2001) The MeCP1 complex represses transcription through preferential binding, remodeling, and deacetylating methylated nucleosomes. Genes Dev 15:827–832

Frank D, Keshet I, Shani M, Levine A, Razin A, Cedar H (1991) Demethylation of CpG islands in embryonic cells. Nature 351:239–241

Fujita N, Takebayashi S, Okumura K, Kudo S, Chiba T, Saya H, Nakao M (1999) Methylation-mediated transcriptional silencing in euchromatin by methyl-CpG binding protein MBD1 isoforms. Mol Cell Biol 19:6415–6426

Fuks F, Burgers WA, Godin N, Kasai M, Kouzarides T (2001) Dnmt3a binds deacetylases and is recruited by a sequence-specific repressor to silence transcription. EMBO J 20:2536–2544

Gibbons RJ, McDowell TL, Raman S, O'Rourke DM, Garrick D, Ayyub H, Higgs DR (2000) Mutations in ATRX, encoding a SWI/SNF-like protein, cause diverse changes in the pattern of DNA methylation. Nat Genet 24:368–371

Goto K, Numata M, Komura JI, Ono T, Bestor TH, Kondo H (1994) Expression of DNA methyltransferase gene in mature and immature neurons as well as proliferating cells in mice. Differentiation 56:39–44

Gruenbaum Y, Cedar H, Razin A (1982) Substrate and sequence specificity of a eukaryotic DNA methylase. Nature 292:620–622

Gruenbaum Y, Szyf M, Cedar H, Razin A (1983) Methylation of replicating and post-repli-cated mouse L-cell DNA. Proc Natl Acad Sci USA 80:4919-4921

Hansen RS, Wijmenga C, Luo P, Stanek AM, Canfield TK, Weemaes CM, Gartler SM (1999) The DNMT3B DNA methyltransferase gene is mutated in the ICF immunodeficiency syndrome. Proc Natl Acad Sci USA 96:14412-14417

Hata K, Okano M, Lei H, Li E (2002) Dnmt3L cooperates with the Dnmt3 family of de novo DNA methyltransferases to establish maternal imprints in mice. Development 129:1983-1993

Hendrich B, Bird A (1998) Identification and characterization of a family of mammalian methyl-CpG binding proteins. Mol Cell Biol 18:6538-6547

Hendrich B, Guy J, Ramsahoye B, Wilson VA, Bird A (2001) Closely related proteins MBD2 and MBD3 play distinctive but interacting roles in mouse development. Genes Dev 15:710-723

Hermann R, Hoeveler A, Doerfler W (1989) Sequence-specific methylation in a downstream region of the late E2A promoter of adenovirus type 2 DNA prevents protein binding. J Mol Biol 210:411-415

Holler M, Westin G, Jiricney J, Schaffner W (1988) Sp1 transcription factor binds DNA and activates transcription even when the binding site is CpG methylated. Genes Dev 2:1127-1135

Holliday R, Pugh JE (1975) DNA modification mechanisms and gene activity during devel-opment. Science 187:226-232

Jones PA (1984) Gene activation by 5-azacytidine. Springer, Berlin Heidelberg New York

Jones PL, Veenstra GJC, Wade PA, Vermaak D, Kass SU, Landsberg N, Strouboulis J, Wolffe AP (1998) Methylated DNA and MeCP2 recruit histone deacetylase to repress transcription. Nature Genet 19:187-191

Kafri T, Ariel M, Brandeis M, Shemer R, Urven L, McCarrey J, Cedar H, Razin A (1992) Devel-opmental pattern of gene-specific DNA methylation in the mouse embryo and germline. Genes Dev 6:705-714

Kafri T, Gao X, Razin A (1993) Mechanistic aspects of genome-wide demethylation in the preimplantation mouse embryo. Proc Natl Acad Sci USA 90:10558-10562

Kantor B, Razin A (2001) DNA methylation, histone deacetylase repressory complexes and development. Gene Funct Dis 2:69-75

Kantor B, Makedonski K, Shemer R, Razin A (2003) Expression and localization of compo-nents of the histone deacetylases multiprotein repressory complexes in the mouse preimplantation embryo. Gene Exp Pattern 3:697-702

Kass SU, Landsberger N, Wolffe AP (1997) DNA methylation directs a time-dependent repression of transcription initiation. Curr Biol 7:157-165

Keshet I, Yisraeli J, Cedar H (1985) Effect of hybrid methylation on gene transcription. Proc Natl Acad Sci USA 82:2560-2564

Keshet I, Lieman-Hurwitz J, Cedar H (1986) DNA methylation affects the formation of active chromatin. Cell 44:535-543

Kondo T, Bobek MP, Kuick R, Lamb B, Zhu X, Narayan A, Bourc'his D, Viegas-Pequignot E, Ehrlich M, Hanash SM (2000) Whole-genome methylation scan in ICF syndrome: hypomethylation of non-satellite DNA repeats D4Z4 and NBL2. Hum Mol Genet 9:597-604

Lewis JD, Meehan RR, Henzel WJ, Maurer-Fogy I, Jeppesen P, Klein F, Bird A (1992) Purifi-cation, sequence, and cellular localization of a novel chromosomal protein that binds to methylated DNA. Cell 69:905-914

Lyle R, Watanabe D, te Vruchte D, Lerchner W, Smrzka OW, Wutz A, Schageman J, Hahner L, Davies C, Barlow DP (2000) The imprinted antisense RNA at the Igf2r locus overlaps but does not imprint Mas1. Nat Genet 25:19-21

McCarrey JR (1993) Development of the germ cell. In: Desjardins C, Ewing LL (eds) Cell and molecular biology of the testis. Oxford University Press, Oxford, pp 58-89

McGrath J, Solter D (1984) Complementation of mouse embryogenesis requires both maternal and paternal genomes. Cell 37:179–183

Meehan RR, Lewis JD, McKay S, Kleiner EL, Bird AP (1989) Identification of a mammalian protein that binds specifically to DNA containing methylated CpGs. Cell 58:499–507

Meehan RR, Lewis JD, Bird AP (1992) Characterization of MeCP2, a vertebrate DNA binding protein with affinity for methylated DNA. Nucleic Acids Res 20:5085–5092

Miniou P, Jeanpierre M, Blanquet V, Sibella V, Bonneau D, Herbelin C, Fischer A, Niveleau A, Viegas-Pequignot E (1994) Abnormal methylation pattern in constitutive and facultative (X inactive chromosome) heterochromatin of ICF patients. Hum Mol Genet 3:2093–2102

Monk M, Boubelik M, Lehnert S (1987) Temporal and regional changes in DNA methylation in the embryonic, extraembryonic and germ cell lineages during mouse embryo development. Development 99:371–382

Nan X, Meehan RR, Bird A (1993) Dissection of the methyl-CpG binding domain from the chromosomal protein MeCP2. Nucleic Acids Res 21:4886–4892

Nan X, Tate P, Li E, Bird A (1996) DNA methylation specifies chromosomal localization of MeCP2. Mol Cell Biol 16:414–421

Nan X, Campoy FJ, Bird A (1997) MeCP2 is a transcriptional repressor with abundant binding sites in genomic chromatin. Cell 88:471–481

Nan X, Ng H-H, Johnson CA, Laherty CD, Turner BM, Eisenman RN, Bird A (1998) Transcriptional repression by the methyl-CpG-binding protein MeCP2 involves a histone deacetylase complex. Nature 393:386–389

Ng HH, Zhang Y, Hendrich B, Johnson CA, Turner BM, Erdjument-Bromage H, Tempst P, Reinberg D, Bird A (1999) MBD2 is a transcriptional repressor belonging to the MeCP1 histone deacetylase complex. Nature Genet 23:58–61

Ng HH, Jeppesen P, Bird A (2000) Active repression of methylated genes by the chromosomal protein MBD1. Mol Cell Biol 20:1394–1406

Oberle I, Rousseau F, Heitz D, Kretz C, Devys D, Hanauer A, Boue J, Bertheas MF, Mandel JL (1991) Instability of a 550-base pair DNA segment and abnormal methylation in fragile X syndrome. Science 252:1097–1102

Okano M, Bell DW, Haber DA, Li E (1999) DNA methyltransferases Dnmt3a and Dnmt3b are essential for de novo methylation and mammalian development. Cell 99:247–257

Paulsen M, El-Maarri O, Engemann S, Strodicke M, Franck O, Davies K, Reinhardt R, Reik W, Walter J (2000) Sequence conservation and variability of imprinting in the Beckwith-Wiedemann syndrome gene cluster in human and mouse. Hum Mol Genet 9:1829–1841

Perk J, Lande L, Cedar H, Razin A, Shemer R (2002) The imprinting mechanism of the Prader Willi/Angelman regional control center. EMBO J 21:5807–5814

Razin A (1998) CpG methylation, chromatin structure and gene silencing – a three-way connection. EMBO J 17:4905–4908

Razin A, Cedar H (1977) Distribution of 5-methylcytosine in chromatin. Proc Natl Acad Sci USA 74:2725–2728

Razin A, Riggs AD (1980) DNA methylation and gene function. Science 210:604–610

Razin A, Cedar H (1984) DNA methylation in eukaryotic cells. Int Rev Cytol 92:159–185

Razin A, Kafri T (1994) DNA methylation from embryo to adult. Prog Nucleic Acids Res Mol Biol 48:53–82

Razin A, Cedar H (1994) DNA methylation and genomic imprinting. Cell 77:473–476

Reis A, Dittrich B, Greger V, Buiting K, Lalande M, Gillessen-Kaesbach G, Anvret M, Horsthemke B (1994) Imprinting mutations suggested by abnormal DNA methylation patterns in familial Angelman and Prader-Willi syndromes. Am J Hum Genet 54:741–747

Riggs AD (1975) X inactivation, differentiation, and DNA methylation. Cytogenet Cell Genet 14:9–25

Robertson KD, Ait-Si-Ali S, Yokochi T, Wade PA, Jones PL, Wolffe AP (2000) DNMT1 forms a complex with Rb, E2F1 and HDAC1 and represses transcription from E2F-responsive promoters. Nat Genet 25:338–342

Rougeulle C, Cardoso C, FontÈs M, Colleaux L, Lalande M (1998) An imprinted antisense RNA overlaps UBE3A and a second maternally expressed transcript. Nat Genet 19:15–16

Rountree MR, Bachman KE, Baylin SB (2000) DNMT1 binds HDAC2 and a new co-repressor, DMAP1, to form a complex at replication foci. Nature Genet 25:269–277

Scarano E (1971) The control of gene function in cell differentiation and in embryogenesis. Adv Cytopharmacol 1:13–24

Shemer R, Eisenberg S, Breslow JL, Razin A (1991) Methylation patterns of the human apoAI-CIII-AIV gene cluster in adult and embryonic tissue suggest dynamic changes in methylation during development. J Biol Chem 266:23676–23681

Shemer R, Birger Y, Riggs AD, Razin A (1997) Structure of the imprinted mouse Snrpn gene and establishment of its parental-specific methylation pattern. Proc Natl Acad Sci USA 94:10267–10272

Shemer R, Hershko AY, Perk J, Mostoslavsky R, Tsuberi B-Z, Cedar H, Buiting K, Razin A (2000) The imprinting box of the Prader-Willi/Angelman syndrome domain. Nat Genet 26:440–443

Sinsheimer RL (1955) The action of pancreatic deoxyribonuclease. II. Isomeric dinucleotides. J Biol Chem 215:579–583

Smilinich NJ, Day CD, Fitzpatrick GV, Caldwell GM, Lossie AC, Cooper PR, Smallwood AC, Joyce JA, Schofield PN, Reik W, Nicholls RD, Weksberg R, Driscoll DJ, Maher ER, Shows TB, Higgins MJ (1999) A maternally methylated CpG island in KvLQT1 is associated with an antisense paternal transcript and loss of imprinting in Beckwith-Wiedemann syndrome. Proc Natl Acad Sci USA 96:8064–8069

Stein R, Razin A, Cedar H (1982) In vitro methylation of the hamster APRT gene inhibits its expression in mouse L-cells. Proc Natl Acad Sci USA 79:3418–3422

Stoger R, Kubicka P, Liu C-G, Kafri T, Razin A, Cedar H, Barlow DP (1993) Maternal-specific methylation of the imprinted mouse Igf2r locus identifies the expressed locus as carrying the imprinting signal. Cell 73:61–71

Surani MA, Barton SC, Norris ML (1984) Development of mouse eggs suggests imprinting of the genome during gametogenesis. Nature 308:548–550

Sweet RW, Chao MV, Axel R (1982) The structure of the thymidine kinase gene promoter: nuclease hypersensitivity correlates with expression. Cell 31:347–353

Tate PH, Bird AP (1993) Effects of DNA methylation on DNA-binding proteins and gene expression. Curr Opin Genet Dev 3:226–231

Tremblay KD, Saam JR, Ingram RS, Tilghman SM, Bartolomei MS (1995) A paternal-specific methylation imprint marks the alleles of the mouse H19 gene. Nat Genet 9:407–413

Watt F, Molloy PL (1988) Cytosine methylation prevents binding to DNA of a HeLa cell transcription factor required for optimal expression of the adenovirus major late promoter. Genes Dev 2:1136–1143

Yeivin A, Razin A (1993) Gene methylation patterns and expression. In: Jost JP, Saluz HP (eds) DNA methylation: molecular biology and biological significance. Birkhauser, Basel, pp 523–568

Yisraeli J, Frank D, Razin A, Cedar H (1988) Effect of in vitro DNA methylation on β globin gene expression. Proc Natl Acad Sci USA 85:4638–4642

Zhang Y, Ng HH, Erdjument-Bromage H, Tempst P, Bird A, Reinberg D (1999) Analysis of the NuRD subunits reveals a histone deacetylase core complex and a connection with DNA methylation. Genes Dev 13:1924–1935

The Epigenetic Breakdown of Cancer Cells: From DNA Methylation to Histone Modifications

Esteban Ballestar, Manel Esteller

Abstract The recognition of epigenetic defects in all types of cancer has represented a revolutionary achievement in cancer research in recent years. DNA methylation aberrant changes (global hypomethylation and CpG island hypermethylation) were among the first events to be recognized. The overall scenario comprises a network of factors in which deregulation of DNA methyltransferases leads to a cancer-type specific profile of tumor suppressor genes that become epigenetically silenced. Over recent years, a better understanding of the machinery that connects DNA methylation, chromatin and transcriptional activity, in which histone modifications stand in a key position, has been achieved. The identification of these connections has contributed not only to understanding how epigenetic deregulation occurs in cancer but also to developing novel therapies that can reverse epigenetic defects in cancer cells.

1
Introduction

Cells encode their heritable information in two major ways: genetic information, defined by the ordered sequence of nucleotides, and epigenetic modifications, which provide cells with heritable states of gene expression.

For years, research on cancer has been concerned with investigating genetic lesions and their downstream consequences, centered on the analysis of the function of target genes. This was due to the assumption that most of the information encoded by cells is harbored in their genomic sequence. However, the recognition of the importance of epigenetic information has been of enormous consequence in reorienting our efforts and opening new doors in cancer research.

E. Ballestar, M. Esteller
Cancer Epigenetics Laboratory, Spanish National Cancer Centre (CNIO),
Melchor Fernández Almagro 3, 28029 Madrid, Spain, e-mail: mesteller@cnio.es

Progress in Molecular and Subcellular Biology
P. Jeanteur (Ed.)
Epigenetics and Chromatin
© Springer-Verlag Berlin Heidelberg 2005

Epigenetic changes are those heritable modifications that do not involve a change in gene sequence. In particular, these changes are involved in protecting cells from endoparasitic sequences, maintaining the imprinting and X-chromosome expression patterns and the identity of cells within a tissue type. In cancer, epigenetic alterations participate in and determine the loss of the original identity of the cell and are known to play a key role in cancer development and progression.

One of the best-studied epigenetic alterations in cancer is that of the content and distribution of 5-methylcytosine in the genome. In normal cells, methylation of the 5' carbon of the pyrimidine ring of cytosine in DNA is the main epigenetic modification of the genome. In mammals, cytosine methylation occurs in the context of the CpG dinucleotide sequence. CpGs are relatively infrequent in the genome, with the exception of regions of variable length, between 0.5 and 2.0 kb, known as CpG islands. Most CpG islands are coincident with the promoter of protein-coding genes and are normally unmethylated, in contrast with the remaining CpGs, which are methylated.

Early analysis of the role of methylation using tissue-specific genes introduced into mammalian cells by transfection gave rise to a general consensus that DNA methylation directs the formation of nuclease-resistant chromatin, leading to repression of gene activity (Keshet et al. 1986; Cedar 1988; Bird 1992). These conclusions have been greatly refined in recent years and detailed mechanisms will be discussed below.

In cancer, three major events related to the balance of the 5-methylcytosine have been demonstrated to occur: (1) the global reduction of 5-methylcytosine content in the genome. This was the first change in DNA methylation to be observed; (2) hypermethylation of the CpG island of many tumor-suppressor genes associated with their transcriptional silencing; and (3) everything occurs with a general increase in the expression of DNA methyltransferases (DNMTs), the enzymes responsible for maintaining and establishing methylation. Thus, cancer is not a problem of the lack or the overexpression of DNMTs; rather, it is the erroneous targeting of DNMTs to incorrect regions of the genome.

Initially, global hypomethylation was thought to be the only significant methylation change in cancer and it was believed that it might lead to the massive overexpression of oncogenes. This is not the case. All the CpG islands of oncogenes are unmethylated in normal cells and cannot be more 'hypomethylated' in cancer cells. However, there is a global decrease in 5-methylcytosine content, which reflects a heterogeneous change in DNA methylation: hypomethylation occurs in isolated CpGs scattered throughout the genome and those CpG dinucleotides present in CpG islands experience hypermethylation.

Currently, it is accepted that hypomethylation of repetitive and parasitic DNA sequences correlates with a number of adverse outcomes in cancer. For example, decreased methylation of repetitive sequences in the satellite DNA of the pericentric region of chromosomes is associated with increased chromosomal rearrangements, a hallmark of cancer. For instance, the finding that DNMT3b mutations, which occur in ICF syndrome, cause centromeric instability is indicative of how global demethylation destabilizes overall chromatin organization. Furthermore, decreased methylation of proviral sequences can lead to reactivation and increased infectivity. In fact, one primary function of DNA methylation is the suppression of transcription and expansion of parasitic elements such as transposons (e.g., SINES and LINES) (Yoder et al. 1997). The vast majority of methylated CpGs in normal cells reside within repetitive elements; global demethylation contributes to the reactivation of these parasitic sequences by transcription and movement.

The idea of the methylation of CpG islands of tumor-suppressor genes as a mechanism of gene inactivation in cancer was proposed in 1994 (Herman et al. 1994) when methylation-dependent silencing of the Von Hippel-Lindau (VHL) gene was demonstrated to be a mechanism of gene inactivation in renal carcinoma. In the following years, parallel studies in the laboratories of Dr. Stephen Baylin and Dr. Peter A. Jones established that CpG island hypermethylation is a common mechanism of gene inactivation in cancer. Recently, we have demonstrated that the profile of CpG island hypermethylation is specific to the tumor type (Esteller et al. 2001). The analysis of a few selected hypermethylated CpG islands can be so powerful that they classify tumors of unknown origin (Paz et al. 2003a). CpG island hypermethylation of tumor-suppressor genes, which leads to their inactivation, is now considered the major epigenetic alteration in cancer (Esteller 2002).

Our vision of the role of genetic lesions in the development and progression of cancer, accepted for many years, has evolved, and nowadays cancer is accepted as having a double origin in which both genetic and epigenetic defects are responsible for the onset and progression of the disease.

2
What Is Responsible for DNA Methylation and for How Deregulation Occurs?

Our appreciation of how DNA methylation changes can occur in cancer depends on how well the DNA methylation machinery itself is understood. As mentioned above, methylation occurs at the 5′ carbon of cytosine, a relatively unreactive position. The catalytic mechanism of DNA (cytosine-5)-methyltransferases has been proposed as being similar to that of thymidylate synthetase, in which an enzyme cysteine thiolate binds covalently to the 6-position. This pushes electrons to the 5-position to make the carbanion, which can then attack the methyl group of N5,N10-methylenetetrahydrofolate. After methyl transfer, abstraction of a proton from the 5-position may allow reformation of the 5–6 double bond and release of the enzyme.

The first DNA cytosine-methyltransferase identified was revealed by purification and cloning. It remains the sole mammalian DNA methyltransferase to have been identified by biochemical assay (Bestor et al. 1988). This enzyme, now termed DNMT1, is a protein that contains 1,620 amino acids and exhibits a 5- to 30-fold preference for hemimethylated substrates. This property led to the assignment of DNMT1 as the enzyme responsible for maintaining the methylation patterns following DNA replication. However, there is no direct evidence that DNMT1 is not also involved in certain types of de novo methylation, and in fact DNMT1 is involved in most of the de novo methylation activity in embryo lysates.

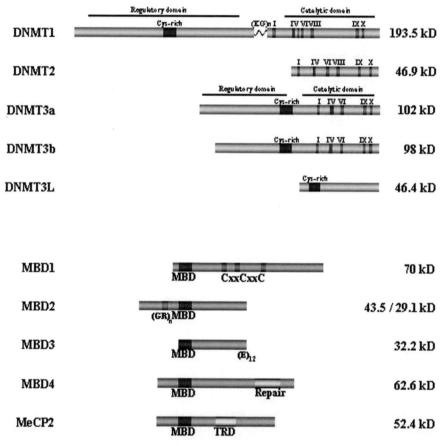

Fig. 1. The DNA methylation system: DNMTs and MBDs. The two families of proteins are represented: *Above* DNMTs. Regulatory and catalytic domains are indicated. DNMT2 contains the full set of sequence motifs that are almost invariably diagnostic of DNA cytosine-methyltransferases but has not been shown to have transmethylase activity by biochemical or genetic tests. It also lacks the N-terminal domain characteristic of eukaryotic DNMTs. The N-terminal regions of DNMT3A and DNMT3B are highly divergent on the N-terminal side of the Cys-rich. DNMT3L lacks canonical DNMT motifs but is otherwise closely related to the C-terminal domain of DNMT3A and DNMT3B. *Below* MBDs. The methyl-CpG binding domain is indicated. In the case of MecP2, its well-defined transcriptional repression domain (TRD) is also depicted. MBD4 possesses a thymine glycosilase domain

The remaining DNA methyltransferases were identified by searches of EST databases. The first of these was DNMT2 (Yoder and Bestor 1998). This lacks the large N-terminal regulatory domain common to other eukaryotic methyl-transferases and does not exhibit comparable DNA methyltransferase activity, although it does seem to have some residual activity in vitro (Hermann et al. 2003).

DNMT3a and DNMT3b were soon identified by searching EST databases (Okano et al. 1998) and were proposed to be the enzymes responsible for de novo methylation (Okano et al. 1999). Mutations in the human *DNMT3B* gene are responsible for ICF syndrome. Figure 1 shows a schematic representation of the DNMT family.

Although DNMTs were originally classified as maintenance or de novo DNA methyltransferases, several strands of evidence indicate that all three DNMTs not only cooperate but also may possess both de novo and maintenance functions in vivo (Rhee et al. 2000; Kim et al. 2002; Chen et al. 2003; Paz et al. 2003b).

Different hypotheses have been advanced to explain the aberrant patterns of DNA methylation in cancer. However, the exact nature of the defects in the methylation machinery in tumor cells is unknown. It has been proposed that an aberrant expression of DNMTs is involved. Failure of the mechanisms that control or prevent methylation in CpG islands in normal cells may also be involved. On the other hand, the defect could occur in the mechanisms that correct aberrantly methylated sequences. Finally, in several systems, methylation of histone H3 methylation has been demonstrated to precede DNA methylation (Bachman et al. 2003). The possibility that defects in the histone methylation machinery are involved in generating aberrant DNA methylation patterns, while, at the same time, DNA methylation changes shift the histone code, cannot be ruled out.

Although no satisfactory answers to this intriguing epigenetic feature of tumoral tissues have been found, the fact that the methylation profile is specific to the tumor type suggests that the group of altered genes is important in the development of each tumor type (Esteller et al. 2001). In fact, a Darwinian evolutionary explanation for the specific profile of aberrant methylation has also been invoked (Esteller 2002). It is possible that indiscriminate aberrant methylation of CpG islands occurs, and that only the methylation and, therefore, the inactivation of certain genes confer the cells with some evolutionary advantage over their neighbors, allowing proliferation and further selection of those hypermethylated CpG islands.

3
Is Methylation Specific to the Tumor Type?

Several lines of evidence imply an active role of hypermethylation of tumor-suppressor genes in the development of cancer. In the first place, hypermethylation is an early event in cancer. However, the comprehensive analysis of methylation in many different tumor types and gene promoters provides evidence of the existence of a tumor-type-specific profile. In theory, CpG islands should be the most 'attractive' substrate for DNA methylation, since, by definition, they contain a high concentration of CpG-rich sequences. It has been speculated that there must be some factors that prevent unscheduled methylation at CpG islands. Many questions arise: why do CpG islands become methylated in cancer? Why do certain CpG islands become methylated while others do not?

The identification of CpG islands that become methylated in cancer has relied primarily on a candidate-gene approach. For this purpose, tumor-suppressor genes whose mutations have been associated with cancer have provided a useful source of candidates. This type of approach has served to identify many methylated genes in cancer for which key roles in tumorigenesis had previously been demonstrated.

The availability of genomic information since the almost completion of the human genome sequencing projects has facilitated the development of new strategies intended to identify novel genes that become methylated in cancer. These genome-wide approaches include the use of microarrays (Suzuki et al. 2002), the use of methylation-sensitive restriction enzymes and the use of two-dimensional gels (Costello et al. 2002), amplification of intermethylated sites (Paz et al. 2003a) and the combination of chromatin immunoprecipitation with genomic microarrays (Ballestar et al. 2003).

4
Connecting DNA Methylation Changes with Transcription: Chromatin Mechanisms

The information stored by methylation of CpGs has functional significance only in the context of chromatin. Since its discovery, DNA methylation has been associated with a transcriptionally inactive state of chromatin; however, the mechanisms by which DNA methylation is translated into transcriptionally silent chromatin have only recently started to be unveiled.

Historically, several hypotheses have been proposed to explain the way in which DNA methylation is interpreted by nuclear factors. The first possibility is that DNA methylation inhibits the binding of sequence-specific transcription factors to their binding sites that contain CpG (Tate and Bird 1993; Deng et al. 2001). In this context, a protein with an affinity for unmethylated CpGs has also been identified (Lee et al. 2001) that is associated with actively transcribed regions of the genome (Lee and Slanik 2002). In this case, methylation of CpGs would result in release of this protein. An alternative model proposed that methylation may have direct consequences for nucleosome positioning, for instance, by leading to the assembly of specialized nucleosomal structures on methylated DNA that silence transcription more effectively than does conventional chromatin (Kass et al. 1997). The third possibility is that methylation leads to the recruitment of specialized factors that selectively recognize methylated DNA and either impede binding of other nuclear factors or have a direct effect on repressing transcription. Although there are examples that support all three possibilities, the active recruitment of methyl-CpG binding activities appears to be the most generally accepted mechanism of methylation-dependent repression.

MeCP1 and MeCP2 were the first two methyl-CpG binding activities to be described (Lewis et al. 1992). While MeCP1 was originally identified as a large multiprotein complex, MeCP2 is a single polypeptide with an affinity for a single methylated-CpG. Characterization of MeCP2 in subsequent years led to the identification of the minimum portion with affinity for methylated DNA, i.e., its methyl-CpG binding domain (MBD) (Meehan et al. 1992) and its transcriptional repression domain (TRD).

Database searches led to the identification of additional proteins harboring the MBD, namely MBD1, MBD2, MBD3 and MBD4 (Hendrich and Bird 1998). While mammalian MBD1 and MBD2 are bona fide methylated-DNA binding proteins, MBD3 is able to bind methylated DNA only in certain species (Hendrich and Bird 1998; Wade et al. 1999). In the case of MBD4, this protein binds preferentially to m5CpG × TpG mismatches. The primary product of deamination at methyl-CpGs and the combined specificities of binding and catalysis indicate that this enzyme functions to minimize mutation at methyl-CpGs. Figure 1 presents a schematic depiction of the MBD family.

In 1997, the laboratories of Dr Adrian Bird and Dr Alan Wolffe reported that MeCP2 represses transcription of methylated DNA through the recruitment of a histone deacetylase-containing complex (Jones et al. 1998; Nan et al. 1998). This finding established for the first time a mechanistic connection between DNA methylation and transcriptional repression by the modification of chromatin. Additional reports have established the mechanism by which the remaining MBDs connect DNA methylation and gene silencing (Ng et al. 1999, 2000; Wade et al. 1999). Ng et al. (1999) reported that MBD2 is, in fact, a component of the formerly identified MeCP1 complex, which exhibits histone deacetylase activity. On the other hand, Alan Wolffe's laboratory identified MBD3 as a component of the Mi-2/NURD complex, which exhibits both histone deacetylase and ATPase-dependent nucleosome remodeling activities (Wade et al. 1999).

In order to understand the implications of the connections between DNA methylation and histones, it is important to define the relevance of these post-translational histone modifications to the determination of different chromatin states. Since the first reports of the occurrence of histone modifications (Allfrey et al. 1964), 40 years of chromatin research has resulted in the description of a variety of histone modifications and the specificity of certain modifications under particular physiological conditions. Most histone modifications occur in their protruding N-terminal tails. This specificity in the pattern of modifications under particular conditions led to the proposal of the 'histone code hypothesis', according to which, histone modifications form a code that may be read by nuclear factors (Strahl and Allis 2000; Turner 2002). Several modifications are compatible with gene silencing. In general, histone deacetylation leads to gene silencing. Furthermore, methylation of lysine 9 of histone H3 has been associated with gene silencing (Fig. 2).

Following the finding of the coupling between DNA methylation and histone deacetylation by MBDs, additional connections have been found. On the one hand, DNMTs are known to be able also to recruit histone deacetylases (Robertson et al. 2000; Fuks et al. 2001); while, on the other hand, both DNMTs and MBDs have been reported to recruit histone methyltransferases that modify lysine 9 of histone H3 (Fujita et al. 2003; Fuks et al. 2003a,b).

Moreover, novel methylated-DNA binding repressors have been identified. The second group of methylated DNA binding proteins is composed of a single member, known as Kaiso (Daniel and Reynolds 1999). Despite the absence of a recognizable MBD motif, Kaiso clearly has the capacity to selectively recognize methyl CpG (Prokhortchouk et al. 2001). Kaiso, identified initially as a p120 catenin-interacting protein by a yeast two-hybrid screen (Daniel and Reynolds 1999), contains no signature MBD domain and is a member of the BTB/POZ (Broad complex, Tramtrak, Bric-a-brac/Pox virus, and Zinc finger) family of transcription factors (Daniel and Reynolds 1999). Kaiso requires at least two symmetrically methylated CpG dinucleotides in its recognition sequence and exhibits methylation-dependent repression in transient transfection assays (Prokhortchouk et al. 2001). Although Kaiso may represent one of the constituent subunits of the MeCP1 complex, it may also interact with other factors to mediate methylation-dependent repression, such as N-CoR (Yoon et al. 2003).

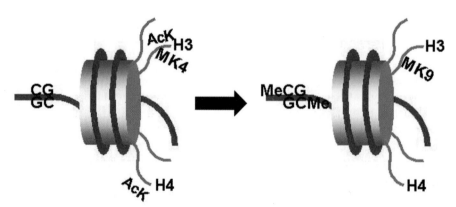

Fig. 2. Interplay between CpG methylation and histone modifications. A nucleosome is shown, where the *gray cylinder* represents the histone octamer and DNA is shown in *red*. CpG methylation is accompanied by loss of acetylation (*Ac*) at both H3 and H4, loss of methylation (*Me*) at lysine 4 of H3 and methylation at K9 of H3

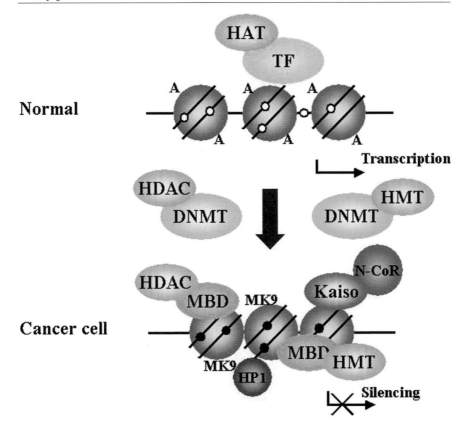

Fig. 3. Multiple factors mediate methylation-dependent silencing of tumor-suppressor genes in cancer. An array of nucleosomes is shown. Histone octamers are represented by *green circles*. DNA is represented as a *black line* in which unmethylated CpG dinucleotides are *white circles*, whereas methylated CpGs are *black circles*. *Above* The situation in an unmethylated promoter in which transcription factors and histone acetyltransferases are recruited and histones are acetylated (*A*). In cancer, DNA methyltransferases (*DNMT*) recruit histone deacetylases (*HDAC*) and histone methyltransferases (*HMTs*). Methylated promoters bind methyl-CpG binding proteins (*MBDs*) that recruit HDACs and HMTs. Histones are deacetylated and histone H3 is methylated at K9 (*MK9*) and facilitates binding of HP1. Kaiso can also bind methylated DNA and recruits the N-CoR co-repressor complex

Therefore, multiple connections are established between hypermethylation of the CpG islands of tumor-suppressor genes in cancer and their transcriptional silencing (Fig. 3). The specificity of these connections and the special circumstances in which these different elements participate for different genes remain to be determined.

In the case of MBD proteins, association with hypermethylated promoters and their involvement in silencing their corresponding genes has now been demonstrated in a number of cases (Magdinier and Wolffe 2001; Nguyen et al.

2001; Ballestar et al. 2003). In fact, MBD proteins appear to be a common feature of the methylated promoter of these genes, and also display a remarkable specificity (Ballestar et al. 2003). Moreover, it has been proposed that fusion proteins such as PML-RAR can contribute to aberrant CpG-island methylation by recruiting DNMTs and HDACs to aberrant sites (di Croce et al. 2002). This latter activity is somewhat controversial, but, in any case, does not seem to be a general mechanism, at least in leukemia patients (Esteller et al. 2002). Research into these issues will lead in the near future to a better understanding of how CpG islands become methylated and how this is precisely translated into specific chromatin structures.

5
Can We Reactivate Epigenetically Silenced Genes?
Towards Epigenetic Therapy

Reversibility of epigenetic events provides a target for chemotherapeutic intervention (Villar-Garea and Esteller 2003). The precise understanding of how epigenetic silencing of tumor-suppressor genes occurs will surely lead to the development of alternative chemotherapies. Nucleoside-analogue inhibitors of DNA methyltransferases, of which 5-aza-2'-deoxycytidine (5-aza-dC) is one of the most common, are able to demethylate DNA and restore silenced gene expression. Unfortunately, the clinical utility of these compounds has not yet been fully realized, mainly because of their side effects. Alternative non-nucleoside inhibitors of DNA methyltransferases have recently been found to function as demethylating agents. These include the antiarrhythmic drug procainamide and the related anesthetic drug procaine (Villar-Garea et al. 2003). Histone deacetylase inhibitors constitute alternative novel targets for epigenetic chemotherapy. Some of these compounds have been demonstrated to have anticancer properties.

Although all these compounds have been shown to reduce tumor growth, alternative and more specific compounds are now being developed. The possibility of designing engineered zinc finger protein (ZFP) transcription factors (TFs) to modify specifically the expression of individual genes is now being investigated (Jamieson et al. 2003; Snowden et al. 2003).

References

Allfrey V, Faulkner RM, Mirsky AE (1964) Acetylation and methylation of histones and their possible role in the regulation of RNA synthesis. Proc Natl Acad Sci USA 51:786–794
Bachman KE, Park BH, Rhee I, Rajagopalan H, Herman JG, Baylin SB, Kinzler KW, Vogelstein B (2003) Histone modifications and silencing prior to DNA methylation of a tumor suppressor gene. Cancer Cell 3:89–95

Ballestar E, Paz MF, Valle L, Wei S, Fraga MF, Espada, J, Cigudosa JC, Huang TH-M, Esteller M (2003) Methyl-CpG binding proteins identify novel sites of epigenetic inactivation in human cancer. EMBO J 22:1–11

Bestor TH, Laudano A, Mattaliano R, Ingram V (1988) Cloning and sequencing of a cDNA encoding DNA methyltransferase of mouse cells. The carboxyl-terminal domain of the mammalian enzyme is related to bacterial restriction methyltransferases. J Mol Biol 203:971–983

Bird A (1992) The essentials of DNA methylation. Cell 70:5–8

Cedar H (1988) DNA methylation and gene activity. Cell 53:3–4

Chen T, Ueda Y, Dodge JE, Wang Z, Li E (2003) Establishment and maintenance of genomic methylation patterns in mouse embryonic stem cells by Dnmt3a and Dnmt3b. Mol Cell Biol 23:5594–5605

Costello JF, Smiraglia DJ, Plass C (2002) Restriction landmark genome scanning. Methods 27:144–149

Daniel JM, Reynolds AB (1999) The catenin p120(ctn) interacts with Kaiso, a novel BTB/POZ domain zinc finger transcription factor. Mol Cell Biol 19:3614–3623

Deng G, Chen A, Pong E, Kim YS (2001) Methylation in hMLH1 promoter interferes with its binding to transcription factor CBF and inhibits gene expression. Oncogene 20:7120–7127

Di Croce L, Raker VA, Corsaro M, Fazi F, Fanelli M, Faretta M, Fuks F, Lo Coco F, Kouzarides T, Nervi C, Minucci S, Pelicci PG (2002) Methyltransferase recruitment and DNA hypermethylation of target promoters by an oncogenic transcription factor. Science 295:1079–1082

Esteller M (2002) CpG island hypermethylation and tumor suppressor genes: a booming present, a brighter future. Oncogene 21:5427–5440

Esteller M, Corn PG, Baylin SB, Herman JG (2001) A gene hypermethylation profile of human cancer. Cancer Res 61:3225–3229

Esteller M, Fraga MF, Paz MF, Campo E, Colomer D, Novo FJ, Calasanz MJ, Galm O, Guo M, Benitez J, Herman JG (2002) Cancer epigenetics and methylation. Science 297:1807–1808

Fujita N, Watanabe S, Ichimura T, Tsuruzoe S, Shinkai Y, Tachibana M, Chiba T, Nakao M (2003) Methyl-CpG binding domain 1 (MBD1) interacts with the Suv39h1-HP1 heterochromatic complex for DNA methylation-based transcriptional repression. J Biol Chem 278:24132–24138

Fuks F, Burgers WA, Godin N, Kasai M, Kouzarides T (2001) Dnmt3a binds deacetylases and is recruited by a sequence-specific repressor to silence transcription. EMBO J 20:2536–2544

Fuks F, Hurd PJ, Deplus R, Kouzarides T (2003a) The DNA methyltransferases associate with HP1 and the SUV39H1 histone methyltransferase. Nucleic Acids Res 31:2305–2312

Fuks F, Hurd PJ, Wolf D, Nan X, Bird AP, Kouzarides T (2003b) The methyl-CpG-binding protein MeCP2 links DNA methylation to histone methylation. J Biol Chem 278:4035–4040

Hendrich B, Bird AP (1998) Identification and characterization of a family of mammalian methyl-CpG binding proteins. Mol Cell Biol 18:6538–6547

Herman JG, Latif F, Weng Y, Lerman MI, Zbar B, Liu S, Samid D, Duan DS, Gnarra JR, Linehan WM, Baylin SB (1994) Silencing of the VHL tumor-suppressor gene by DNA methylation in renal carcinoma. Proc Natl Acad Sci USA 91:9700–9704

Hermann A, Schmitt S, Jeltsch A (2003) The human Dnmt2 has residual DNA-(cytosine-C5) methyltransferase activity. J Biol Chem 278:31717–31721

Jamieson AC, Miller JC, Pabo CO (2003) Drug discovery with engineered zinc-finger proteins. Nat Rev Drug Discov 2:361–368

Jones PL, Veenstra GJ, Wade PA, Vermaak D, Kass SU, Landsberger N, Strouboulis J, Wolffe AP (1998) Methylated DNA and MeCP2 recruit histone deacetylase to repress transcription. Nat Genet 19:187–191

Kass SU, Pruss D, Wolffe AP (1997) How does DNA methylation repress transcription? Trends Genet 13:444–449

Keshet I, Lieman-Hurwitz J, Cedar H (1986) DNA methylation affects the formation of active chromatin. Cell 44:535–545

Kim GD, Ni J, Kelesoglu N, Roberts RJ, Pradhan S (2002) Co-operation and communication between the human maintenance and de novo DNA (cytosine-5) methyltransferases. EMBO J 21:4183–4195

Lee JH, Skalnik DG (2002) CpG-binding protein is a nuclear matrix- and euchromatin-associated protein localized to nuclear speckles containing human trithorax. Identification of nuclear matrix targeting signals. J Biol Chem 277:42259–42267

Lee JH, Voo KS, Skalnik DG (2001) Identification and characterization of the DNA binding domain of CpG-binding protein. J Biol Chem 276:44669–44676

Lewis JD, Meehan RR, Henzel WJ, Maurer-Fogy I, Jeppesen P, Klein F, Bird A (1992) Purification, sequence, and cellular localization of a novel chromosomal protein that binds to methylated DNA. Cell 69:905–914

Magdinier F, Wolffe AP (2001) Selective association of the methyl-CpG binding protein MBD2 with the silent p14/p16 locus in human neoplasia. Proc Natl Acad Sci USA 98:4990–4995

Meehan RR, Lewis JD, Bird AP (1992) Characterization of MeCP2, a vertebrate DNA binding protein with affinity for methylated DNA. Nucleic Acids Res 20:5085–5092

Nan X, Ng HH, Johnson CA, Laherty CD, Turner BM, Eisenman RN, Bird A (1998) Transcriptional repression by the methyl-CpG-binding protein MeCP2 involves a histone deacetylase complex. Nature 393:386–389

Ng HH, Zhang Y, Hendrich B, Johnson CA, Turner BM, Erdjument-Bromage H, Tempst P, Reinberg D, Bird A (1999) MBD2 is a transcriptional repressor belonging to the MeCP1 histone deacetylase complex. Nat Genet 23:58–61

Ng HH, Jeppesen P, Bird A (2000) Active repression of methylated genes by the chromosomal protein MBD1. Mol Cell Biol 20:1394–1406

Nguyen CT, Gonzales FA, Jones PA (2001) Altered chromatin structure associated with methylation-induced gene silencing in cancer cells: correlation of accessibility, methylation, MeCP2 binding and acetylation. Nucleic Acids Res 29:4598–4606

Okano M, Xie S, Li E (1998) Cloning and characterization of a family of novel mammalian DNA (cytosine-5) methyltransferases. Nat Genet 19:219–220

Okano M, Bell DW, Haber DA, Li E (1999) DNA methyltransferases Dnmt3a and Dnmt3b are essential for de novo methylation and mammalian development. Cell 99:247–257

Paz MF, Fraga MF, Avila S, Guo M, Pollan M, Herman JG, Esteller M (2003a) A systematic profile of DNA methylation in human cancer cell lines. Cancer Res 63:1114–1121

Paz MF, Wei S, Cigudosa JC, Rodriguez-Perales S, Peinado MA, Huang TH, Esteller M (2003b) Genetic unmasking of epigenetically silenced tumor suppressor genes in colon cancer cells deficient in DNA methyltransferases. Hum Mol Genet 12:2209–2219

Prokhortchouk A, Hendrich B, Jorgensen H, Ruzov A, Wilm M, Georgiev G, Bird A, Prokhortchouk E (2001) The p120 catenin partner Kaiso is a DNA methylation-dependent transcriptional repressor. Genes Dev 15:1613–1618

Rhee I, Jair KW, Yen RW, Lengauer C, Herman JG, Kinzler KW, Vogelstein B, Baylin SB, Schuebel KE (2000) CpG methylation is maintained in human cancer cells lacking DNMT1. Nature 404:1003–1007

Robertson KD, Ait-Si-Ali S, Yokochi T, Wade PA, Jones PL, Wolffe AP (2000) DNMT1 forms a complex with Rb, E2F1 and HDAC1 and represses transcription from E2F-responsive promoters. Nat Genet 25:338–342

Snowden AW, Zhang L, Urnov F, Dent C, Jouvenot Y, Zhong X, Rebar EJ, Jamieson AC, Zhang HS, Tan S, Case CC, Pabo CO, Wolffe AP, Gregory PD (2003) Repression of vascular endothelial growth factor A in glioblastoma cells using engineered zinc finger transcription factors. Cancer Res 63:8968–8976

Strahl BD, Allis CD (2000) The language of covalent histone modifications. Nature 403:41–45

Suzuki H, Gabrielson E, Chen W, Anbazhagan R, van Engeland M, Weijenberg MP, Herman JG, Baylin SB (2002) A genomic screen for genes upregulated by demethylation and histone deacetylase inhibition in human colorectal cancer. Nat Genet 31:141–149

Tate PH, Bird AP (1993) Effects of DNA methylation on DNA-binding proteins and gene expression. Curr Opin Genet Dev 3:226–231

Turner BM (2002) Cellular memory and the histone code. Cell 111:285–291

Villar-Garea A, Esteller M (2003) DNA demethylating agents and chromatin-remodelling drugs: which, how and why? Curr Drug Metab 4:11–31

Villar-Garea A, Fraga MF, Espada J, Esteller M (2003) Procaine is a DNA-demethylating agent with growth-inhibitory effects in human cancer cells. Cancer Res 63:4984–4989

Wade PA, Gegonne A, Jones PL, Ballestar E, Aubry F, Wolffe AP (1999) Mi-2 complex couples DNA methylation to chromatin remodelling and histone deacetylation. Nat Genet 23:62–66

Yoder JA, Walsh CP, Bestor TH (1997) Cytosine methylation and the ecology of intragenomic parasites. Trends Genet 13:335–340

Yoder JA, Bestor TH (1998) A candidate mammalian DNA methyltransferase related to pmt1p of fission yeast. Hum Mol Genet 7:279–284

Yoon HG, Chan DW, Reynolds AB, Qin J, Wong J (2003) N-CoR mediates DNA methylation-dependent repression through a methyl CpG binding protein Kaiso. Mol Cell 12:723–734

Developmental Regulation of the β-Globin Gene Locus

Lyubomira Chakalova, David Carter, Emmanuel Debrand,
Beatriz Goyenechea, Alice Horton, Joanne Miles, Cameron Osborne,
Peter Fraser

Abstract The β-globin genes have become a classical model for studying regulation of gene expression. Wide-ranging studies have revealed multiple levels of epigenetic regulation that coordinately ensure a highly specialised, tissue- and stage-specific gene transcription pattern. Key players include *cis*-acting elements involved in establishing and maintaining specific chromatin conformations and histone modification patterns, elements engaged in the transcription process through long-range regulatory interactions, *trans*-acting general and tissue-specific factors. On a larger scale, molecular events occurring at the locus level take place in the context of a highly dynamic nucleus as part of the cellular epigenetic programme.

1 Introduction

Our picture of gene regulation at the transcriptional level, and particularly the functional relationships with higher-order chromatin and nuclear structure, remains largely incomplete. Any given gene is subject to many regulatory echelons that ultimately control when and where it is transcribed as well as the level of transcription. These include the transcription factor environment, regional and distal controlling elements, local chromatin conformation, chromatin modifications, as well as higher-order folding and nuclear organization. Many of these parameters encode or have the potential to be influenced by epigenetic information. Though the β-globin locus has long served as a paradigm in the study of many of these regulatory levels, the recognition of epigenetic control of developmental β-globin gene expression has only recently emerged. This chapter is not intended to be an exhaustive review of the literature on globin gene regulation, but will instead focus on some of the key elements of potential epigenetic control.

L. Chakalova, D. Carter, E. Debrand, B. Goyenechea, A. Horton, J. Miles, C. Osborne, P. Fraser
Laboratory of Chromatin and Gene Expression, The Babraham Institute, Cambridge, CB2 4AT, UK, e-mail: peter.fraser@bbsrc.ac.uk

Progress in Molecular and Subcellular Biology
P. Jeanteur (Ed.)
Epigenetics and Chromatin
© Springer-Verlag Berlin Heidelberg 2005

2 The β-Globin Clusters and Their Ontogeny

The α- and β-like globin genes are very highly expressed in erythroid cells, making up approximately 90 % of the total poly(A) RNA in mature reticulocytes (Hunt 1974). This is due in part to the exceptionally high levels of globin gene transcription. The β-like genes are clustered on chromosome 7 in mice and chromosome 11 in humans. The temporal and spatial expression of the genes is tightly regulated during development, providing an appropriate and unique haemoglobin type for each stage. The structure and regulation of the mouse and human β-globin loci are similar in many aspects (Fig. 1).

The mouse β-globin cluster contains four genes (εy, βH1, βmaj, and βmin), while the human locus contains five genes (ε, $^G\gamma$, $^A\gamma$, δ, and β). In both cases the linear arrangement of the genes along the chromosome reflects the developmental order of expression. This phenomenon, often referred to as co-linearity, has been observed in other gene clusters such as the hox genes (Kmita and Duboule 2003). During human development ε-globin is expressed from 6–8 weeks of gestation in primitive nucleated erythroid cells derived mainly from the embryonic blood islands of the yolk sac (Collins and Weissman 1984). As the major site of erythropoiesis changes to the fetal liver, the ε gene is silenced and transcription of the γ-globin genes is switched on. At around the time of birth the main site of erythropoiesis changes again to the adult bone marrow concomitant with a further 'switch' in gene expression. Erythroid cells from the bone marrow express predominantly the β-globin gene (95 %); however, low levels of δ- and γ-globin expression are detectable. Though the γ genes are normally considered to be silenced in adult erythroid cells, low-level γ expression is the result of transient expression in the early stages of adult erythroid differentiation (Pope et al. 2000; Wojda et al. 2002;

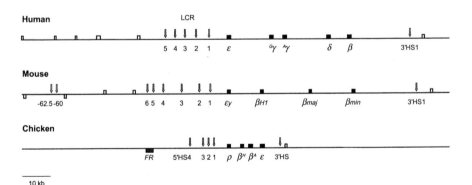

Fig. 1. Maps of the human, mouse, and chicken β-globin loci. Genes are presented as *boxes*: *black boxes* represent globin genes; *open boxes* olfactory receptor genes; *grey box* the chicken folate receptor gene. Genes transcribed in the sense direction, such as all globin genes, are located *above the line*; genes transcribed in the opposite direction are *below the line*. *Vertical arrows* indicate DNase-hypersensitive sites

Chakalova et al., unpubl. observ.). The δ-globin is normally expressed at low levels throughout adult life (2.5% of the β-globin-like protein), which may reflect deficiencies in its promoter sequence.

In mice, red blood cell formation begins at day 8 in yolk sac blood islands and by day 9 primitive, nucleated erythrocytes are released into circulation from the yolk sac. These cells express primarily the embryonic εy and βH1 genes, although low levels of βmaj and βmin transcription are detectable (Brotherton et al. 1979; Trimborn et al. 1999). Interestingly, the transcriptional output of each gene is inversely proportional to its distance from the upstream locus control region (Trimborn et al. 1999; see also discussion below). The fetal liver becomes the main site of erythropoiesis after day 10, giving rise to definitive erythroid cells. The embryonic genes εy and βH1 are completely silenced and the adult βmaj and βmin genes are transcriptionally upregulated to full activity which persists into adult life (Brotherton et al. 1979; Trimborn et al. 1999).

The human β-globin locus has also been studied by inserting part, or all, of the locus into transgenic mice. Transgenic mice containing the entire human globin locus express ε-globin and γ-globin in primitive erythrocytes derived from the yolk sac (Strouboulis et al. 1992; Peterson et al. 1993). The switch to expression in the fetal liver occurs after day 10 and is accompanied by complete silencing of ε and persistent but reduced transcription of the γ genes, concomitant with increased δ- and β-gene transcription (Strouboulis et al. 1992; Peterson et al. 1993; Fraser et al. 1998). By day E16.5 γ-gene silencing is nearly complete, whereas δ- and β-gene transcription continues through adult life. Though there are differences that may be attributable to the dramatically shortened gestation period, the consistent observation that spatial and temporal globin switching of the human genes can be retained in a transgenic mouse highlights the degree of conservation between the human and mouse β-globin loci.

3
Models for Studying the β-Globin Locus

Over the years many experimental systems have been employed to study the β-globin genes. The best methods involve looking directly at the organism in question. To this end, chicken and mouse have been widely used as model systems to study developmental regulation. Though the literature is rich in the characterization of specific mutations that affect expression in the human locus (Stamotoyannopoulos and Grosveld 2001), developmental analysis presents more of a challenge for ethical and logistic reasons; so various alternatives have been sought. As described above, the human locus can be cloned and inserted into mice. In such transgenic mice many of the features of the β-globin locus, including tissue and developmental specificity, are retained. Thus, transgenic mice provide a reasonable and useful method to assay function in the human locus, although caution must be exercised when interpreting the results. An alternative to studying the organisms in vivo is to use cell lines. Cell lines have been isolated that appear to reflect some aspects of erythroid and developmental specificity. For example, murine erythroleukaemia (MEL) cells, immortalized with the Friend leukaemia virus (Friend et al. 1966; Marks and Rifkind 1978), are thought to represent a murine pro-erythroblast stage cell. They can proliferate in a predifferentiated state, or be induced to differentiate with a variety of chemical compounds into cells that mimic the adult stage of murine erythroid development, expressing βmaj and βmin. Erythroid cell lines from chicken, human, and mouse have been very useful in studying various aspects of globin gene regulation; however, it should be noted that none of them reproduces the magnitude of gene expression dynamics seen in the globin locus in vivo nor fully recapitulates the epigenetic changes associated with development.

More recently, attention has focused on manipulation of the endogenous mouse locus by homologous recombination as a means of functional analysis of the locus. This approach represents the most definitive system for functional analyses of a complex gene locus.

4
The LCR Is Required for High-Level Expression

The globin genes are transcribed at exceptionally high levels in erythroid cells. Initial attempts to characterize the DNA sequence elements responsible for this high-level activation concentrated on the regions immediately flanking the gene (Wright et al. 1984; Townes et al. 1985; Kollias et al. 1986; Behringer et al. 1987; Antoniou et al. 1988). Many β-globin transgenic mice were generated with various promoter and downstream sequences (Chada et al. 1986; Kollias et al. 1986, 1987). Though much useful information was gained regarding gene-proximal regulatory elements, the level of expression generated in such mice was highly variable, and in most cases one or two orders of magnitude lower than the endogenous mouse globin genes. The variability of expression in different transgenic lines was attributed to genomic position effects (PE) at the site of integration. PE can be either positive or negative presumably depending on the chromatin flanking the transgene. Transgene integration in or near constitutive heterochromatin such as pericentromeric regions is now generally recognized as one of the causes of some types of transgene silencing in mammals (Festenstein et al. 1996; Milot et al. 1996). Position effect variegation (PEV), which was initially recognized in yeast and *Drosophila* and has now been seen for a number of genes in vertebrates, results in clonal heritable silencing of a transgene in a subpopulation of cells of a tissue. PEV has also been observed in non-centromeric transgene integration sites (Savelier et al. 2003). For globin transgenes affected by PEV it is worth noting that whilst overall expression is low due to the reduced number of expressing cells, erythroid specificity and developmental timing is often retained (Milot et al. 1996). Collectively, these results suggest that the sequences immediately flanking the globin gene contribute to tissue- and developmental-specific expression but are insufficient to overcome the effects of the surrounding chromatin to ensure high-level expression in all cells.

Clues that distant elements were involved in globin gene regulation came from naturally occurring deletions in humans that led to deregulation of the β-globin genes. Patients with β-thalassaemia have severely reduced HbA (adult hemoglobin) due to downregulation of β-globin gene transcription. Several thalassaemia mutations have been identified and characterized; the most informative in this case are the large deletions involving regions upstream of the β gene such as the Dutch thalassaemia (van der Ploeg et al. 1980; Kioussis et al. 1983; Harteveld et al. 2003), English thalassaemia (Curtin et al. 1985; Curtin and Kan 1988), and Hispanic thalassaemia deletions (Driscoll et al. 1989; Forrester et al. 1990). In the search for potential distant regulatory elements, DNase I-hypersensitive sites (HS) in the β-globin locus were mapped. Several erythroid-specific HS were found in a 15-kb region upstream of the ε gene (Tuan et al. 1985; Forrester et al. 1986, 1987; Grosveld et al. 1987). Collectively known as the locus control region (LCR) each HS contains a core with several binding sites for erythroid and ubiquitous transcription factors (Philipsen et al. 1990; Talbot et al. 1990; Pruzina et al. 1991; Strauss and Orkin 1992; Furukawa et al. 1995; Goodwin et al. 2001). When the LCR was linked to the β-globin gene and inserted into transgenic mice, the level of β-gene expression was found to be equivalent to the endogenous mouse globin genes (Grosveld et al. 1987). The LCR was able to drive high-level expression of the gene in all transgenic mice regardless of the integration site, and the level of expression was directly proportional to the number of LCR–gene constructs integrated. Thus the LCR was functionally defined as a sequence that confers integration-independent, copy number-dependent, high-level expression upon a linked transgene. The LCR's ability to drive expression regardless of integration site was postulated to be due to a positive chromatin opening activity that allowed it to 'open' the chromatin of the adjacent transgene regardless of the site of integration. Indeed, analyses of a number of different LCRs linked to transgenes have demonstrated that even the repressive effects of pericentromeric heterochromatin can be overcome by a complete LCR (Festenstein et al. 1996; Milot et al. 1996; Kioussis and Festenstein 1997; Fraser and Grosveld 1998). The LCR was also shown to be capable of reprogramming heterologous genes, through its ability to drive erythroid-specific expression of non-erythroid genes (Blom van Assendelft et al. 1989). These findings led to suggestions that the LCR was responsible for opening the entire β-globin locus in erythroid cells. However, conclusive evidence showing that the LCR is not necessary for chromatin opening was provided by seminal experiments in which the mouse and human LCRs were deleted by homologous recombination (Epner et al. 1998; Reik et al. 1998; Alami et al. 2000). Globin gene expression was virtually abolished in the absence of the LCR in the human locus whereas β-gene expression in the mouse locus was greatly reduced but still developmentally specific. However, the erythroid-specific open chromatin structure and histone hyperacetylation of the locus were maintained in the absence of the LCR (Schubeler et al. 2000). Collectively, these results show that the LCR is required for high-level transcription of the globin genes, but they also demonstrate that other sequence elements in the locus control opening and maintenance of an active chromatin structure.

5
The Role of Individual HS

The functional contribution of the different HS has been dissected in many experiments. The initial approaches assayed gene expression in small constructs containing one or all of the LCR HS(s) linked to a gene. As the technology for larger transgenes became available, individual HS deletions were assayed in what were often described as 'full locus' constructs. Human HS1, HS2, HS3, and HS4 can all increase the level of expression of a linked gene depending on the experimental system. Interestingly, HS2 is the only LCR HS with classical enhancer activity (Tuan et al. 1989; Moon and Ley 1991), which by definition is a sequence element that can enhance expression of a linked transgene in a transient transfection assay in either orientation, upstream or downstream relative to the gene. All of the other HS show little activity in this type of assay. However, when integrated into chromatin, as occurs in stably transfected cells, both HS2 and HS3 were able to drive increased expression levels, indicating that the sites are functionally distinct (Collis et al. 1990). In transgenic mice, Human HS1, HS2, HS3, and HS4 were able to significantly increase transgene expression levels depending on the developmental stage assayed (Ryan et al. 1989; Fraser et al. 1990, 1993; Jackson et al. 1996). HS2 and HS3 appeared to provide the bulk of the activity at most developmental stages (Fraser et al. 1990, 1993). In general, expression of linked genes increased in an additive rather than a synergistic fashion when multiple HS were included in the same construct (Collis et al. 1990; Fraser et al. 1990). Interestingly, in multi-gene constructs the developmental pattern of γ- versus β-gene expression varied depending on the HS used, suggesting distinct developmental specificities (Fraser et al. 1993). HS3 was the only site capable of driving γ-gene expression in the fetal liver, while HS4 activity peaked during β-gene expression in adult cells. Deletion of single HS from YAC transgenes appears to support the findings of developmental or gene specificities for some of the LCR HS in the human locus (Peterson et al. 1996; Navas et al. 1998, 2001, 2003). However, similar phenomena were not as apparent in knockouts of single HS in the endogenous mouse locus (Fiering et al. 1995; Hug et al. 1996; Bender et al. 2001). When the core of HS2, HS3, or HS4 was removed, expression of the globin genes was severely reduced (Bungert et al. 1995, 1999). Likewise, substitution of one HS core for another resulted in greatly reduced expression, again suggesting that each HS plays a unique role. Deletion by homologous targeting of HS1, HS2, HS3 or HS4 lead to modest reductions in adult globin expression consistent with the additive effect on transcription as observed in transgene constructs (Fiering et al. 1995; Hug et al. 1996; Bender et al. 2001). Deletion of HS2 had the largest effect in adult erythroid cells, reducing gene expression levels to 65 % of wild type. On the other hand, deletion of HS5 and HS6 (Bender et al. 1998; Farrell et al. 2000) had little or no apparent effect on gene expression levels in adult erythroid cells. Collectively, these findings are consistent with the assertion that HS1–4 all contribute to increasing gene expression levels but appear to achieve this via distinct mechanisms. Human HS2 can act as a classical enhancer, whilst HS1, HS3, and HS4 may increase the level of expression through chromatin-mediated, gene-specific, developmental or structural mechanisms.

6
Gene Competition and the LCR Holocomplex

An unusual phenomenon occurred when the adult β-globin gene was linked to the LCR in transgenic mice. Expression was seen in fetal and adult cells as expected, but abnormal expression of the β gene was also observed in embryonic cells (Enver et al. 1990; Hanscombe et al. 1991). Interestingly, placement of a γ gene between the β gene and LCR restored correct developmental expression of the β gene. In such LCR-γ/-β constructs γ was expressed normally in embryonic and early fetal cells while β was expressed in fetal and adult cells. Reversal of the gene order, i.e. LCR-β/-γ, led to co-expression of both genes in embryonic cells followed by fetal silencing of the γ gene. These experiments were interpreted as evidence of gene competition for the LCR and suggested a looping mechanism in which the LCR interacted preferentially with the nearest gene promoter in embryonic cells, thereby suppressing transcription of a downstream gene. One prediction of the gene competition hypothesis was that the LCR could activate only one gene in cis at any given moment. This was tested in single cells by RNA FISH (fluorescence in-situ hybridization) to detect γ- and β-primary transcripts in early fetal cells of transgenic mice carrying a single copy of a 70-kb construct (Wijgerde et al. 1995). At this stage nearly all cells co-express both the γ- and β-genes as shown by immunofluorescent detection of γ- and β-polypeptides (Fraser et al. 1993). However, RNA FISH showed that in the vast majority of cells only a single primary transcript signal was detected per locus. Since primary transcripts have a very short half-life, detection is indicative of ongoing or very recent transcription. Interestingly, a significant number of cells in homozygotes showed transcription of the γ-gene on one allele and transcription of the β-gene on the other, showing that each locus could respond independently and differently to the same *trans*-acting factor environment. Cells transcribing only the γ-gene were observed containing large amounts of the developmentally late gene product β mRNA in the cytoplasm, demonstrating that transcription of the genes alternated or switched back and forth. These results supported the gene competition hypothesis and lead to the LCR flip-flop model in which it was proposed that the LCR formed semi-stable interactions with individual genes to activate transcription (Wijgerde et al. 1995). The competition mechanism was extended to suggest that developmental regulation of the entire cluster was accomplished through polar gene competition for the LCR. In the early stages, sequestration of the LCR by the ε- and γ-genes was implicated in the prevention of β-gene transcription. Silencing of the ε-gene in fetal cells would then allow the LCR to occasionally interact with the β-gene in competition with the γ-genes, and, finally, γ-gene silencing in bone marrow-derived cells led to sole expression of the β-gene by default. This scenario was further supported by transgenic experiments in which the ε-gene and flanking sequences were deleted and replaced with a marked β-gene (Dillon et al. 1997). In these mice the marked β-gene was transcribed in embryonic cells and partially suppressed γ-gene transcription. In fetal cells, γ expression was completely suppressed by the LCR-proximal β-gene as was fetal and adult transcription of the wild-type β-gene in its normal downstream position. Although it is clear from several experiments that gene competition for the LCR occurs, the suggestion that it is responsible for developmental regulation of the locus requires caution since in nearly every case the genes were removed from their normal epigenetic context (see below). Nevertheless these and other data led to the postulation of the LCR holocomplex theory (Ellis et al. 1996). This idea proposes that all the HS in the LCR interact to form a univalent nucleoprotein structure (the holocomplex) capable of interacting with and activating transcription of a single globin gene at one time.

7
The β-Globin Locus Resides in a Region of Tissue-Specific Open Chromatin

The β-globin loci of human and mouse are embedded in clusters of olfactory receptor genes (*Org*) that are transcriptionally silent in erythroid tissues (Bulger et al. 1999, 2000). The chicken β-globin locus is flanked on the 3' side by *Org*s but has an erythroid-expressed folate receptor (*FR*) gene upstream (Litt et al. 2001a). General DNase I sensitivity of the β-globin locus has been assayed at various resolutions in the mouse, human and chicken. The chicken locus which is the most highly characterized in terms of DNase I sensitivity has four genes. The chicken LCR and globin genes lie in an approximately 30-kb region of open chromatin (Felsenfeld 1993; Litt et al. 2001b). Directly upstream of the LCR is a 16-kb stretch of relatively closed chromatin followed by the *FR* gene that is expressed in erythroid cells prior to β-gene activation (Prioleau et al. 1999). The human locus is also in a large region of open chromatin, but appears to be further divided into developmentally controlled sub-domains of increased sensitivity to DNase I digestion (Gribnau et al. 2000). The LCR and active genes lie in regions of hyperaccessible chromatin, while the developmentally inactive globin genes are surrounded by chromatin of intermediate sensitivity compared to a non-erythroid gene. This pattern changes during development concurrent with the gene expression pattern. The mouse locus also resides in a relatively open chromatin region that spans approximately 150 kb from upstream of the –62.5 HS to downstream of 3′HS1 (Bulger et al. 2003). Some evidence from MEL cells suggests that the mouse locus is also divided into subdomains of differential sensitivity (Smith et al. 1984). The open region upstream of the mouse LCR contains a number of *Org*, which raises questions as to how these genes are kept silent in erythroid cells especially in view of their proximity to the LCR, which has been shown capable of activating heterologous genes.

8
The Role of Insulators

Insulators are sequence elements capable of enhancer blocking and/or chromatin barrier activity (Bell and Felsenfeld 1999). Several insulator elements have been described so far, but the most highly characterized vertebrate insulator is the chicken β-globin LCR element HS4. HS4 resides at the very 5′ end of the β-globin locus in chicken and marks the transition between the highly condensed chromatin upstream and the open chromatin of the globin locus (Hebbes et al. 1994; Prioleau et al. 1999). When HS4 flanks a transgene, the expression of that gene in transformed cells is stably maintained (Pikaart et al. 1998; Mutskov et al. 2002). In the absence of the insulators the transgene is silenced soon after transfection, a process associated with increased levels of H3/K9 methylation. Flanking HS4 insulators thus appear to possess a chromatin barrier function, which is thought to prevent the spread or influence of nearby repressive chromatin (Litt et al. 2001b; Mutskov et al. 2002). HS4 is also able to block the action of an enhancer when placed between the gene and enhancer in transfection assays. The enhancer blocking activity is mediated by the CTCF protein, and is separable from the chromatin barrier function (Recillas-Targa et al. 2002). Potential CTCF binding sites have been identified by sequence analysis in both human and mouse HS5 and 3′HS1 (Farrell et al. 2002). These sites have been assessed for CTCF binding in vitro and enhancer-blocking activity. Mouse 3′HS1 showed CTCF-binding activity comparable to chicken HS4, and only slightly lower enhancer-blocking activity (Farrell et al. 2002). Consistent with this data, it has been shown that the 3′ chromatin boundary of the mouse locus resides near 3′HS1 (Bulger et al. 2003). Human 3′HS1, and both human and mouse HS5, have an intermediate affinity for CTCF, and intermediate to low enhancer-blocking activity (Farrell et al. 2002). However, deletion of human and mouse HS5 from their endogenous positions had little or no effect on chromatin structure, or globin, or *Org* expression (Reik et al. 1998; Farrell et al. 2000; Bulger et al. 2003), suggesting that these HS do not play a major role in locus organization or preventing LCR activation of the *Org* genes. Chromatin immunoprecipitation using antibodies against CTCF in mouse suggests that it may bind at the upstream –62.5 HS (Bulger et al. 2003). These HS did not show enhancer-blocking activity, but the transition of closed to open chromatin was mapped to a region closely upstream, suggesting they may be involved in the formation of a chromatin domain boundary (Bulger et al. 2003). Thus the role of insulators in the human and mouse globin loci is not entirely clear nor is the function of chicken HS4 in its endogenous position. It is possible that these loci may not be completely analogous and, therefore, it remains to be seen what role HS4 plays in the organization of the chicken β-globin domain and whether it prevents inappropriate activation of the upstream FR gene as suggested by its chromatin barrier and enhancer-blocking activities seen in transfection experiments.

9
Intergenic Transcription

Intergenic transcripts were first described in the β-globin locus by Imaizumi et al. (1973) as long RNA species encompassing both intergenic and gene sequences. These giant transcripts were interpreted as examples of eukaryotic polycistronic pre-mRNAs, similar to the prokaryotic polycistronic RNAs that had recently been discovered. The discovery of β-globin primary transcripts and intron splicing led to the rejection of the vertebrate polycistronic transcript theory, and intergenic transcripts were largely dismissed as an artefact. However, β-globin intergenic transcripts re-emerged in the 1990s (Tuan et al. 1992; Ashe et al. 1997; Kong et al. 1997; Gribnau et al. 2000) and were shown to be rare, long, erythroid-specific, nuclear-restricted RNA molecules, encompassing both the LCR and intergenic regions. The direction of intergenic transcription in the locus is the same sense direction as gene transcription (Ashe et al. 1997). Subsequent work showed that intergenic transcripts in the human β-globin locus are developmentally specific (Gribnau et al. 2000). At least three subdomains have been defined in the human β-globin locus in transgenic mice on the basis of intergenic transcript abundance. The LCR subdomain, which is devoid of genes, is transcribed throughout development, while transcription of the embryonic subdomain containing the ε- and γ-genes, is five- to ten-fold higher in embryonic cells compared to adult cells. Transcription of the adult subdomain encompassing the δ- and β-genes is very low in embryonic cells and five- to ten-fold higher in fetal and adult cells (Gribnau et al. 2000). The same adult stage-specific pattern of intergenic transcripts seen in the human transgene locus is also found in the endogenous human locus, in ex vivo cultured adult erythroid precursors (Goyenechea et al., unpubl. observ.; Miles and Fraser, unpubl. observ.). Similar developmental changes in intergenic transcription are also detected in the mouse locus (Chakalova et al., unpubl. observ.). The most intriguing aspect of the intergenic transcription pattern was its precise correlation with areas of increased general DNase I sensitivity. Intergenic transcripts delineate the highly accessible chromatin domains that surround the active genes and the LCR HS (Gribnau et al. 2000). Recently, several groups have focused their attention on defining the start sites of intergenic transcription.

10
Intergenic Promoters

The intergenic transcription initiation site for the adult subdomain is located approximately 3 kb upstream of the δ-gene in the human β-globin locus in transgenic mice by 5′ RACE (Gribnau et al. 2000). The upstream region (referred to as the δβ promoter) does not contain a canonical TATAA element or typical gene promoter elements. The δβ promoter behaves like a weak promoter when linked to a promoter-less EGFP reporter gene in stable transfection assays (Debrand et al. 2004). The level of EGFP expression per cell is much lower with the δβ promoter compared to a gene promoter; however, a much larger percentage of cells express EGFP under the δβ promoter in comparison to a gene promoter. In the β locus, non-coding transcripts initiated from the δβ promoter presumably extend through the entire δβ domain since removal of the element leads to extinction of all intergenic transcription throughout the subdomain (Gribnau et al. 2000; Debrand et al. 2004).

Intergenic promoters in the εγ domain, active in embryonic red cells, have not yet been pinpointed. Initiation in the LCR subdomain appears to be much more complicated. Multiple sites within the LCR are capable of initiating transcription (Tuan et al. 1992; Kong et al. 1997; Leach et al. 2001; Routledge and Proudfoot 2002). Human HS2 gives rise to sense intergenic transcripts in erythroleukaemia K562 cells (Kong et al. 1997). Moreover, HS2 enhancer activity appears dependent on its ability to produce intergenic transcripts (Tuan et al. 1992). Intrinsic intergenic promoter activity has also been shown for human HS3 (Leach et al. 2001; Routledge and Proudfoot 2002). Upstream of the LCR, transcripts are initiated from a human endogenous retroviral LTR element located approximately 1 kb upstream of HS5 (Long et al. 1998; Plant et al. 2001). The properties of the LTR element have been partially dissected by transient transfection analysis. The LTR fires long non-coding transcripts that read through a downstream gene promoter, and that correlate with high-level synthesis of the coding gene mRNA. Reversal of the direction of intergenic transcription away from the gene drastically reduces initiation from the gene promoter (Long et al. 1998), showing that the LTR possesses an intrinsic, directional, enhancer-like activity associated with transcriptional sense. This conclusion was most elegantly demonstrated by experiments in YAC transgenic mice in which the LCR was inverted with respect to the genes by cre recombinase (Tanimoto et al. 1999). LCR inversion severely reduced expression of all the β-genes.

11
Histone Modification and Developmental Globin Gene Expression

In recent years, much attention has focused on the correlation between post-translational covalent modifications to the amino-terminal tails of histones and gene expression status. The first demonstration of a link between histone hyperacetylation and an 'open', DNase I-sensitive, chromatin structure was in the chicken β-globin chromosomal domain (Hebbes et al. 1994). Chromatin immunoprecipitation (ChIP) revealed a broad pattern of acetylation over the 33 kb of DNase I-sensitive chromatin containing the genes and LCR. Both the transcriptionally active and previously active globin genes were found to be acetylated; however, later analyses suggested that some of the developmentally early genes picked up histone H3 lysine 9 (H3/K9) dimethylation, indicative of inactive chromatin (Litt et al. 2001a). Using antibodies to a range of acetylated histone isoforms and cell lines representing different stages of erythroid maturation, changes in the pattern of acetylation during differentiation of the chicken β-globin locus and the neighbouring FR gene were demonstrated (Litt et al. 2001b). Condensed chromatin and developmentally inactive genes maintained the lowest acetylation, whilst activation of genes correlated with a dramatic increase in acetylation (Litt et al. 2001b). Using antibodies to dimethylated H3/K4 revealed a perfect correlation to the pattern of histone acetylation. Dimethylation of H3/K9, however, was inversely correlated.

Tissue-specific histone acetylation has also been demonstrated for the mouse and human β-globin loci. Chromatin immunoprecipitation of the murine β-globin locus in both primary tissue (fetal liver) and MEL cells revealed an enrichment of histone H3 and H4 acetylation over the locus control region (LCR) and transcriptionally active βmaj and βmin gene promoters, whilst the region encompassing the silenced embryonic genes εy and βH1 was relatively hypoacetylated (Forsberg et al. 2000). A higher resolution analysis of histone modification across the murine β-globin locus in adult anaemic spleen demonstrated the presence of subdomains of acetylation and histone H3/K4 dimethylation over the LCR and the active βmajor and βminor genes (Bulger et al. 2003). It was observed that in contrast to the chicken β-globin locus, the pattern of histone acetylation and H3/4 dimethylation did not precisely correlate with nuclease-sensitivity. There appeared to be a large 150-kb domain of nuclease sensitivity.

The histone H3 and H4 acetylation pattern of the human β-globin locus has, until recently, only been analysed on human chromosomes that have been transferred into MEL cells (Schubeler et al. 2000). Assessing a limited number of sites, the authors detected basal H3 and H4 acetylation throughout the locus, with peaks of H3 acetylation at the LCR and the active β-globin gene promoter. The peak of H3 acetylation at the promoter was lost in a mutant locus with a deletion of HS2–5 of the LCR; however, the general pattern of H3 and H4 acetylation was maintained.

High-resolution ChIP analysis of the human β-globin locus during development in transgenic mice and in human adult primary erythroid cells has revealed subdomains of histone H3 and H4 acetylation and also H3 K4 di- and trimethylation (Miles and Fraser, unpubl. observ.). In embryonic blood the enriched domains comprise the LCR and the transcriptionally active ε- and γ-genes. These modifications correlate with the occurrence of intergenic transcription and increased DNase I general sensitivity. In adult cells two clear domains of modified chromatin exist surrounding the LCR and the δ- and β-genes, again matching precisely the pattern of intergenic transcription and increased general DNase I sensitivity. Thus a tight correlation between 'active' histone modifications, intergenic transcription, and nuclease sensitivity demarcate developmentally regulated subdomains in the human β-globin locus.

12
The Role of Intergenic Transcription

The question of the functional significance of intergenic transcription has been addressed by analyzing a number of deletions of the δβ promoter. A 2.5-kb deletion was first studied in transgenic mice (Gribnau et al. 2000). As expected, intergenic transcription in the adult domain is lost in the mutant locus. More importantly, the adult domain fails to open to a DNase I hyperaccessible structure in definitive erythroid cells, showing that the deleted region is essential for chromatin remodelling of the adult domain. Consequently, transcriptional activation of the β-gene is also abolished. In contrast, developmentally specific transcription of the ε- and γ-genes is normal, indicating that the defect is specific only for the adult domain. To exclude the possibility that other elements in the deleted region contribute to the observed phenotype, a specific 300-bp deletion was engineered with cre/lox technology that includes the minimal δβ promoter. The resultant phenotype exactly matches that of the 2.5-kb deletion. The ε- and γ-genes are highly transcribed in all embryonic erythroid cells, but β gene transcription in fetal and adult stages is dramatically reduced. In addition, the adult domain does not acquire the developmentally specific histone modifications normally present across intergenic and gene regions (H3/K4 di- and trimethylation, and H3 and H4 acetylation). These results show that domain-wide chromatin remodelling is dependent on the δβ promoter, strongly suggesting that transcription leads to remodelling. This concept is further strengthened by biochemical studies, which show that ATP-dependent remodelling complexes, HATs, and histone methyltransferases are associated with the elongating RNA polymerase II complex (Orphanides and Reinberg 2000). Furthermore, transcription through chromatin by RNAP II has been shown to be highly disruptive, leading to partial disassembly of nucleosomes in the wake of the polymerase complex (Studitsky et al. 2004). This presents an ideal opportunity to modify existing nucleosomal tails and/or insert replacement histones associated with active chromatin. Thus intergenic transcription appears to be part of the mechanism of an essential, processive, remodelling machine, which disrupts chromatin fibres and modifies histones through the specific activities associated with the RNAP II complex.

13
The Cell Cycle Connection

Intergenic transcription is not continuous in erythroid cells. Approximately 15–25 % of globin loci show intergenic signals in a non-synchronized population of erythroid cells assessed by RNA FISH. A number of assays have been used to show that intergenic transcription is cell cycle regulated, occurring primarily in G1 phase cells and also in early S phase (Gribnau et al. 2000). Some evidence suggests that the G1 phase intergenic transcription occurs directly after cells exit mitosis, while S phase-specific intergenic transcripts are detectable on newly replicated globin alleles in early S phase. These are major potential control points in chromatin remodelling and reorganization in the cell cycle. In the first hour after mitosis, active gene loci are rapidly decondensed and correct nuclear positioning is established (Dimitrova and Gilbert 1999; Thomson et al. 2004). After DNA replication of the globin locus in early S phase, epigenetic marks need to be re-established on newly replicated daughter alleles. Thus cell type-specific chromatin states could be implemented and/or existing epigenetic conformations maintained through this process. A replication-independent chromatin assembly pathway has been proposed, which includes transcription-coupled histone replacement. This process is thought to occur in active chromatin through replacement of histone H3 with the variant histone H3.3 during transcription (Ahmad and Henikoff 2002a,b). Interestingly, H3.3 is highly enriched in modifications associated with transcriptionally active chromatin, such as methylated H3/K4, and deficient in heterochromatin-specific marks such as H3/K9 methylation (McKittrick et al. 2004). In *Drosophila* cells nearly all H3/K4 methylation is found on H3.3, suggesting that detection of H3/K4 methylation (as seen in intergenic transcription domains in the globin locus) is indicative of the replication-independent chromatin assembly pathway. Nucleosome replacement also provides a mechanism for the activation of domains that are silenced by presumably stable histone methylation marks.

14
The Corfu Deletion

Over the years the many naturally occurring mutations in the β-globin locus have provided a rich source of information regarding the effects of various sequence elements on globin gene regulation (Collins and Weissman 1984; Stamatoyannopoulos and Grosveld 2001). The Corfu mutation involves a 7.2-kb deletion that removes the 5' part of the δ-globin gene and several kilobases of upstream sequence (Kulozik et al. 1988) including the δβ intergenic promoter responsible for transcription of the adult subdomain (Fig. 2). This is the smallest known deletion which removes a critical 1-kb region that appears to be the minimal region of difference between deletions that cause HPFH (hereditary persistence of fetal haemoglobin) or β thalassaemia (Collins and Weissman 1984). HPFH is a relatively benign condition characterized by elevated pancellular expression of fetal haemoglobin (HbF; product of the γ genes) in adult cells, while β thalassaemia can be a life-threatening anemia resulting from abnormally low β-gene expression coupled with normal low expression of the γ genes (Stamatoyannopoulos and Grosveld 2001). Deletions that remove the δ- and β-genes result in thalassaemia, whereas similar mutations that also delete the critical 1-kb region, located approximately 3 kb upstream of the δ-gene, often result in an HPFH phenotype. The Corfu mutation has interesting phenotypic consequences depending on gene dosage. Corfu heterozygotes have moderately reduced β-gene expression and normal low-level γ expression (~2%). Paradoxically, Corfu homozygotes exhibit greatly elevated HbF (80–90%) and barely detectable HbA (product of the β-gene). We have shown that the Corfu deletion which removes the intergenic δβ promoter results in high-level transcription of the fetal γ-genes on the affected chromosome in both heterozygotes and homozygotes by RNA FISH (Chakalova et al., unpubl. observ.). Post-transcriptional regulation of γ mRNA appears to be responsible for the low level of γ expression in heterozygotes

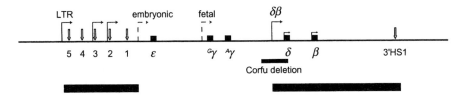

Fig. 2. Features of the human β-globin locus in adult erythroid cells. Active chromatin subdomains are represented by *large black boxes*. These domains show high levels of intergenic transcription, nuclease hyperaccessibility, and histone modifications indicative of active chromatin, such as H3 and H4 hyperacetylation, H3/K4 methylation. Active δ- and β-gene promoters are denoted by *small arrows*. *Large arrows* represent intergenic promoters across the locus: LTR, HS3, HS2 and δβ promoter. Putative embryonic and fetal intergenic promoters are shown as *dashed arrows*. The Corfu deletion is shown as a *thick line below the map*

which seems to be de-repressed in homozygotes. Intergenic transcription analysis by RNA FISH reveals that intergenic transcripts in the δβ domain of the deleted allele appear to initiate approximately 16 kb upstream of the breakpoint, 5′ of the γ-genes (Goyenechea et al., unpubl. observ.). This entire 16-kb region which is normally in a closed conformation and unmodified in adult cells exhibits increased DNase I sensitivity, hyperacetylation of H3 and H4, as well as increased H3/K4 trimethylation. Thus it appears that deletion of the normal 3′ boundary of the γ domain results in a fusion of the fetal and adult domains and persistent activation of the intergenic promoter upstream of the fetal γ-genes. This persistent intergenic transcription through the γ domain is thought to unmask the γ-genes, allowing greater access to the adult transcription factor environment, and perhaps equally importantly recognition and interaction with the upstream LCR leading to high-level γ-gene transcription in adult cells in competition with the β-gene.

15
Higher Order Folding and Long-Range Regulation

High-level transcription of the globin genes is absolutely dependent on the LCR. Although much indirect evidence suggested direct physical contact between the LCR and genes as the mechanism of transcriptional activation, direct evidence was lacking. Recently, two different techniques have been used which have come to the same general conclusion: the LCR functions through direct contact with the globin genes. The first method, RNA TRAP (tagging and recovery of associated proteins), is a modification of the RNA FISH method (Carter et al. 2002). Briefly, hapten-labelled RNA FISH probes are hybridized to nascent transcripts of an actively transcribing globin gene. Antibodies coupled to horseradish peroxidase (HRP) recognize and bind the haptens on the RNA FISH probes, concentrating HRP activity to the site of the transcribing gene. The HRP is used to catalyze the activation of a biotin-tyramide molecule, which covalently attaches and thus deposits biotin on chromatin proteins in the immediate vicinity of the transcribing gene. Biotinylated proteins and associated DNA are then purified and the enrichment of various DNA sequences measured. RNA TRAP directed towards either the βmaj or βmin genes reveals that the classical enhancer component of the LCR, HS2, is in proximity to both genes when they are actively transcribed. The other HS (HS2, HS3 and HS4) appear to be more peripheral, or only transiently associated with the gene, suggesting a major role for HS2 in transcription. This is consistent with individual HS deletion studies, showing that deletion of HS2 leads to the largest drop in expression compared to the other HS (Fiering et al. 1995; Hug et al. 1996; Farrell et al. 2000; Bender et al. 2001). The only HS found in proximity to the active genes are HS1-4, totally consistent with functional studies, showing that individual deletion of these HS has an effect on gene expression (Fiering et al. 1995; Hug et al. 1996; Bender et al. 2001). Deletion of HS5 and HS6, which were not found in proximity to the genes by RNA TRAP, have been shown to have no measurable effect on adult β-globin expression when deleted (Farrell et al. 2000). The HS at −62, −60 and 3′HS1 also were never found in proximity to the transcribing gene, suggesting no direct role in transcription. The spatial arrangement of the different HS may give insight into their function in vivo. For example, the proximity of HS2 to the active genes may represent a direct interaction with the transcriptional apparatus, whilst the more peripheral positions of HS1, HS3 and HS4 could represent a different role in adding to the level of transcription, perhaps by modulating chromatin or stabilizing the interaction through binding to flanking sequences. However, it is also possible that HS1, HS3 and HS4 are closely associated with the active gene (as is HS2) but are only required transiently for activation.

The CCC (chromosome conformation capture) method involves crosslinking chromatin in cells followed by restriction digestion and ligation under dilute conditions, which favour intramolecular ligation (Dekker et al. 2002). Distant genomic regions that are often in contact, and thus have high crosslinking frequencies, are then detectable by PCR across the novel ligation junctions. Though the resolution of CCC is somewhat poor compared to RNA TRAP, the active β-globin genes were seen to ligate to the LCR at a higher frequency than the inactive genes, again providing direct evidence for LCR–gene contact (Tolhuis et al. 2002). However, unlike RNA TRAP, the CCC data suggested contact between many other sequence elements such as the HS at −62 and −60 (approximately 35 kb upstream of the LCR) as well as HS5 and HS6 of the LCR and 3′HS1 located approximately 25 kb downstream of the βmin gene. These data were interpreted as evidence of an active chromatin hub consisting of a cluster of all the HS over the 150-kb region studied playing a role in initiating and maintaining a structure conducive to transcription.

16
Nuclear Organization

As is the case with other genes, nuclear compartmentalization and organization appears to impact β-globin gene regulation. Both the human locus on chromosome 11 and the mouse locus on chromosome 7 reside in regions of high gene density. Bickmore and colleagues have demonstrated a correlation between the gene density of regions in which a particular gene resides and its nuclear position relative to the chromosomal territory (Mahy et al. 2002b). They found that genes on human chromosome 11 and mouse chromosome 7, in the gene-rich region containing the globin genes, tend to be located on the edge of the territory, or indeed outside the territory, presumably looped out from the intensely stained territorial mass. In contrast, genes located in relatively gene-poor regions tend to reside within the bulk of the territory. Although transcription may be able to occur within the central regions of the territory, extraterritorial positioning appears also to relate to transcriptional status or potential (Mahy et al. 2002a). Treatment of cells with transcriptional inhibitors results in movement of extraterritorial regions to more territory-proximal positions. Whereas the β-globin locus has been observed in a slightly internal position relative to the chromosome territory and adjacent to centromeric heterochromatin in lymphoid cells (Brown et al. 2001), it is often in an extraterritorial position in MEL cells (Ragoczy et al. 2003).

The LCR is likely to contribute to functional positioning of the β-globin locus. The human locus does not adopt an extraterritorial position in MEL cell hybrids when the LCR has been deleted from chromosome 11 (Ragoczy et al. 2003). Furthermore, the Hispanic thalassaemia deletion, which removes the LCR and approximately 30 kb of upstream sequences including the LTR, causes the locus to locate to centromeric heterochromatin (Schubeler et al. 2000), whereas loci with deletions that remove only parts of the LCR remain positioned away from centromeric heterochromatin. This suggests that positioning away from repressive chromatin environments may be linked to sequences outside the LCR.

17
Summary Model

Obviously, there is still much to learn regarding the gene activation pathway, and though there are many gaps we are not deterred from proposing a general hierarchy of events. The initial pre-activation of the β-globin locus domain during erythroid differentiation in human and mouse most likely involves chromatin modifications over a region of approximately 150 kb beginning tens of kilobases upstream of the LCR and extending down to 3'HS1. This could probably best be described as a poised conformation in which the entire locus is relatively decondensed compared to a non-erythroid locus, but not quite as open as an active subdomain during gene expression. The HS at –62 and –60 in mouse and the potential human homologue at approximately –110 kb may play a role in this initial opening along with other elements in the locus (Bulger et al. 1999; Palstra et al. 2003). Intergenic transcription may be involved in setting up this initial domain through very long, S phase-specific intergenic transcription through the entire region. These changes may be involved in preventing the globin locus from associating with repressive centromeric heterochromatin. As erythroid differentiation proceeds and the globin genes become activated, non-S-phase intergenic transcription in the LCR and the developmentally appropriate active gene subdomains may promote histone replacement with variant histone H3.3 and hyperacetylation of H3, H4 and H3/K4 methylation. These modifications may allow increased factor access and additional modifications of histones in promoter regions, for example. This would promote LCR recognition of, and dynamic interaction with, the appropriately modified gene(s) while ignoring intervening, distal, or upstream genes lacking the proper conformation or modifications. The

LCR–gene complex may promote extension of the locus beyond the confines of the chromosome territory as a result of, or in order to facilitate, highly efficient transcription of the globin genes.

References

Ahmad K, Henikoff S (2002a) Histone H3 variants specify modes of chromatin assembly. Proc Natl Acad Sci USA 99 (Suppl 4):16477–16484

Ahmad K, Henikoff S (2002b) The histone variant H3.3 marks active chromatin by replication-independent nucleosome assembly. Mol Cell 9(6):1191–1200

Alami R, Bender MA, Feng YQ, Fiering SN, Hug BA, Ley TJ, Groudine M, Bouhassira EE (2000) Deletions within the mouse beta-globin locus control region preferentially reduce beta(min) globin gene expression. Genomics 63(3):417–424

Antoniou M, deBoer E, Habets G, Grosveld F (1988) The human beta-globin gene contains multiple regulatory regions: identification of one promoter and two downstream enhancers. EMBO J 7(2):377–384

Ashe HL, Monks J, Wijgerde M, Fraser P, Proudfoot NJ (1997) Intergenic transcription and transinduction of the human beta-globin locus. Genes Dev 11(19):2494–2509

Behringer RR, Hammer RE, Brinster RL, Palmiter RD, Townes TM (1987) Two 3′ sequences direct adult erythroid-specific expression of human beta-globin genes in transgenic mice. Proc Natl Acad Sci USA 84(20):7056–7060

Bell AC, Felsenfeld G (1999) Stopped at the border: boundaries and insulators. Curr Opin Genet Dev 9(2):191–198

Bender MA, Reik A, Close J, Telling A, Epner E, Fiering S, Hardison R, Groudine M (1998) Description and targeted deletion of 5′ hypersensitive site 5 and 6 of the mouse beta-globin locus control region. Blood 92(11):4394–4403

Bender MA, Roach JN, Halow J, Close J, Alami R, Bouhassira EE, Groudine M, Fiering SN (2001) Targeted deletion of 5′HS1 and 5′HS4 of the beta-globin locus control region reveals additive activity of the DNaseI hypersensitive sites. Blood 98(7):2022–2027

Blom van Assendelft G, Hanscombe O, Grosveld F, Greaves DR (1989) The beta-globin dominant control region activates homologous and heterologous promoters in a tissue-specific manner. Cell 56(6):969–977

Brotherton TW, Chui DH, Gauldie J, Patterson M (1979) Hemoglobin ontogeny during normal mouse fetal development. Proc Natl Acad Sci USA 76(6):2853–2857

Brown KE, Amoils S, Horn JM, Buckle VJ, Higgs DR, Merkenschlager M, Fisher AG (2001) Expression of alpha- and beta-globin genes occurs within different nuclear domains in haemopoietic cells. Nat Cell Biol 3(6):602–606

Bulger M, van Doorninck JH, Saitoh N, Telling A, Farrell C, Bender MA, Felsenfeld G, Axel R, Groudine M, von Doorninck JH (1999) Conservation of sequence and structure flanking the mouse and human beta-globin loci: the beta-globin genes are embedded within an array of odorant receptor genes. Proc Natl Acad Sci USA 96(9):5129–5134

Bulger M, Bender MA, van Doorninck JH, Wertman B, Farrell CM, Felsenfeld G, Groudine M, Hardison R (2000) Comparative structural and functional analysis of the olfactory receptor genes flanking the human and mouse beta-globin gene clusters. Proc Natl Acad Sci USA 97(26):14560–14565

Bulger M, Schubeler D, Bender MA, Hamilton J, Farrell CM, Hardison RC, Groudine M (2003) A complex chromatin landscape revealed by patterns of nuclease sensitivity and histone modification within the mouse beta-globin locus. Mol Cell Biol 23(15):5234–5244

Bungert J, Dave U, Lim KC, Lieuw KH, Shavit JA, Liu Q, Engel JD (1995) Synergistic regulation of human beta-globin gene switching by locus control region elements HS3 and HS4. Genes Dev 9(24):3083–3096

Bungert J, Tanimoto K, Patel S, Liu Q, Fear M, Engel JD (1999) Hypersensitive site 2 specifies a unique function within the human beta-globin locus control region to stimulate globin gene transcription. Mol Cell Biol 19(4):3062–3072

Carter D, Chakalova L, Osborne CS, Dai YF, Fraser P (2002) Long-range chromatin regulatory interactions in vivo. Nat Genet 32(4):623–626

Chada K, Magram J, Costantini F (1986) An embryonic pattern of expression of a human fetal globin gene in transgenic mice. Nature 319(6055):685–689

Collins FS, Weissman SM (1984) The molecular genetics of human hemoglobin. Prog Nucleic Acid Res Mol Biol 31:315–462

Collis P, Antoniou M, Grosveld F (1990) Definition of the minimal requirements within the human beta-globin gene and the dominant control region for high level expression. EMBO J 9(1):233–240

Curtin P, Pirastu M, Kan YW, Gobert-Jones JA, Stephens AD, Lehmann H (1985) A distant gene deletion affects beta-globin gene function in an atypical gamma delta beta-thalassemia. J Clin Invest 76(4):1554–1558

Curtin PT, Kan YW (1988) The inactive beta globin gene on a gamma delta beta thalassemia chromosome has a normal structure and functions normally in vitro. Blood 71(3):766–770

Dekker J, Rippe K, Dekker M, Kleckner N (2002) Capturing chromosome conformation. Science 295(5558):1306–1311

Dillon N, Trimborn T, Strouboulis J, Fraser P, Grosveld F (1997) The effect of distance on long-range chromatin interactions. Mol Cell 1(1):131–139

Dimitrova DS, Gilbert DM (1999) The spatial position and replication timing of chromosomal domains are both established in early G1 phase. Mol Cell 4(6):983–993

Driscoll MC, Dobkin CS, Alter BP (1989) Gamma delta beta-thalassemia due to a de novo mutation deleting the 5′ beta-globin gene activation-region hypersensitive sites. Proc Natl Acad Sci USA 86(19):7470–7474

Ellis J, Tan-Un KC, Harper A, Michalovich D, Yannoutsos N, Philipsen S, Grosveld F (1996) A dominant chromatin-opening activity in 5′ hypersensitive site 3 of the human beta-globin locus control region. EMBO J 15(3):562–568

Enver T, Raich N, Ebens AJ, Papayannopoulou T, Costantini F, Stamatoyannopoulos G (1990) Developmental regulation of human fetal-to-adult globin gene switching in transgenic mice. Nature 344(6264):309–313

Epner E, Reik A, Cimbora D, Telling A, Bender MA, Fiering S, Enver T, Martin DI, Kennedy M, Keller G, Groudine M (1998) The beta-globin LCR is not necessary for an open chromatin structure or developmentally regulated transcription of the native mouse beta-globin locus. Mol Cell 2(4):447–455

Farrell CM, West AG, Felsenfeld G (2002) Conserved CTCF insulator elements flank the mouse and human beta-globin loci. Mol Cell Biol 22(11):3820–3831

Farrell CM, Grinberg A, Huang SP, Chen D, Pichel JG, Westphal H, Felsenfeld G (2000) A large upstream region is not necessary for gene expression or hypersensitive site formation at the mouse beta-globin locus. Proc Natl Acad Sci USA 97(26):14554–14559

Felsenfeld G (1993) Chromatin structure and the expression of globin-encoding genes. Gene 135(1–2):119–124

Festenstein R, Tolaini M, Corbella P, Mamalaki C, Parrington J, Fox M, Miliou A, Jones M, Kioussis D (1996) Locus control region function and heterochromatin-induced position effect variegation. Science 271(5252):1123–1125

Fiering S, Epner E, Robinson K, Zhuang Y, Telling A, Hu M, Martin DI, Enver T, Ley TJ, Groudine M (1995) Targeted deletion of 5′HS2 of the murine beta-globin LCR reveals that it is not essential for proper regulation of the beta-globin locus. Genes Dev 9(18):2203–2213

Forrester WC, Thompson C, Elder JT, Groudine M (1986) A developmentally stable chromatin structure in the human beta-globin gene cluster. Proc Natl Acad Sci USA 83(5):1359–1363

Forrester WC, Takegawa S, Papayannopoulou T, Stamatoyannopoulos G, Groudine M (1987) Evidence for a locus activation region: the formation of developmentally stable hypersensitive sites in globin-expressing hybrids. Nucleic Acids Res 15(24):10159–10177

Forrester WC, Epner E, Driscoll MC, Enver T, Brice M, Papayannopoulou T, Groudine M (1990) A deletion of the human beta-globin locus activation region causes a major alteration in chromatin structure and replication across the entire beta-globin locus. Genes Dev 4(10):1637–1649

Forsberg EC, Downs KM, Christensen HM, Im H, Nuzzi PA, Bresnick EH (2000) Developmentally dynamic histone acetylation pattern of a tissue-specific chromatin domain. Proc Natl Acad Sci USA 97(26):14494–14499

Fraser P, Grosveld F (1998) Locus control regions, chromatin activation and transcription. Curr Opin Cell Biol 10(3):361–365

Fraser P, Hurst J, Collis P, Grosveld F (1990) DNaseI hypersensitive sites 1, 2 and 3 of the human beta-globin dominant control region direct position-independent expression. Nucleic Acids Res 18(12):3503–3508

Fraser P, Pruzina S, Antoniou M, Grosveld F (1993) Each hypersensitive site of the human beta-globin locus control region confers a different developmental pattern of expression on the globin genes. Genes Dev 7(1):106–113

Fraser P, Gribnau J, Trimborn T (1998) Mechanisms of developmental regulation in globin loci. Curr Opin Hematol 5(2):139–144

Friend C, Patuleia MC, de Harven E (1966) Erythrocytic maturation in vitro of murine (Friend) virus-induced leukemic cells. Natl Cancer Inst Monogr 22:505–522

Furukawa T, Navas PA, Josephson BM, Peterson KR, Papayannopoulou T, Stamatoyannopoulos G (1995) Coexpression of epsilon, G gamma and A gamma globin mRNA in embryonic red blood cells from a single copy beta-YAC transgenic mouse. Blood Cells Mol Dis 21(2):168–178

Goodwin AJ, McInerney JM, Glander MA, Pomerantz O, Lowrey CH (2001) In vivo formation of a human beta-globin locus control region core element requires binding sites for multiple factors including GATA-1, NF-E2, erythroid Kruppel-like factor, and Sp1. J Biol Chem 276(29):26883–26892

Gribnau J, Diderich K, Pruzina S, Calzolari R, Fraser P (2000) Intergenic transcription and developmental remodeling of chromatin subdomains in the human beta-globin locus. Mol Cell 5(2):377–386

Grosveld F, van Assendelft GB, Greaves DR, Kollias G (1987) Position-independent, high-level expression of the human beta-globin gene in transgenic mice. Cell 51(6):975–985

Hanscombe O, Whyatt D, Fraser P, Yannoutsos N, Greaves D, Dillon N, Grosveld F (1991) Importance of globin gene order for correct developmental expression. Genes Dev 5(8):1387–1394

Harteveld CL, Osborne CS, Peters M, van der Werf S, Plug R, Fraser P, Giordano PC (2003) Novel 112 kb (epsilon G gamma A gamma) delta beta-thalassaemia deletion in a Dutch family. Br J Haematol 122(5):855–858

Hebbes TR, Clayton AL, Thorne AW, Crane-Robinson C (1994) Core histone hyperacetylation co-maps with generalized DNase I sensitivity in the chicken beta-globin chromosomal domain. EMBO J 13(8):1823–1830

Hug BA, Wesselschmidt RL, Fiering S, Bender MA, Epner E, Groudine M, Ley TJ (1996) Analysis of mice containing a targeted deletion of beta-globin locus control region 5' hypersensitive site 3. Mol Cell Biol 16(6):2906–2912

Hunt JA (1974) Rate of synthesis and half-life of globin messenger ribonucleic acid. Rate of synthesis of globin messenger ribonucleic acid calculated from data of cell haemoglobin content. Biochem J 138(3):499–510

Imaizumi T, Diggelmann H, Scherrer K (1973) Demonstration of globin messenger sequences in giant nuclear precursors of messenger RNA of avian erythroblasts. Proc Natl Acad Sci USA 70(4):1122–1126

Jackson JD, Petrykowska H, Philipsen S, Miller W, Hardison R (1996) Role of DNA sequences outside the cores of DNase hypersensitive sites (HSs) in functions of the beta-globin locus control region. Domain opening and synergism between HS2 and HS3. J Biol Chem 271(20):11871–11878

Kioussis D, Festenstein R (1997) Locus control regions: overcoming heterochromatin-induced gene inactivation in mammals. Curr Opin Genet Dev 7(5):614–619

Kioussis D, Vanin E, deLange T, Flavell RA, Grosveld F (1983) Beta-globin gene inactivation by DNA translocation in gamma beta-thalassaemia. Nature 306(5944):662–666

Kmita M, Duboule D (2003) Organizing axes in time and space; 25 years of colinear tinkering. Science 301(5631):331–333

Kollias G, Wrighton N, Hurst J, Grosveld F (1986) Regulated expression of human A gamma-, beta-, and hybrid gamma beta-globin genes in transgenic mice: manipulation of the developmental expression patterns. Cell 46(1):89–94

Kollias G, Hurst J, deBoer E, Grosveld F (1987) The human beta-globin gene contains a downstream developmental specific enhancer. Nucleic Acids Res 15(14):5739–5747

Kong S, Bohl D, Li C, Tuan D (1997) Transcription of the HS2 enhancer toward a cis-linked gene is independent of the orientation, position, and distance of the enhancer relative to the gene. Mol Cell Biol 17(7):3955–3965

Kulozik AE, Yarwood N, Jones RW (1988) The Corfu delta beta zero thalassemia: a small deletion acts at a distance to selectively abolish beta globin gene expression. Blood 71(2):457–462

Leach KM, Nightingale K, Igarashi K, Levings PP, Engel JD, Becker PB, Bungert J (2001) Reconstitution of human beta-globin locus control region hypersensitive sites in the absence of chromatin assembly. Mol Cell Biol 21(8):2629–2640

Litt MD, Simpson M, Gaszner M, Allis CD, Felsenfeld G (2001a) Correlation between histone lysine methylation and developmental changes at the chicken beta-globin locus. Science 293(5539):2453–2455

Litt MD, Simpson M, Recillas-Targa F, Prioleau MN, Felsenfeld G (2001b) Transitions in histone acetylation reveal boundaries of three separately regulated neighboring loci. EMBO J 20(9):2224–2235

Long Q, Bengra C, Li C, Kutlar F, Tuan D (1998) A long terminal repeat of the human endogenous retrovirus ERV-9 is located in the 5′ boundary area of the human beta-globin locus control region. Genomics 54(3):542–555

Mahy NL, Perry PE, Gilchrist S, Baldock RA, Bickmore WA (2002a) Spatial organization of active and inactive genes and noncoding DNA within chromosome territories. J Cell Biol 157(4):579–589

Mahy NL, Perry PE, Bickmore WA (2002b) Gene density and transcription influence the localization of chromatin outside of chromosome territories detectable by FISH. J Cell Biol 159(5):753–763

Marks PA, Rifkind RA (1978) Erythroleukemic differentiation. Annu Rev Biochem 47:419–448

McKittrick E, Gafken PR, Ahmad K, Henikoff S (2004) Histone H3.3 is enriched in covalent modifications associated with active chromatin. Proc Natl Acad Sci USA 101(6):1525–1530

Milot E, Strouboulis J, Trimborn T, Wijgerde M, de Boer E, Langeveld A, Tan-Un K, Vergeer W, Yannoutsos N, Grosveld F, Fraser P (1996) Heterochromatin effects on the frequency and duration of LCR-mediated gene transcription. Cell 87(1):105–114

Moon AM, Ley TJ (1991) Functional properties of the beta-globin locus control region in K562 erythroleukemia cells. Blood 77(10):2272–2284

Mutskov VJ, Farrell CM, Wade PA, Wolffe AP, Felsenfeld G (2002) The barrier function of an insulator couples high histone acetylation levels with specific protection of promoter DNA from methylation. Genes Dev 16(12):1540–1554

Navas PA, Peterson KR, Li Q, Skarpidi E, Rohde A, Shaw SE, Clegg CH, Asano H, Stamatoyannopoulos G (1998) Developmental specificity of the interaction between the locus control region and embryonic or fetal globin genes in transgenic mice with an HS3 core deletion. Mol Cell Biol 18(7):4188–4196

Navas PA, Peterson KR, Li Q, McArthur M, Stamatoyannopoulos G (2001) The 5'HS4 core element of the human beta-globin locus control region is required for high-level globin gene expression in definitive but not in primitive erythropoiesis. J Mol Biol 312(1):17–26

Navas PA, Swank RA, Yu M, Peterson KR, Stamatoyannopoulos G (2003) Mutation of a transcriptional motif of a distant regulatory element reduces the expression of embryonic and fetal globin genes. Hum Mol Genet 12(22):2941–2948

Orphanides G, Reinberg D (2000) RNA polymerase II elongation through chromatin. Nature 407(6803):471–475

Palstra RJ, Tolhuis B, Splinter E, Nijmeijer R, Grosveld F, de Laat W (2003) The beta-globin nuclear compartment in development and erythroid differentiation. Nat Genet 35(2):190–194

Peterson KR, Clegg CH, Huxley C, Josephson BM, Haugen HS, Furukawa T, Stamatoyannopoulos G (1993) Transgenic mice containing a 248-kb yeast artificial chromosome carrying the human beta-globin locus display proper developmental control of human globin genes. Proc Natl Acad Sci USA 90(16):7593–7597

Peterson KR, Clegg CH, Navas PA, Norton EJ, Kimbrough TG, Stamatoyannopoulos G (1996) Effect of deletion of 5'HS3 or 5'HS2 of the human beta-globin locus control region on the developmental regulation of globin gene expression in beta-globin locus yeast artificial chromosome transgenic mice. Proc Natl Acad Sci USA 93(13):6605–6609

Philipsen S, Talbot D, Fraser P, Grosveld F (1990) The beta-globin dominant control region: hypersensitive site 2. EMBO J 9(7):2159–2167

Pikaart MJ, Recillas-Targa F, Felsenfeld G (1998) Loss of transcriptional activity of a transgene is accompanied by DNA methylation and histone deacetylation and is prevented by insulators. Genes Dev 12(18):2852–2862

Plant KE, Routledge SJ, Proudfoot NJ (2001) Intergenic transcription in the human beta-globin gene cluster. Mol Cell Biol 21(19):6507–6514

Pope SH, Fibach E, Sun J, Chin K, Rodgers GP (2000) Two-phase liquid culture system models normal human adult erythropoiesis at the molecular level. Eur J Haematol 64(5):292–303

Prioleau MN, Nony P, Simpson M, Felsenfeld G (1999) An insulator element and condensed chromatin region separate the chicken beta-globin locus from an independently regulated erythroid-specific folate receptor gene. EMBO J 18(14):4035–4048

Pruzina S, Hanscombe O, Whyatt D, Grosveld F, Philipsen S (1991) Hypersensitive site 4 of the human beta globin locus control region. Nucleic Acids Res 19(7):1413–1419

Ragoczy T, Telling A, Sawado T, Groudine M, Kosak ST (2003) A genetic analysis of chromosome territory looping: diverse roles for distal regulatory elements. Chromosome Res 11(5):513–525

Recillas-Targa F, Pikaart MJ, Burgess-Beusse B, Bell AC, Litt MD, West AG, Gaszner M, Felsenfeld G (2002) Position-effect protection and enhancer blocking by the chicken beta-globin insulator are separable activities. Proc Natl Acad Sci USA 99(10):6883–6888

Reik A, Telling A, Zitnik G, Cimbora D, Epner E, Groudine M (1998) The locus control region is necessary for gene expression in the human beta-globin locus but not the maintenance of an open chromatin structure in erythroid cells. Mol Cell Biol 18(10):5992–6000

Routledge SJ, Proudfoot NJ (2002) Definition of transcriptional promoters in the human beta globin locus control region. J Mol Biol 323(4):601–611

Ryan TM, Behringer RR, Townes TM, Palmiter RD, Brinster RL (1989) High-level erythroid expression of human alpha-globin genes in transgenic mice. Proc Natl Acad Sci USA 86(1):37–41

Saveliev A, Everett C, Sharpe T, Webster Z, Festenstein R (2003) DNA triplet repeats mediate heterochromatin-protein-1-sensitive variegated gene silencing. Nature 422(6934):909–913

Schubeler D, Francastel C, Cimbora DM, Reik A, Martin DI, Groudine M et al (2000) Nuclear localization and histone acetylation: a pathway for chromatin opening and transcriptional activation of the human beta-globin locus. Genes Dev 14(8):940–950

Smith RD, Yu J, Seale RL (1984) Chromatin structure of the beta-globin gene family in murine erythroleukemia cells. Biochemistry 23(4):785–790

Stamatoyannopoulos G, Grosveld F (2001) Hemoglobin switching. In: Stamatoyannopoulos G, Majerus PW, Perlmutter RM, Varmus H (eds) The molecular basis of blood diseases. Saunders, Philadelphia, pp 135–182

Strauss EC, Orkin SH (1992) In vivo protein–DNA interactions at hypersensitive site 3 of the human beta-globin locus control region. Proc Natl Acad Sci USA 89(13):5809–5813

Strouboulis J, Dillon N, Grosveld F (1992) Developmental regulation of a complete 70-kb human beta-globin locus in transgenic mice. Genes Dev 6(10):1857–1864

Studitsky VM, Walter W, Kireeva M, Kashlev M, Felsenfeld G (2004) Chromatin remodeling by RNA polymerases. Trends Biochem Sci 29(3):127–135

Talbot D, Philipsen S, Fraser P, Grosveld F (1990) Detailed analysis of the site 3 region of the human beta-globin dominant control region. EMBO J 9(7):2169–2177

Tanimoto K, Liu Q, Bungert J, Engel JD (1999) Effects of altered gene order or orientation of the locus control region on human beta-globin gene expression in mice. Nature 398(6725):344–348

Thomson I, Gilchrist S, Bickmore WA, Chubb JR (2004) The radial positioning of chromatin is not inherited through mitosis but is established de novo in early G1. Curr Biol 14(2):166–172

Tolhuis B, Palstra RJ, Splinter E, Grosveld F, de Laat W (2002) Looping and interaction between hypersensitive sites in the active beta-globin locus. Mol Cell 10(6):1453–1465

Townes TM, Lingrel JB, Chen HY, Brinster RL, Palmiter RD (1985) Erythroid-specific expression of human beta-globin genes in transgenic mice. EMBO J 4(7):1715–1723

Trimborn T, Gribnau J, Grosveld F, Fraser P (1999) Mechanisms of developmental control of transcription in the murine alpha- and beta-globin loci. Genes Dev 13(1):112–124

Tuan D, Solomon W, Li Q, London IM (1985) The 'beta-like-globin' gene domain in human erythroid cells. Proc Natl Acad Sci USA 82(19):6384–6388

Tuan D, Solomon WB, London IM, Lee DP (1989) An erythroid-specific, developmental-stage-independent enhancer far upstream of the human 'beta-like globin' genes. Proc Natl Acad Sci USA 86(8):2554–2558

Tuan D, Kong S, Hu K (1992) Transcription of the hypersensitive site HS2 enhancer in erythroid cells. Proc Natl Acad Sci USA 89(23):11219–11223

Van der Ploeg LH, Konings A, Oort M, Roos D, Bernini L, Flavell RA (1980) Gamma-beta-thalassaemia studies showing that deletion of the gamma- and delta-genes influences beta-globin gene expression in man. Nature 283(5748):637–642

Wijgerde M, Grosveld F, Fraser P (1995) Transcription complex stability and chromatin dynamics in vivo. Nature 377(6546):209–213

Wojda U, Noel P, Miller JL (2002) Fetal and adult hemoglobin production during adult erythropoiesis: coordinate expression correlates with cell proliferation. Blood 99(8):3005–3013

Wright S, Rosenthal A, Flavell R, Grosveld F (1984) DNA sequences required for regulated expression of beta-globin genes in murine erythroleukemia cells. Cell 38(1):265–273

Epigenetic Regulation of Mammalian Imprinted Genes: From Primary to Functional Imprints

Michaël Weber, Hélène Hagège, Nathalie Aptel, Claude Brunel, Guy Cathala, Thierry Forné

Abstract Parental genomic imprinting was discovered in mammals some 20 years ago. This phenomenon, crucial for normal development, rapidly became a key to understanding epigenetic regulation of mammalian gene expression. In this chapter we present a general overview of the field and describe in detail the 'imprinting cycle'. We provide selected examples that recapitulate our current knowledge of epigenetic regulation at imprinted loci. These epigenetic mechanisms lead to the stable repression of imprinted genes on one parental allele by interfering with 'formatting' for gene expression that usually occurs on expressed alleles. From this perspective, genomic imprinting remarkably illustrates the complexity of the epigenetic mechanisms involved in the control of gene expression in mammals.

1
Introduction

From the classical point of view of Mendelian genetics, inheritance of each parental allele of an autosomal gene in a diploid cell is considered as equivalent. However, the discovery of parental genomic imprinting in mammals, some 20 years ago, radically changed this conception.

Indeed, the two parental genomes do not equally contribute to embryonic development. Mouse embryos issued from eggs containing two female pronuclei (gynogenotes) or two male pronuclei (androgenotes) have developmental abnormalities and die before birth (McGrath and Solter 1984; Surani et al. 1984). Gynogenotes have placental defects, whereas androgenotes are characterized by an overgrowth of extra-embryonic tissues. These phenotypes are very similar to those observed in humans in the case of fertilization of an enucleated oocyte by one or two spermatozoa, or spontaneous activation of an

M. Weber, H. Hagège, N. Aptel, C. Brunel, G. Cathala, T. Forné
Institut de Génétique Moléculaire de Montpellier, UMR5535 CNRS-UMII, IFR122,
1919, Route de Mende, 34293 Montpellier Cedex 5, France, e-mail: forne@igm.cnrs-mop.fr

Progress in Molecular and Subcellular Biology
P. Jeanteur (Ed.)
Epigenetics and Chromatin
© Springer-Verlag Berlin Heidelberg 2005

oocyte after duplication of the maternal genome. Thus, the paternal and the maternal genomes have complementary roles and are both required for normal embryonic development.

The notion of genomic imprinting was also suggested by studies of mouse embryos with uniparental disomies (i.e. mice that have inherited both copies of a portion of a chromosome from the same parent). Such embryos show strong developmental defects and, interestingly, it appeared that paternal or maternal disomies have opposite effects on embryonic development (Cattanach and Kirk 1985). Initially, ten distinct imprinted regions distributed over six chromosomes were identified in the mouse genome (Cattanach and Jones 1994) and some of them showed phenotypes similar to those observed in human uniparental disomies (Ledbetter and Engel 1995).

Altogether, these observations supported the idea that some mammalian genes might behave differently depending on the parental origin of the allele. This hypothesis was demonstrated a few years later with the identification of the first imprinted genes: *H19* (Bartolomei et al. 1991), *Igf2r* (insulin-like growth factor 2 receptor) (Barlow et al. 1991) and *Igf2* (insulin-like growth factor 2) genes (DeChiara et al. 1991). Indeed, these so-called imprinted genes are stably repressed on one allele depending on its parental origin. By convention, the repressed allele is defined as the one bearing the 'imprint'. Nearly 80 imprinted genes have now been described in human and mouse. Over the last decade, genomic imprinting acquired an important place in the field of developmental biology, as it turned out to be essential for normal embryonic development in mammals and to be altered in numerous human pathologies (Lalande 1996; Falls et al. 1999).

2
Imprinting Evolution

2.1
Conservation of Parental Genomic Imprinting in Therian Mammals

The term 'imprinting' was first used to describe a phenomenon observed in insects (*Sciara* sp.) leading to the inactivation of the paternal set of chromosomes (Crouse 1960). However, parental genomic imprinting, i.e. the repression of one allele according to its parental origin, has only been described in mammals and in flowering plants (Baroux et al. 2002). One should also mention that a form of epigenetic regulation depending on the 'parental' origin also exists in yeast (Allshire and Bickmore 2000).

So far, investigations into imprinted genes have been mostly restricted to human and mouse, two mammals belonging to the group of eutherians (placental mammals). To better understand the evolution of imprinting, studies have been performed in metatherians (marsupials), prototherians (monotremes) and birds. Based on the analysis of the *Igf2* and *Igf2r* genes, genomic

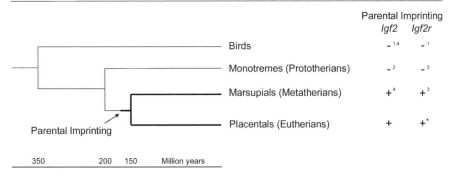

Fig. 1. Parental genomic imprinting and evolution of therian mammals. The *Igf2* and *Igf2r* genes are imprinted in placental mammals and in marsupials, but not in monotremes or in birds. This suggests that genomic imprinting appeared some 150–200 million years ago in the branch leading to therian mammals (placental and marsupial). The alternative theory of mammalian evolution regroups the monotremes and the marsupials in the same lineage excluding the placental mammals. In this case, one should imagine that imprinting appeared independently in placental mammals and marsupials, or that it has been specifically lost in monotremes. * *Igf2r* is not imprinted in the primates (Killian et al. 2001b). References: *1* Nolan et al. (2001); *2* Killian et al. (2001a); *3* Killian et al. (2000); *4* O'Neill et al. (2000)

imprinting appears to be conserved in marsupials but not in monotremes or in birds (Killian et al. 2000, 2001a; O'Neill et al. 2000; Nolan et al. 2001). These data favour the hypothesis that genomic imprinting appeared some 150 million years ago at the time when monotremes and therian mammals (placental mammals and marsupials) diverged (Fig. 1).

However, the evolutionary history of some imprinted genes, like the *Igf2r* gene, seems to be more complex. In contrast to what has been shown in rodents, the *IGF2R* gene is not imprinted in humans (Kalscheuer et al. 1993; Ogawa et al. 1993), although *IGF2R* imprinting has been found sporadically in some individuals (Xu et al. 1993). A systematic analysis of imprinting of this gene in several mammalian species confirmed that it is imprinted in rodents and artiodactyls (sheep, cow, pig), but not in primates. This suggests that imprinting of the *IGF2R* gene was lost some 75 million years ago in the primate lineage (Killian et al. 2001b).

2.2
Theories on the Evolution of Parental Genomic Imprinting

By silencing one allele of several autosomous genes, genomic imprinting seems to be in contradiction with the evolutionary advantage of diploidy in mammals. Therefore, it has become a fundamental challenge for mammalian developmental biology to understand the functional significance of genomic imprinting (Wilkins and Haig 2003).

2.2.1
The Parental Conflict Theory

One key to better understanding the emergence of genomic imprinting is to point out that most imprinted genes are involved in the control of embryonic development. Indeed, several imprinted genes code for factors involved in the insulin pathway (*Igf2, Igf2r, Ins2, Rasgrf1* and *Grb10*). Others code for factors regulating placental development (*Mash2, Igf2* and *Ipl*) (Guillemot et al. 1995; Georgiades et al. 2001; Constância et al. 2002; Frank et al. 2002) or are involved in brain development and maternal or neonate behaviour (Lefebvre et al. 1998; Li et al. 1999). Of course, some imprinted genes appear to be unrelated to any of the above-mentioned functions, but one has to keep in mind that some genes may be imprinted only because they are located within imprinted domains.

Strikingly, paternally expressed genes tend to favour placental development and/or fetal growth (*Igf2, Ins2, Peg1/Mest, Peg3, Rasgrf1, Gtl2, Gnasxl*), whereas maternally expressed genes, with a few exceptions such as the *Mash2* gene (Guillemot et al. 1994, 1995), exhibit an opposite effect (*Igf2r, H19, Grb10, Cdkn1c, Gnas, Ipl*). This is coherent with the phenotype of both androgenotes, which are large embryos, and gynogenotes, which present growth retardation.

These observations, together with the phylogenic distribution of genomic imprinting (see Sect. 1.1), support the so-called parental conflict theory (Moore and Haig 1991). According to this theory, genomic imprinting results from a conflict between the paternal and maternal genomes concerning the amount of nutrients allocated to the embryo. The interest of the father would be to favour the growth of the embryos in order to maximize their survival rates, even at the expense of future progenies of the mother. Indeed, these latter have little chance to develop since mammals are essentially polygamous. Conversely, the interest of the mother would be to limit fetal growth in order to save its own resources for future progenies. Therefore, in the embryo, paternally inherited genes are selected to stimulate the extraction of resources from the mother, whereas maternally transmitted genes are selected to limit the extraction of resources from the mother.

Such a conflict would give a selective advantage to imprinted alleles versus non-imprinted alleles for the genes involved in development. This model predicts that genomic imprinting should not appear in oviparous species (monotremes and birds) since the amount of nutrients allocated to the embryo is predetermined by the mother (Haig 1999). This theory also implies that imprinted genes are not necessarily the same in each species, and, in this respect, it has been shown that the *Igf2r, U2af-rs1* and *Tssc4* genes are imprinted in mouse but not in human (Kalscheuer et al. 1993; Pearsall et al. 1996; Paulsen et al. 2000), whereas the *Ltrpc5* gene is imprinted in human but not in mouse (Paulsen et al. 2000). Similarly, the *Rasgrf1* gene is imprinted in rat and mouse, but not in *Peromyscus*, another rodent species (Pearsall et al. 1999).

Following this logic, one could argue that genomic imprinting does not present any intrinsic role in mammalian development, but only reflects a 'parental conflict' leading to monoallelic expression. From this point of view, the phenotypes associated with imprinting defects in human and mouse would not reflect a crucial role of genomic imprinting in development, but would rather be due to the perturbation of gene dosage at imprinted loci (Jaenisch 1997).

2.2.2
Alternative Theories

Several other theories have been proposed to explain the evolution of parental genomic imprinting in mammals. Among these, one of the oldest suggests that genomic imprinting evolved to prevent parthenogenesis (Solter 1988). This hypothesis also inspired the 'ovarian time-bomb' theory, which proposes that genomic imprinting has been selected to reduce the risks of trophoblast invasion following the parthenogenetic activation of unfecundated oocytes (Varmuza and Mann 1994). However, this theory is controversial (Haig 1994; Moore 1994; Solter 1994).

A recent and original hypothesis proposes that the monoallelic expression of some genes is only the consequence of a selective pressure maintaining distinct chromatin structures on paternal and maternal chromosomes in order to facilitate pairing of homologous chromosomes, DNA recombination and repair (Pardo-Manuel de Villena et al. 2000). However, this theory does not explain why imprinting evolved only in therian mammals and why it mostly affects genes involved in fetal development.

3
Characteristics of Mammalian Imprinted Genes

Most mammalian imprinted genes are clustered into large chromosomal domains encompassing hundreds of kilobase pairs. These domains contain both paternally and maternally expressed genes, which are often regulated by a common imprinting centre. They may also contain genes that escape imprinting and are biallelically expressed. This organization implies that genomic imprinting has evolved at the level of large chromatin domains rather than at the level of individual genes.

There are as many paternally as maternally expressed genes and, so far, no sequences or structures that would always and only specify imprinting have been identified. However, it has been established that the proportion of imprinted genes associated with a CpG island is higher than for the other genes, suggesting that the presence of a CpG island could be involved in the establishment or the maintenance of parental imprints (Paulsen et al. 2000).

Numerous imprinted genes also contain tandem repeats (Neumann et al. 1995). These repeats do not share any homology but can be differentially methylated in male and female germ lines and are frequently located close to differentially methylated regions (DMRs) (Smrzka et al. 1995; Tremblay et al. 1995; Moore et al. 1997; Pearsall et al. 1999), suggesting that they may be involved in the establishment of differential methylation at imprinted loci. This latter observation has led to the hypothesis that imprinting has evolved from a host defence mechanism against parasitic repeat sequences (Barlow 1993). In the case of the *Rasgrf1* gene, it has been clearly shown that repeats located 3' of the DMR (41 nucleotides repeated 40 times) are required for the establishment of the paternal methylation and the monoallelic expression of the gene (Yoon et al. 2002). Remarkably, these repeats are missing in rodent species in which this gene is not imprinted (Pearsall et al. 1999). In contrast, for some other imprinted genes like *Igf2* (Moore et al. 1997), *H19* (Stadnick et al. 1999) and *Grb10* (Arnaud et al. 2003), repeats found in the mouse are not conserved in human despite conservation of the adjacent DMRs, and functional studies could not reveal any obvious role in genomic imprinting (Stadnick et al. 1999; Reed et al. 2001; Thorvaldsen et al. 2002).

Finally, it has been shown in human and mouse that imprinted chromosomal domains show significantly reduced numbers of short interspersed transposable elements (SINEs) (Greally 2002; Ke et al. 2002). One interpretation is that these elements are excluded from imprinted loci because they are targeted by DNA methylation and could alter the regulation of imprinting. These studies also revealed that maternally expressed genes show a higher G/C and CpG content than paternally expressed genes.

4
Epigenetic Control of Imprinted Genes

How can both alleles of an imprinted gene be differentially expressed in the same cellular context despite having an identical nucleotidic sequence? The answer is that parental genomic imprinting involves *epigenetic modifications*, i.e. heritable modifications that affect gene expression without changes in the DNA sequence. These parental-specific marks are initiated in the germ lines and are maintained during embryonic development to distinguish the parental origin of the alleles.

4.1
DNA Methylation

Methylation of cytosine residues is a major epigenetic modification in eukaryotic genomes. In mammals, cytosine methylation occurs predominantly at CpG dinucleotides and is dispersed over the whole genome. These

CpG dinucleotides are globally underrepresented in the genome, but some sequences of several hundreds of base pairs, the so-called CpG islands, harbour a high CpG dinucleotide content. Paradoxically, these CpG islands have the general property to remain usually unmethylated, and they often co-localize with promoters of constitutively active genes (Larsen et al. 1992). However, hypermethylation of promoter-associated CpG islands in tumour suppressor genes has been described in many tumours and is probably involved in the tumourigenic process (Baylin and Herman 2000; Esteller 2002). Finally, it has also been proposed that CpG islands could play a role in DNA replication (Delgado et al. 1998).

Over the last decade, several DNA methyltransferases (Dnmts) have been identified in mammals. *Dnmt1* was the first methyltransferase to be cloned in mouse (Bestor et al. 1988), and then in human (Yen et al. 1992). DNMT1 has a higher affinity for double-stranded hemimethylated DNA than for unmethylated DNA, and is responsible for the maintenance of DNA methylation after each cycle of replication. Thereafter, studies with *Dnmt1*-deficient cells suggested that other enzymes account for the de novo DNA methylation activities (Lei et al. 1996), and three candidates have been identified by sequence comparisons: *Dnmt2*, *Dnmt3a* and *Dnmt3b* (Okano et al. 1998a; Yoder and Bestor 1998). DNMT3a and DNMT3b turned out to be the major de novo DNA methyltransferases in mammals, establishing genomic methylation patterns during early development (Okano et al. 1999). Moreover, a mutation of the human form of *DNMT3b* is associated with ICF syndrome (immunodeficiency, centromere instability, facial abnormalities), characterized by hypomethylation of the satellite sequences and by centromeric instability (Okano et al. 1999; Xu et al. 1999). In addition, the *Dnmt3* family members seem also to be involved in the maintenance of DNA methylation patterns in cooperation with Dnmt1 (Liang et al. 2002; Rhee et al. 2002). In contrast, the biological function of *Dnmt2* still remains elusive. Early attempts failed to detect any catalytic activity for DNMT2, despite the presence of a catalytic domain (Okano et al. 1998b), but recent studies have suggested that DNMT2 could have residual cytosine-5 DNA methyltransferase activity in vivo (Hermann et al. 2003; Liu et al. 2003; Tang et al. 2003).

DNA methylation is essential for embryonic development. Indeed, *Dnmt1* inactivation in the mouse is embryonically lethal as a consequence of a global demethylation of the genome (Li et al. 1992), whereas its partial inactivation leads to chromosomal instability and increased tumour frequency (Gaudet et al. 2003). Similarly, deficiency in *Dnmt3a* or *Dnmt3b* is lethal during embryonic development or shortly after birth, whereas double mutants die at the gastrulation stage (Okano et al. 1999).

DNA methylation was the first epigenetic mark to be identified as playing a role in genomic imprinting in mammals. As early as 1987, several studies showed that the expression of a transgene could be regulated by the establishment of distinct methylation profiles in parental gametes (Reik et al. 1987; Sapienza et al. 1987; Swain et al. 1987). Thereafter, the function of DNA methylation was reinforced by the discovery that the *H19*, *Igf2* and *Igf2r* genes contain regions that are differentially methylated on the parental alleles (DMRs) (Sasaki et al. 1992; Bartolomei et al. 1993; Brandeis et al. 1993; Ferguson-Smith et al. 1993). Indeed, DMRs have now been described in most of the imprinted genes.

However, the main argument in favour of a role of DNA methylation in genomic imprinting comes from the observation that imprinted expression of numerous genes is lost in *Dnmt1* –/– mouse embryos (homozygous deletion of the *Dnmt1* gene) (Li et al. 1993; Caspary et al. 1998). Additionally, the overexpression of *Dnmt1* induces loss of imprinting of some genes because of biallelic methylation of regulatory sequences (Biniszkiewicz et al. 2002). The sole exception is the *Mash2* gene that maintains monoallelic expression even in the absence of *Dnmt1* (Caspary et al. 1998; Tanaka et al. 1999). Finally, the use of 5-azacytidine, a DNA methylation inhibitor, induces loss of imprinting of numerous genes both ex vivo in cultured cells (El Kharroubi et al. 2001) and in vivo in the mouse (Hu et al. 1997).

These works emphasize the fundamental role of DNA methylation in the maintenance of genomic imprinting in mammals.

4.2
Histone Modifications

Other epigenetic modifications are associated with genomic imprinting in mammals. DNase sensitivity experiments and chromatin immunoprecipitation assays (ChIPs) performed on the *Igf2/H19* (Bartolomei et al. 1993; Ferguson-Smith et al. 1993; Hark and Tilghman 1998; Hu et al. 1998; Szabo et al. 1998; Khosla et al. 1999; Pedone et al. 1999; Kanduri et al. 2000; Grandjean et al. 2001), *U2AF1-rs1* (Shibata et al. 1996; Feil et al. 1997; Gregory et al. 2001), *Igf2r* (Hu et al. 2000) and *Snrpn* (Schweizer et al. 1999; Saitoh and Wada 2000; Fulmer-Smentek and Francke 2001; Gregory et al. 2001) loci have demonstrated that imprinted genes harbour different chromatin structures and histone acetylation patterns on each parental allele (Feil and Khosla 1999). As a general rule, the repressed methylated allele is characterized by compact chromatin and histone deacetylation, whereas the unmethylated expressed allele is characterized by nuclease sensitivity and histone acetylation. Different profiles of histone H3 methylation have also been described: H3 lysine 9 methylation is associated with the repressed methylated allele, and H3 lysine 4 methylation is associated with the active unmethylated allele (Xin et al. 2001; Fournier et al. 2002).

However, the importance of histone modifications in the establishment of imprinting remains to be clarified. The fundamental question is to know whether histone modifications can act as primary imprinting marks in the gametes, or if they are established after CpG methylation to maintain and strengthen the repressed state of the chromatin. Indeed, recent studies in mammalian systems suggest that DNA methylation can direct histone modifications (Hashimshony et al. 2003) through the recruitment of histone deacetylases (HDACs) and histone methyltransferases (against lysine 9 of histone H3) by the MBD proteins (methyl-CpG binding proteins) (Wade 2001; Fujita et al. 2003; Fuks et al. 2003).

However, at the *Snrpn* locus, chromatin modifications seem to represent imprinting marks that are independent of DNA methylation. In human, the imprinting centre AS-SRO maintains differential allelic chromatin structure in the absence of differential DNA methylation (Perk et al. 2002). Similarly, in mouse, the preferential H3 lysine 9 methylation of the maternal sequence of the PWS-IC is maintained in the absence of DNA methylation and appears to be sufficient for *Snrpn* gene repression (Xin et al. 2001).

Finally, the repressed allele of numerous imprinted genes can be reactivated by some DNA methylation inhibitors, but not by histone deacetylase inhibitors (Saitoh and Wada 2000; El Kharroubi et al. 2001; Fulmer-Smentek and Francke 2001; Grandjean et al. 2001; Lynch et al. 2002), suggesting that DNA methylation is the major epigenetic mark for most imprinted genes.

4.3
Asynchronous DNA Replication Timing

Another characteristic of imprinted genes is that they replicate asynchronously during the S-phase of the cell cycle. The initial observations were made by FISH experiments on the mouse *H19*, *Igf2* and *Snrpn* genes (Kitsberg et al. 1993) and on the human PWS/AS locus (Knoll et al. 1994; Gunaratne et al. 1995). At the time, the common rule was that imprinted genes replicate earlier on the paternal allele than on the maternal allele, independent of whether they are paternally or maternally expressed. This suggested that early replication is a paternal characteristic of imprinted loci that is regulated at the level of higher-order chromatin architecture (Greally et al. 1998; Kagotani et al. 2002).

However, one should be careful with the interpretations of these FISH data. In these experiments, a replicating sequence is visualized as two close dots and the asynchrony is revealed by a high number of cells presenting a characteristic pattern of three dots (i.e. only one allele has been replicated), but this could possibly reflect a difference in the kinetics of chromatid segregation rather than a delay in chromosomal replication (Shuster et al. 1998; Gribnau et al. 2003). Moreover, the FISH results could not be confirmed with alternative techniques like BrdU incorporation followed by S-phase fractionation (Kawame et al. 1995). More recently, it has been shown in mouse ES cells that, in disagreement with the FISH results, the maternal allele of the *Igf2/H19* locus replicates before the paternal allele (Gribnau et al. 2003).

Despite these reserves, asynchronous replication seems to be directly linked to genomic imprinting, since it is abolished at the human PWS/AS locus or at the mouse *Igf2/H19* locus when their respective imprinting centres are deleted (Gunaratne et al. 1995; Greally et al. 1998). Furthermore, it has been shown at some imprinted loci that asynchronous replication timings are set up in the gametes (Simon et al. 1999) and can be maintained in the absence of DNA methylation (Gribnau et al. 2003), suggesting that they might be involved in the establishment of genomic imprinting in the gametes.

4.4
Chromatin Architecture

Genomic imprinting may also establish a different higher-order chromatin architecture on each parental allele, as suggested by the existence of allele-specific patterns of association to the nuclear matrix at the *Igf2* locus (Weber et al. 2003). The existence of a differential chromatin organization at imprinted chromosomal domains is also suggested by the observation that the human *PWS/AS* and *IGF2/H19* loci show different meiotic recombination frequencies in the male and female germ lines (Paldi et al. 1995; Robinson and Lalande 1995). However, it remains to be clarified how these large-scale properties of imprinted domains are related to local epigenetic features at the imprinting centres.

5
The Parental Genomic Imprinting Cycle

Parental imprints undergo a cycle during the life of the organism that allows their reprogramming at each generation. Imprinting marks are inherited from the parental gametes and are then maintained and 'read' in the somatic cells of the individual before being erased in the germ line and re-established according to the sex of the individual for the next generation (Fig. 2).

5.1
Erasure

Parental epigenetic imprints are erased in primordial germ cells (PGCs) before being reprogrammed according to the type of the gamete (Szabo and Mann 1995; Villar et al. 1995). The genome of primordial germ cells under-goes a global demethylation, leading to the erasure of methylation at imprinted loci between the 10th and 12th days of mouse embryonic develop-ment (Szabo and Mann 1995; Davis et al. 2000; Hajkova et al. 2002; Lee et al. 2002; also see Chap. 6, this Vol.). Nuclear transplantation experiments have confirmed that functional imprints, i.e. epigenetic features required for a sta-ble monoallelic repression in the embryo (see below), have been erased at this

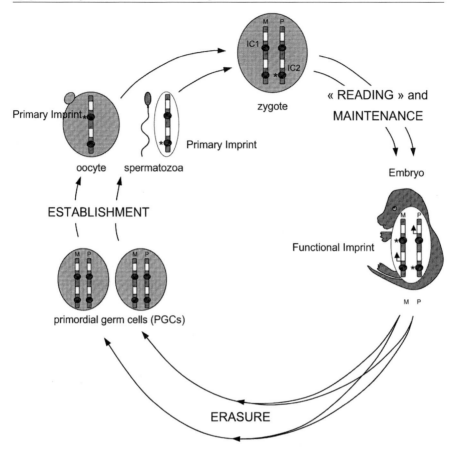

Fig. 2. The parental genomic imprinting cycle. Genomic imprinting results from epigenetic modifications established in the male and female germ lines, maintained and 'read' during cell divisions in the embryo to constitute functional imprints leading to stable monoallelic repression. *Above* Example of a paternally and maternally repressed gene following the apposition of epigenetic marks at imprinting centres (*IC*) in the oocyte or sperm. Parental imprints are erased during early gametogenesis and then re-established to be transmitted to the next generation according to sex of the individual. *M* Maternal chromosome; *P* paternal chromosome

stage (Obata et al. 1998; Kato et al. 1999; Lee et al. 2002). The mechanism of this demethylation is still unknown, but the demethylation activity appears dominant when the germ cells are fused with somatic cells (Tada et al. 1997).

Conceptually, it is interesting to note that the allele inherited from the parent with the same sex is reprogrammed in the germ cells of the embryo rather than being transmitted without modifications to the next generation.

5.2
Establishment

5.2.1
Primary Imprinting Marks

The so-called germ line or *primary imprinting marks* are those epigenetic modifications that are established in the gametes (when the two alleles are physically separated) and that keep the epigenetic 'memory' of the parental origin of the alleles in somatic cells throughout development (Fig. 2). Establishment of imprinting is defined as the acquisition of such primary imprinting marks in the germ lines between the erasure step and fertilization.

DNA methylation (or protection from DNA methylation) probably represents such a primary imprinting mark for most of the imprinted genes identified so far. Indeed, numerous imprinted genes acquire differential methylation patterns in the male and female gametes (these are called 'primary methylation imprints') that are subsequently maintained in the embryo (Brandeis et al. 1993; Stoger et al. 1993; Tremblay et al. 1995; Tucker et al. 1996; Shemer et al. 1997; Ueda et al. 2000). Establishment of primary methylation imprints occurs during late differentiation of the gametes (Szabo and Mann 1995; Obata and Kono 2002). In contrast to other autosomal genes, passage through the germ line is required to establish methylation patterns at imprinted loci (Tucker et al. 1996), suggesting that they are targeted by de novo methylation activities specific to the germ cells. A major breakthrough came from the identification of the DNMT3L factor (Dnmt3-like), a protein belonging to the DNMT3 family (Aapola et al. 2000, 2001). DNMT3L is the first factor known to be involved in the establishment of primary imprints during gametogenesis, a class of factor that could be named 'imprinters'. *Dnmt3L* is expressed in the male and female germ lines precisely at the time when the parental imprints are established. The homozygous deletion of *Dnmt3L* in the mouse inhibits differentiation of spermatozoa in males, whereas embryos issued from *Dnmt3L* –/– females show deregulation of numerous imprinted genes leading to premature death. Indeed, the *Dnmt3L* deletion induces an absence of methylation at imprinted sequences that are usually methylated in the oocytes without affecting methylation patterns in the rest of the genome (Bourc'his et al. 2001; Hata et al. 2002). Therefore, the *Dnmt3L* gene is specifically required to establish maternal methylation imprints during oogenesis. Surprisingly, the DNMT3L protein lacks the DNA methyltransferase catalytic domain and it has been proposed that it could act indirectly via interactions with other members of the *Dnmt3* family (Chedin et al. 2002; Hata et al. 2002). Indeed, it has been shown that the establishment of methylation imprints does not require *Dnmt1o* (the *Dnmt1* isoform expressed during oogenesis) (Howell et al. 2001).

Finally, it remains unclear how male and female germ cells can establish different methylation patterns. This could result from the existence of different chromatin organizations in the male and female germ cells or from the expression of germ-line-specific 'imprinter' factors. Among these factors, one could mention the zinc-finger protein CTCF which may protect the *H19* ICR from DNA methylation in the female germ line (Fedoriw et al. 2004), even if contradictory results have been obtained (Schoenherr et al. 2003).

5.2.2
Imprinting Centres

Over the past 10 years, it has become clear that most imprinted genes are not regulated individually, but in a coordinate fashion at the level of structured chromatin domains. Numerous imprinted loci contain an imprinting centre (IC), a sequence of a few kilobases carrying the 'primary imprinting marks' established in the gametes and maintained in the embryo. These ICs regulate the imprinting of several genes over long distances, as demonstrated by deletion experiments in the mouse.

The existence of an IC was first proposed at the human *15q11–13* locus after observation that microdeletions in Angelman and Prader-Willi syndrome patients are associated with disruption of imprinting of several genes dispersed throughout 2 Mb of the locus (Buiting et al. 1995). This IC is a bipartite regulatory element (Fig. 3D). The AS-SRO element, located 35 kb upstream of the *SNRPN* gene, is deleted in Angelman syndrome patients and regulates the establishment of maternal imprints in the germ cells. The PWS-SRO element, which includes exon 1 of the *SNRPN* gene, is deleted in Prader-Willi syndrome patients and controls the establishment of paternal imprints (Dittrich et al. 1996; Saitoh et al. 1996; Yang et al. 1998; Shemer et al. 2000; Bressler et al. 2001; Perk et al. 2002). In addition, the PWS-SRO element is also required to maintain imprinting on the paternal chromosome during embryonic development (Bielinska et al. 2000).

An IC was also identified at the *Igf2r* locus, located on chromosome 17 in the mouse (Fig. 3E). This intronic element, methylated on the maternal allele, was initially described as being required for *Igf2r* imprinting (Stoger et al. 1993; Wutz et al. 1997), then it turned out that it acts on the paternal allele as a bidirectional repressor for several genes dispersed throughout a 400-kb domain (Wutz et al. 2001; Zwart et al. 2001).

Finally, two ICs have been identified in the distal part of mouse chromosome 7, the orthologue of the human *11p15.5* locus, which includes the *Cdkn1c*, *Igf2* and *H19* genes. The KvDMR1, located in intron 10 of the *Kvlqt1* (*Kcnq1*) gene, is specifically methylated on the maternal allele and induces the repression of several genes of the locus on the paternal allele (Fig. 3F) (Horike et al. 2000; Fitzpatrick et al. 2002; Kanduri et al. 2002). This region is probably

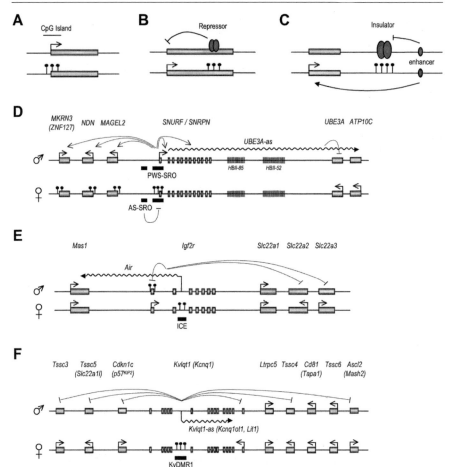

Fig. 3. Regulatory mechanisms of imprinted gene expression. **A** Gene promoter is repressed by DNA methylation (*lollipops*). **B** Binding of a *cis*-acting repressor is regulated by DNA methylation. **C** Insulator activity, which blocks promoter access to enhancers, is regulated by DNA methylation. **D–F** Imprinting centres and antisense RNAs. **D** The human *15q11–13* locus, involved in Angelman (*AS*) and Prader-Willi (*PWS*) syndromes (orthologue of the central part of chromosome 7 in the mouse). PWS is associated with a deficiency in paternally expressed genes, whereas AS results from a deficiency in maternally expressed genes (*UBE3A* is the main candidate gene for AS). At this locus, all paternally expressed genes are associated with a 5'DMR. The PWS-SRO element, which includes the promoter and exon 1 of the *SNRPN* gene, contains a DMR whose maternal methylation is inherited from oocytes in mouse but is established after fertilization in human (El-Maarri et al. 2001). PWS-SRO is required on the paternal allele to activate paternally expressed genes and represses *UBE3A*, whereas AS-SRO is required on the maternal allele to repress paternally expressed genes and activates *UBE3A* (Dittrich et al. 1996; Saitoh et al. 1996; Yang et al. 1998; Bressler et al. 2001). The AS-SRO element does not show any differential methylation but displays a specific chromatin structure on the maternal allele that inactivates the PWS-SRO element and prevents gene activation in *cis* (Shemer et al. 2000; Perk et al. 2002). The *UBE3A* gene is subjected to imprinting only in the brain (Albrecht et al. 1997; Rougeulle et al. 1997; Vu and Hoffman 1997). According to the current model, the PWS-SRO element

involved in imprinting disorders leading to the human Beckwith-Wiedemann syndrome (Lee et al. 1999; Smilinich et al. 1999).

Imprinting of the *Igf2* and *H19* genes depends on a second IC located 2 kb upstream of the *H19* gene. This sequence, called the imprinting-control region (ICR), has an insulator activity that is regulated by DNA methylation (Fig. 3C). On the maternal allele, the CTCF protein, a factor involved in boundary function in vertebrates, binds the unmethylated ICR and blocks the access of *Igf2* promoters to enhancers located downstream of *H19*. On the paternal allele, binding of CTCF is inhibited by DNA methylation.

Establishment of primary methylation marks at ICs may also require additional *cis*-regulatory elements. This is the case at the *Rasgrf1* locus where a repeat sequence controls the establishment of paternal methylation on the IC during spermatogenesis (Yoon et al. 2002).

5.3
Maintenance

Immediately after fertilization, a global demethylation wave affects the genome. The paternal genome is actively demethylated in the zygote, whereas the maternal genome undergoes a passive demethylation during cell divisions due to the absence of the maintenance methyltransferase until the blastocyst stage (Rougier et al. 1998; Reik et al. 2001). Remarkably, the sequences carrying primary methylation imprints have the unique property to resist this

---

upregulates an antisense RNA (UBE3A-as) in the brain that represses *UBE3A* on the paternal allele (Rougeulle et al. 1998; Chamberlain and Brannan 2001; Runte et al. 2001). This RNA is also the host of brain-specific snoRNAs that are repeated in tandem in numerous copies (HPII-85 and 52) (Cavaillé et al. 2000; Runte et al. 2001). E The *Igf2r/Air* locus on mouse chromosome 17 (orthologue of the human *6q* locus). The maternally expressed *Igf2r* gene contains two DMRs: one in the promoter region, which harbours a paternal methylation that is established after fertilization, and a sequence located in intron 2 that is methylated in the oocytes and corresponds to the imprinting control element (*ICE*) of this locus (Stoger et al. 1993; Wutz et al. 1997). The *Slc22a2* and *Slc22a3* genes are maternally expressed in the placenta despite a stable hypomethylation of both parental alleles (Zwart et al. 2001). Unmethylated ICE corresponds to the active promoter of the 108-kb-long Air (Antisense *Igf2r* RNA) transcript (Lyle et al. 2000). Deletion of the ICE (Wutz et al. 2001; Zwart et al. 2001) or expression of a truncated Air RNA (Sleutels et al. 2002) on the paternal allele releases the repression of the *Igf2r*, *Slc22a2* and *Slc22a3* genes, suggesting a direct role for Air RNA in the mechanism of bidirectional repression on the paternal chromosome. F The *Kvlqt1* locus on the distal part of mouse chromosome 7 (orthologue of the human *11p15.5* locus). Intronic KvDMR1 is methylated on the maternal allele in somatic and germ cells and corresponds to the Kvlqt1-as promoter on the paternal allele (Lee et al. 1999; Smilinich et al. 1999; Engemann et al. 2000; Yatsuki et al. 2002). Deletion of KvDMR1 on the paternal allele induces reactivation of numerous genes in *cis*. (Horike et al. 2000; Fitzpatrick et al. 2002)

demethylation wave and maintain their gametic methylation patterns throughout development (Brandeis et al. 1993; Shemer et al. 1997; Tremblay et al. 1997). This property is crucial to maintain distinct marks between the parental genomes, because methylation imprints could not be re-established in the somatic cells of the embryo (Tucker et al. 1996).

The mechanism by which imprinted domains resist demethylation after fertilization is still poorly understood. One can imagine that paternally methylated regions resist active demethylation in the zygote by acquiring some specific chromatin features. DNMT1o is a DNMT1 isoform issued from an alternative promoter (Mertineit et al. 1998) that replaces DNMT1 in the embryo until the implantation stage. It specifically maintains the methylation at imprinted loci during the fourth replication cycle at the transition from an 8- to a 16-cell embryo (Howell et al. 2001). However, except at the eight-cell stage, DNMT1o is sequestered in the cytoplasm and the factors responsible for the maintenance of methylation imprints during the other divisions until the implantation stage are unknown.

The maintenance of methylation imprints in the embryo also implies that unmethylated sequences resist the de novo methylation. This property is illustrated by the fact that imprinted regions can be de novo methylated in the germ cells but not in the somatic cells of the embryo (Tucker et al. 1996). The molecular bases of this resistance are still unknown, but probably involve proteic factors that protect unmethylated sequences from the methylation machinery (Birger et al. 1999), as it has been proposed for the CTCF protein at the *H19* ICR (Pant et al. 2003; Schoenherr et al. 2003).

5.4
Monoallelic Expression of Imprinted Genes

Primary imprinting marks established in the gametes lead to stable monoallelic expression of imprinted genes in the embryo. This monoallelic expression is the consequence of distinct cascades of epigenetic events occurring on the parental alleles during development. The active allele undergoes all the required chromatin reorganization leading to gene expression, a process that could be named '*genome formatting*' (Paro 2000). In contrast, the imprinted allele acquires a *functional imprint* that impairs the 'formatting' for gene expression on this allele and maintains a stable silent state.

The acquisition of functional imprints, which is often called 'imprint reading', involves the establishment of '*secondary imprinting marks*', i.e. epigenetic modifications established in *cis* after fertilization from the primary imprinting marks. These secondary marks include additional DMRs that are possibly cell-type specific and this hierarchy in the establishment of DMRs is crucial for the mechanisms of genomic imprinting (Weber et al. 2001; Lopes et al. 2003).

5.4.1
Formatting for Gene Expression

On the non-imprinted allele, 'formatting' of the locus leads to a proper spatio-temporal regulation of gene expression. This 'formatting' consists of a combination of local and higher-order chromatin reorganizations allowing activation and modulation of transcription during development, and requires specific regulatory elements. Although not being directly involved in the imprinting process, understanding how these elements are controlled is crucial since, on the repressed allele, functional imprints are constituted to block their activity. One striking example is provided by the DMR2/MAR2 endoderm-specific regulatory element in the mouse *Igf2* gene. The DMR2 is not required to maintain *Igf2* imprinting on the maternal allele, but the methylated DMR2 is required for high *Igf2* expression levels on the paternal allele (Murrell et al. 2001). This DMR controls the activity of a neighbouring matrix attachment region (MAR2) that would itself favour long-range interaction with distal enhancers (Weber et al. 2003). On the repressed allele, the functional imprint keeps the DMR2 unmethylated, possibly to prevent activation of MAR2 and *Igf2* expression.

This example reveals how some secondary imprinting marks can reflect functional imprints without being necessarily involved in the imprinting process, i.e. the repression of the imprinted allele.

5.4.2
Acquisition of Functional Imprints

There is no single model for monoallelic repression that would be common to all known imprinted genes. Indeed, as most imprinted genes are clustered and share regulatory elements that can be tissue-specific, the acquisition of functional imprints is subject to various spatio-temporal constraints during development. Therefore, the mechanisms of imprinted expression may differ not only from one imprinted gene to another, but also from one cell type to another. Moreover, even if it is commonly accepted that repression of imprinted genes is determined at the transcriptional level, one cannot exclude a post-transcriptional contribution as suggested for the *H19* and *Igf2* genes (Jouvenot et al. 1999).

Functional imprints in the embryo result from multiple mechanisms that involve secondary DMRs, allele-specific binding of proteic factors as well as non-coding antisense RNAs (ncRNAs).

Promoter Methylation

One common mechanism that gives rise to stable monoallelic repression is differential allelic methylation of promoters (Fig. 3A). This allele-specific methylation can either be established directly in the gametes or, as is the case for most imprinted genes, be acquired secondarily in the embryo from a *cis*-IC. It is the case at the *H19* locus, for example, where the IC located 2 kb upstream of the gene induces promoter methylation on the paternal allele during early development (Srivastava et al. 2000).

DNA methylation can lead to transcriptional repression notably by preventing the binding of transcription factors, or through the binding of MBD proteins (Wade 2001). These factors induce a local compaction of chromatin by recruiting histone deacetylase or histone methyltransferase complexes. Indeed, MeCP2 forms a complex with HDACs and the Sin3A co-repressor (Jones et al. 1998; Nan et al. 1998), whereas MBD2 recruits the histone deacetylase complex Mi-2/NurD (Ng et al. 1999; Wade et al. 1999; Zhang et al. 1999). Similarly, MeCP2 and MBD1 recruit methyltransferase activities targeted against lysine 9 of histone H3 (Fujita et al. 2003; Fuks et al. 2003).

Modulation of Regulatory Sequences

The observation that some imprinted genes harbour methylated sequences on the active allele led to the hypothesis that these regions have repressor activities that are inhibited by DNA methylation (Fig. 3B; Sasaki et al. 1992; Brandeis et al. 1993; Stoger et al. 1993; Feil et al. 1994). This model has been demonstrated in the case of the *Igf2* gene where the DMR1, an intragenic DMR bearing a secondary imprinting mark, acts as a repressor on the maternal allele and is inactivated by DNA methylation on the paternal allele in mesodermic tissues (Constância et al. 2000; Eden et al. 2001). A similar model could be proposed for the *Dlk1/Gtl2* locus where repression of the maternal *Dlk1* allele depends on the unmethylated intergenic germ-line DMR (Ig-DMR) (Lin et al. 2003).

Additionally, functional imprints may also involve indirect mechanisms like the modulation of enhancer or silencer sequences that are located far away from the gene. For example, functional imprint of the *Igf2* gene involves the CTCF/ICR insulator that impairs access to endodermic enhancers (Fig. 3C).

Antisense Transcripts

Numerous imprinted genes are associated with oppositely imprinted genes encoding untranslated antisense RNAs. The best-characterized examples are the *Igf2r/Air*, *Ube3A/Ube3A-as* and *Kvlqt1/Kvlqt1-as* loci (see Fig. 3D–F). However, the function of these RNAs in the establishment and the maintenance of genomic imprinting remains to be clarified (Rougeulle and Heard 2002). These RNAs are very large and are paternally expressed from unmethylated ICs. The *Air* promoter is located in the intronic IC identified within the *Igf2r* gene, and it has been demonstrated that the Air transcript is necessary for paternal repression of several genes in the locus (Sleutels et al. 2002). This was the first demonstration that a non-coding RNA can be directly involved in the regulation of genomic imprinting. However, the mechanism by which the Air RNA affects imprinting is still unknown. A puzzling observation is that it overlaps only one imprinted gene (*Igf2r*) while being involved in the imprinting of several other genes of the locus. One can imagine that the Air RNA initiates repression at the *Igf2r* gene that subsequently propagates to neighbouring genes, or that it covers a portion of the chromosome like the Xist RNA does on the X chromosome (Clemson et al. 1996). Possibly, the Kvlqt1-as transcript could play a similar role in paternal gene repression at the *Kvlqt1* locus.

6
Conclusion

After almost complete sequencing of the mouse and human genomes, most protein-coding genes have now been identified. However, the dynamics of genome organization that governs gene expression remains a mystery. It is clear, though, that, during development, genes undergo a 'formatting' for gene expression and that genomic imprinting interferes with this formatting to obtain a stable repression on the imprinted allele.

By convention, parental genomic imprinting is defined as the stable *repression* of one allele depending on its parental origin. The term 'imprint' can be used for 'primary' imprints, which keep the memory of parental origins, 'secondary' imprints, these allele-specific epigenetic modifications established after fertilization, or 'functional' imprints, which include all the genomic and epigenetic features involved in stable repression of the imprinted allele. These intricate concepts can easily be misleading and therefore it is always very useful to use a rigorous and consensual terminology. For example, the somatic epigenetic reprogramming that occurs during reproductive cloning (cloning of animals from somatic cells) probably involves 'reformatting' of functional imprints but certainly not the reprogramming of primary imprints that would require passage through the germ line.

In addition, the constitution of functional imprints depends on the genomic context and on cell-type-specific constraints, and thus it is perhaps not so surprising that imprinting mechanisms are so different from one gene to another. Furthermore, regulatory elements involved in imprinting or gene expression are shared between several genes in imprinted loci and one regulatory element involved in the imprinting of one gene may be required for the expression of another gene. Therefore, when speaking about the function of such regulatory elements, we should always specify from which gene we adopt the point of view. Finally, the function of a given regulatory element can also vary for the same gene during the imprinting cycle. For example, the ICR/CTCF sequence at the *Igf2/H19* locus is involved in the maintenance of the *Igf2* primary imprint (Schoenherr et al. 2003) but also as an insulator in the functional imprint of *Igf2*.

Mammalian genomes are highly and dynamically structured and research during recent years has revealed how much local histone modifications or DNA methylation are crucial for gene expression. However, how these local epigenetic events are first brought into specific regulatory sequences remains unclear (Orphanides and Reinberg 2002). Studies on parental genomic imprinting indicate that some features of mammalian genomes, like the timing of DNA replication and the imprinting process itself, are controlled over large distances. Obviously, investigating higher-order chromatin architecture at the level of several tens of kilobase pairs will become crucial to understanding how gene expression and genomic imprinting are regulated. This domain remains largely unexplored and this is essentially due to the lack of techniques that would allow us to investigate chromatin organization at that scale. The development of such methodologies, while emerging as a central preoccupation for researchers in the field (Tolhuis et al. 2002), remains a major technological challenge if we want to unfold the mystery hidden behind epigenetic control of eukaryotic gene expression.

Acknowledgements. We thank colleagues from the Institut de Génétique Moléculaire de Montpellier (IGMM) for reading the manuscript, and the Association pour la Recherche contre le Cancer (ARC contract no. 3279), as well as the Fond National de la Science (ACI jeune chercheur) for their support and grants given to T. Forné. M. Weber and H. Hagège were supported by PhD fellowship Allocations de Moniteur Normalien (Ministère de l'Education Nationale, de la Recherche et de la Technologie).

References

Aapola U, Kawasaki K, Scott HS, Ollila J, Vihinen M, Heino M, Shintani A, Minoshima S, Krohn K, Antonarakis SE et al. (2000) Isolation and initial characterization of a novel zinc finger gene, DNMT3L, on 21q22.3, related to the cytosine-5-methyltransferase 3 gene family. Genomics 65:293–298

Aapola U, Lyle R, Krohn K, Antonarakis SE, Peterson P (2001) Isolation and initial characterization of the mouse Dnmt3l gene. Cytogenet Cell Genet 92:122–126

Albrecht U, Sutcliffe JS, Cattanach BM, Beechey CV, Armstrong D, Eichele G, Beaudet AL (1997) Imprinted expression of the murine Angelman syndrome gene, Ube3a, in hippocampal and Purkinje neurons. Nat Genet 17:75–78

Allshire R, Bickmore W (2000) Pausing for thought on the boundaries of imprinting. Cell 102:705–708

Arnaud P, Monk D, Hitchins M, Gordon E, Dean W, Beechey CV, Peters J, Craigen W, Preece M, Stanier P et al. (2003) Conserved methylation imprints in the human and mouse GRB10 genes with divergent allelic expression suggests differential reading of the same mark. Hum Mol Genet 12:1005–1019

Barlow DP (1993) Methylation and imprinting: from host defense to gene regulation? Science 260:309–310

Barlow DP, Stoger R, Herrmann BG, Saito K, Schweifer N (1991) The mouse insulin-like growth factor type-2 receptor is imprinted and closely linked to the Tme locus. Nature 349:84–87

Baroux C, Spillane C, Grossniklaus U (2002) Genomic imprinting during seed development. Adv Genet 46:165–214

Bartolomei MS, Zemel S, Tilghman SM (1991) Parental imprinting of the mouse H19 gene. Nature 351:153–155

Bartolomei MS, Webber AL, Brunkow ME, Tilghman SM (1993) Epigenetic mechanisms underlying the imprinting of the mouse H19 gene. Genes Dev 7:1663–1673

Baylin SB, Herman JG (2000) DNA hypermethylation in tumorigenesis: epigenetics joins genetics. Trends Genet 16:168–174

Bestor T, Laudano A, Mattaliano R, Ingram V (1988) Cloning and sequencing of a cDNA encoding DNA methyltransferase of mouse cells. The carboxyl-terminal domain of the mammalian enzymes is related to bacterial restriction methyltransferases. J Mol Biol 203:971–983

Bielinska B, Blaydes SM, Buiting K, Yang T, Krajewska-Walasek M, Horsthemke B, Brannan CI (2000) De novo deletions of SNRPN exon 1 in early human and mouse embryos result in a paternal to maternal imprint switch. Nat Genet 25:74–78

Biniszkiewicz D, Gribnau J, Ramsahoye B, Gaudet F, Eggan K, Humpherys D, Mastrangelo MA, Jun Z, Walter J, Jaenisch R (2002) Dnmt1 overexpression causes genomic hypermethylation, loss of imprinting, and embryonic lethality. Mol Cell Biol 22:2124–2135

Birger Y, Shemer R, Perk J, Razin A (1999) The imprinting box of the mouse Igf2r gene. Nature 397:84–88

Bourc'his D, Xu GL, Lin CS, Bollman B, Bestor TH (2001) Dnmt3L and the establishment of maternal genomic imprints. Science 294:2536–2539

Brandeis M, Kafri T, Ariel M, Chaillet JR, McCarrey J, Razin A, Cedar H (1993) The ontogeny of allele-specific methylation associated with imprinted genes in the mouse. EMBO J 12:3669–3677

Bressler J, Tsai TF, Wu MY, Tsai SF, Ramirez MA, Armstrong D, Beaudet AL (2001) The SNRPN promoter is not required for genomic imprinting of the Prader-Willi/Angelman domain in mice. Nat Genet 28:232–240

Buiting K, Saitoh S, Gross S, Dittrich B, Schwartz S, Nicholls RD, Horsthemke B (1995) Inherited microdeletions in the Angelman and Prader-Willi syndromes define an imprinting centre on human chromosome 15. Nat Genet 9:395–400

Caspary T, Cleary MA, Baker CC, Guan XJ, Tilghman SM (1998) Multiple mechanisms regulate imprinting of the mouse distal chromosome 7 gene cluster. Mol Cell Biol 18:3466–3474

Cattanach BM, Jones J (1994) Genetic imprinting in the mouse: implications for gene regulation. J Inherit Metab Dis 17:403–420

Cattanach BM, Kirk M (1985) Differential activity of maternally and paternally derived chromosome regions in mice. Nature 315:496–498

Cavaillé J, Buiting K, Kiefmann M, Lalande M, Brannan CI, Horsthemke B, Bachellerie JP, Brosius J, Huttenhofer A (2000) Identification of brain-specific and imprinted small nucleolar RNA genes exhibiting an unusual genomic organization. Proc Natl Acad Sci USA 97:14311–14316

Chamberlain SJ, Brannan CI (2001) The Prader-Willi syndrome imprinting center activates the paternally expressed murine Ube3a antisense transcript but represses paternal Ube3a. Genomics 73:316–322

Chedin F, Lieber MR, Hsieh CL (2002) The DNA methyltransferase-like protein DNMT3L stimulates de novo methylation by Dnmt3a. Proc Natl Acad Sci USA 99:16916–16921

Clemson CM, McNeil JA, Willard HF, Lawrence JB (1996) XIST RNA paints the inactive X chromosome at interphase: evidence for a novel RNA involved in nuclear/chromosome structure. J Cell Biol 132:259–275

Constância M, Dean W, Lopes S, Moore T, Kelsey G, Reik W (2000) Deletion of a silencer element in Igf2 results in loss of imprinting independent of H19. Nat Genet 26:203–206

Constância M, Hemberger M, Hughes J, Dean W, Ferguson-Smith A, Fundele R, Stewart F, Kelsey G, Fowden A, Sibley C, Reik W (2002) Placental-specific IGF-II is a major modulator of placental and fetal growth. Nature 417:945–948

Crouse HV (1960) The controlling element in sex chromosome behavior in *Sciara*. Genetics 45:1429–1443

Davis TL, Yang GJ, McCarrey JR, Bartolomei MS (2000) The H19 methylation imprint is erased and re-established differentially on the parental alleles during male germ cell development. Hum Mol Genet 9:2885–2894

DeChiara TM, Robertson EJ, Efstratiadis A (1991) Parental imprinting of the mouse insulin-like growth factor II gene. Cell 64:849–859

Delgado S, Gomez M, Bird A, Antequera F (1998) Initiation of DNA replication at CpG islands in mammalian chromosomes. EMBO J 17:2426–2435

Dittrich B, Buiting K, Korn B, Rickard S, Buxton J, Saitoh S, Nicholls RD, Poustka A, Winterpacht A, Zabel B, Horsthemke B (1996) Imprint switching on human chromosome 15 may involve alternative transcripts of the SNRPN gene. Nat Genet 14:163–170

Eden S, Constância M, Hashimshony T, Dean W, Goldstein B, Johnson AC, Keshet I, Reik W, Cedar H (2001) An upstream repressor element plays a role in Igf2 imprinting. EMBO J 20:3518–3525

El Kharroubi A, Piras G, Stewart CL (2001) DNA demethylation reactivates a subset of imprinted genes in uniparental mouse embryonic fibroblasts. J Biol Chem 276:8674–8680

El-Maarri O, Buiting K, Peery EG, et al. (2001) Maternal methylation imprints on human chromosome 15 are established during or after fertilization. Nat Genet 27:341–344

Engemann S, Strodicke M, Paulsen M, Franck O, Reinhardt R, Lane N, Reik W, Walter J (2000) Sequence and functional comparison in the Beckwith-Wiedemann region: implications for a novel imprinting centre and extended imprinting. Hum Mol Genet 9:2691–2706

Esteller M (2002) CpG island hypermethylation and tumor suppressor genes: a booming present, a brighter future. Oncogene 21:5427–5440

Falls JG, Pulford DJ, Wylie AA, Jirtle RL (1999) Genomic imprinting: implications for human disease. Am J Pathol 154:635–647

Fedoriw A, Stein P, Svoboda P, Schultz R, Bartolomei M (2004) Transgenic RNAi reveals essential function for CTCF in H19 gene imprinting. Science 303:238–240

Feil R, Khosla S (1999) Genomic imprinting in mammals: an interplay between chromatin and DNA methylation? Trends Genet 15:431–435

Feil R, Walter J, Allen ND, Reik W (1994) Developmental control of allelic methylation in the imprinted mouse Igf2 and H19 genes. Development 120:2933–2943

Feil R, Boyano MD, Allen ND, Kelsey G (1997) Parental chromosome-specific chromatin conformation in the imprinted U2af1-rs1 gene in the mouse. J Biol Chem 272:20893–20900

Ferguson-Smith AC, Sasaki H, Cattanach BM, Surani MA (1993) Parental-origin-specific epigenetic modification of the mouse H19 gene. Nature 362:751–755

Fitzpatrick GV, Soloway PD, Higgins MJ (2002) Regional loss of imprinting and growth deficiency in mice with a targeted deletion of KvDMR1. Nat Genet 32:426–431

Fournier C, Goto Y, Ballestar E, Delaval K, Hever AM, Esteller M, Feil R (2002) Allele-specific histone lysine methylation marks regulatory regions at imprinted mouse genes. EMBO J 21:6560–6570

Frank D, Fortino W, Clark L, Musalo R, Wang W, Saxena A, Li CM, Reik W, Ludwig T, Tycko B (2002) Placental overgrowth in mice lacking the imprinted gene Ipl. Proc Natl Acad Sci USA 99:7490–7495

Fujita N, Watanabe S, Ichimura T, Tsuruzoe S, Shinkai Y, Tachibana M, Chiba T, Nakao M (2003) Methyl-CpG binding domain 1 (MBD1) interacts with the Suv39h1-HP1 heterochromatic complex for DNA methylation-based transcriptional repression. J Biol Chem 278:24132–24138

Fuks F, Hurd PJ, Wolf D, Nan X, Bird AP, Kouzarides T (2003) The methyl-CpG-binding protein MeCP2 links DNA methylation to histone methylation. J Biol Chem 278:4035–4040

Fulmer-Smentek SB, Francke U (2001) Association of acetylated histones with paternally expressed genes in the Prader-Willi deletion region. Hum Mol Genet 10:645–652

Gaudet F, Hodgson JG, Eden A, Jackson-Grusby L, Dausman J, Gray JW, Leonhardt H, Jaenisch R (2003) Induction of tumors in mice by genomic hypomethylation. Science 300:489–492

Georgiades P, Watkins M, Burton GJ, Ferguson-Smith AC (2001) Roles for genomic imprinting and the zygotic genome in placental development. Proc Natl Acad Sci USA 98:4522–4527

Grandjean V, O'Neill L, Sado T, Turner B, Ferguson-Smith A (2001) Relationship between DNA methylation, histone H4 acetylation and gene expression in the mouse imprinted Igf2-H19 domain. FEBS Lett 488:165–169

Greally JM (2002) Short interspersed transposable elements (SINEs) are excluded from imprinted regions in the human genome. Proc Natl Acad Sci USA 99:327–332

Greally JM, Starr DJ, Hwang S, Song L, Jaarola M, Zemel S (1998) The mouse H19 locus mediates a transition between imprinted and non-imprinted DNA replication patterns. Hum Mol Genet 7:91–95

Gregory RI, Randall TE, Johnson CA, Khosla S, Hatada I, O'Neill LP, Turner BM, Feil R (2001) DNA methylation is linked to deacetylation of histone H3, but not H4, on the imprinted genes Snrpn and U2af1-rs1. Mol Cell Biol 21:5426–5436

Gribnau J, Hochedlinger K, Hata K, Li E, Jaenisch R (2003) Asynchronous replication timing of imprinted loci is independent of DNA methylation, but consistent with differential subnuclear localization. Genes Dev 17:759–773

Guillemot F, Nagy A, Auerbach A, Rossant J, Joyner AL (1994) Essential role of Mash-2 in extraembryonic development. Nature 371:333–336

Guillemot F, Caspary T, Tilghman SM, Copeland NG, Gilbert DJ, Jenkins NA, Anderson DJ, Joyner AL, Rossant J, Nagy A (1995) Genomic imprinting of Mash2, a mouse gene required for trophoblast development. Nat Genet 9:235–242

Gunaratne PH, Nakao M, Ledbetter DH, Sutcliffe JS, Chinault AC (1995) Tissue-specific and allele-specific replication timing control in the imprinted human Prader-Willi syndrome region. Genes Dev 9:808–820

Haig D (1994) Refusing the ovarian time bomb. Trends Genet 10:346–347; author reply 348–349

Haig D (1999) Multiple paternity and genomic imprinting. Genetics 151:1229–1231

Hajkova P, Erhardt S, Lane N, Haaf T, El-Maarri O, Reik W, Walter J, Surani M (2002) Epigenetic reprogramming in mouse primordial germ cells. Mech Dev 117:15

Hark AT, Tilghman SM (1998) Chromatin conformation of the H19 epigenetic mark. Hum Mol Genet 7:1979–1985

Hashimshony T, Zhang J, Keshet I, Bustin M, Cedar H (2003) The role of DNA methylation in setting up chromatin structure during development. Nat Genet 34:187–192

Hata K, Okano M, Lei H, Li E (2002) Dnmt3L cooperates with the Dnmt3 family of de novo DNA methyltransferases to establish maternal imprints in mice. Development 129:1983–1993

Hermann A, Schmitt S, Jeltsch A (2003) The human Dnmt2 has residual DNA-(cytosine-C5) methyltransferase activity. J Biol Chem 278:31717–31721

Horike S, Mitsuya K, Meguro M, Kotobuki N, Kashiwagi A, Notsu T, Schulz TC, Shirayoshi Y, Oshimura M (2000) Targeted disruption of the human LIT1 locus defines a putative imprinting control element playing an essential role in Beckwith-Wiedemann syndrome. Hum Mol Genet 9:2075–2083

Howell CY, Bestor TH, Ding F, Latham KE, Mertineit C, Trasler JM, Chaillet JR (2001) Genomic imprinting disrupted by a maternal effect mutation in the Dnmt1 gene. Cell 104:829–838

Hu JF, Nguyen PH, Pham NV, Vu TH, Hoffman AR (1997) Modulation of Igf2 genomic imprinting in mice induced by 5-azacytidine, an inhibitor of DNA methylation. Mol Endocrinol 11:1891–1898

Hu JF, Oruganti H, Vu TH, Hoffman AR (1998) The role of histone acetylation in the allelic expression of the imprinted human insulin-like growth factor II gene. Biochem Biophys Res Commun 251:403–408

Hu JF, Pham J, Dey I, Li T, Vu TH, Hoffman AR (2000) Allele-specific histone acetylation accompanies genomic imprinting of the insulin-like growth factor II receptor gene. Endocrinology 141:4428–4435

Jaenisch R (1997) DNA methylation and imprinting: why bother? Trends Genet 13:323–329

Jones PL, Veenstra GJ, Wade PA, Vermaak D, Kass SU, Landsberger N, Strouboulis J, Wolffe AP (1998) Methylated DNA and MeCP2 recruit histone deacetylase to repress transcription. Nat Genet 19:187–191

Jouvenot Y, Poirier F, Jami J, Paldi A (1999) Biallelic transcription of Igf2 and H19 in individual cells suggests a post-transcriptional contribution to genomic imprinting. Curr Biol 9:1199–1202

Kagotani K, Takebayashi S, Kohda A, Taguchi H, Paulsen M, Walter J, Reik W, Okumura K (2002) Replication timing properties within the mouse distal chromosome 7 imprinting cluster. Biosci Biotechnol Biochem 66:1046–1051

Kalscheuer VM, Mariman EC, Schepens MT, Rehder H, Ropers HH (1993) The insulin-like growth factor type-2 receptor gene is imprinted in the mouse but not in humans. Nat Genet 5:74–78

Kanduri C, Holmgren C, Pilartz M, Franklin G, Kanduri M, Liu L, Ginjala V, Ulleras E, Mattsson R, Ohlsson R (2000) The 5′ flank of mouse H19 in an unusual chromatin conformation unidirectionally blocks enhancer–promoter communication. Curr Biol 10:449–457

Kanduri C, Fitzpatrick G, Mukhopadhyay R, Kanduri M, Lobanenkov V, Higgins M, Ohlsson R (2002) A differentially methylated imprinting control region within the Kcnq1 locus harbors a methylation-sensitive chromatin insulator. J Biol Chem 277:18106–18110

Kato Y, Rideout WM III, Hilton K, Barton SC, Tsunoda Y, Surani MA (1999) Developmental potential of mouse primordial germ cells. Development 126:1823–1832

Kawame H, Gartler SM, Hansen RS (1995) Allele-specific replication timing in imprinted domains: absence of asynchrony at several loci. Hum Mol Genet 4:2287–2293

Ke X, Thomas SN, Robinson DO, Collins A (2002) The distinguishing sequence characteristics of mouse imprinted genes. Mamm Genome 13:639–645

Khosla S, Aitchison A, Gregory R, Allen ND, Feil R (1999) Parental allele-specific chromatin configuration in a boundary-imprinting-control element upstream of the mouse H19 gene. Mol Cell Biol 19:2556–2566

Killian JK, Byrd JC, Jirtle JV, Munday BL, Stoskopf MK, MacDonald RG, Jirtle RL (2000) M6P/IGF2R imprinting evolution in mammals. Mol Cell 5:707–716

Killian JK, Nolan CM, Stewart N, Munday BL, Andersen NA, Nicol S, Jirtle RL (2001a) Monotreme IGF2 expression and ancestral origin of genomic imprinting. J Exp Zool 291:205–212

Killian JK, Nolan CM, Wylie AA, Li T, Vu TH, Hoffman AR, Jirtle RL (2001b) Divergent evolution in M6P/IGF2R imprinting from the Jurassic to the Quaternary. Hum Mol Genet 10:1721–1728

Kitsberg D, Selig S, Brandeis M, Simon I, Keshet I, Driscoll DJ, Nicholls RD, Cedar H (1993) Allele-specific replication timing of imprinted gene regions. Nature 364:459–463

Knoll JH, Cheng SD, Lalande M (1994) Allele specificity of DNA replication timing in the Angelman/Prader-Willi syndrome imprinted chromosomal region. Nat Genet 6:41–46

Lalande M (1996) Parental imprinting and human disease. Annu Rev Genet 30:173–195

Larsen F, Gundersen G, Lopez R, Prydz H (1992) CpG islands as gene markers in the human genome. Genomics 13:1095–1107

Ledbetter DH, Engel E (1995) Uniparental disomy in humans: development of an imprinting map and its implications for prenatal diagnosis. Hum Mol Genet 4:1757–1764

Lee J, Inoue K, Ono R, Ogonuki N, Kohda T, Kaneko-Ishino T, Ogura A, Ishino F (2002) Erasing genomic imprinting memory in mouse clone embryos produced from day 11.5 primordial germ cells. Development 129:1807–1817

Lee MP, DeBaun MR, Mitsuya K, Galonek HL, Brandenburg S, Oshimura M, Feinberg AP (1999) Loss of imprinting of a paternally expressed transcript, with antisense orientation to KVLQT1, occurs frequently in Beckwith-Wiedemann syndrome and is independent of insulin-like growth factor II imprinting. Proc Natl Acad Sci USA 96:5203–5208

Lefebvre L, Viville S, Barton SC, Ishino F, Keverne EB, Surani MA (1998) Abnormal maternal behaviour and growth retardation associated with loss of the imprinted gene Mest. Nat Genet 20:163–169

Lei H, Oh SP, Okano M, Juttermann R, Goss KA, Jaenisch R, Li E (1996) De novo DNA cytosine methyltransferase activities in mouse embryonic stem cells. Development 122:3195–3205

Li E, Beard C, Jaenisch R (1993) Role for DNA methylation in genomic imprinting. Nature 366:362–365

Li E, Bestor TH, Jaenisch R (1992) Targeted mutation of the DNA methyltransferase gene results in embryonic lethality. Cell 69:915–926

Li L, Keverne EB, Aparicio SA, Ishino F, Barton SC, Surani MA (1999) Regulation of maternal behavior and offspring growth by paternally expressed Peg3. Science 284:330–333

Liang G, Chan MF, Tomigahara Y, Tsai YC, Gonzales FA, Li E, Laird PW, Jones PA (2002) Cooperativity between DNA methyltransferases in the maintenance methylation of repetitive elements. Mol Cell Biol 22:480–491

Lin SP, Youngson N, Takada S, Seitz H, Reik W, Paulsen M, Cavaille J, Ferguson-Smith AC (2003) Asymmetric regulation of imprinting on the maternal and paternal chromosomes at the Dlk1-Gtl2 imprinted cluster on mouse chromosome 12. Nat Genet 35:97–102

Liu K, Wang YF, Cantemir C, Muller MT (2003) Endogenous assays of DNA methyltransferases: evidence for differential activities of DNMT1, DNMT2, and DNMT3 in mammalian cells in vivo. Mol Cell Biol 23:2709–2719

Lopes S, Lewis A, Hajkova P, Dean W, Oswald J, Forné T, Murrell A, Constância M, Bartolomei M, Walter J, Reik W (2003) Epigenetic modifications in an imprinting cluster are controlled by a hierarchy of DMRs suggesting long-range chromatin interactions. Hum Mol Genet 12:295–305

Lyle R, Watanabe D, te Vruchte D, Lerchner W, Smrzka OW, Wutz A, Schageman J, Hahner L, Davies C, Barlow DP (2000) The imprinted antisense RNA at the Igf2r locus overlaps but does not imprint Mas1. Nat Genet 25:19–21

Lynch CA, Tycko B, Bestor TH, Walsh CP (2002) Reactivation of a silenced H19 gene in human rhabdomyosarcoma by demethylation of DNA but not by histone hyperacetylation. Mol Cancer 1:2

McGrath J, Solter D (1984) Completion of mouse embryogenesis requires both the maternal and paternal genomes. Cell 37:179–183

Mertineit C, Yoder JA, Taketo T, Laird DW, Trasler JM, Bestor TH (1998) Sex-specific exons control DNA methyltransferase in mammalian germ cells. Development 125:889–897

Moore T (1994) Refusing the ovarian time bomb. Trends Genet 10:347–349

Moore T, Haig D (1991) Genomic imprinting in mammalian development: a parental tug-of-war. Trends Genet 7:45–49

Moore T, Constância M, Zubair M, Bailleul B, Feil R, Sasaki H, Reik W (1997) Multiple imprinted sense and antisense transcripts, differential methylation and tandem repeats in a putative imprinting control region upstream of mouse Igf2. Proc Natl Acad Sci USA 94:12509–12514

Murrell A, Heeson S, Bowden L, Constância M, Dean W, Kelsey G, Reik W (2001) An intragenic methylated region in the imprinted Igf2 gene augments transcription. EMBO Rep 2:1101–1106

Nan X, Ng HH, Johnson CA, Laherty CD, Turner BM, Eisenman RN, Bird A (1998) Transcriptional repression by the methyl-CpG-binding protein MeCP2 involves a histone deacetylase complex. Nature 393:386–389

Neumann B, Kubicka P, Barlow DP (1995) Characteristics of imprinted genes. Nat Genet 9:12–13

Ng HH, Zhang Y, Hendrich B, Johnson CA, Turner BM, Erdjument-Bromage H, Tempst P, Reinberg D, Bird A (1999) MBD2 is a transcriptional repressor belonging to the MeCP1 histone deacetylase complex. Nat Genet 23:58–61

Nolan CM, Killian JK, Petitte JN, Jirtle RL (2001) Imprint status of M6P/IGF2R and IGF2 in chickens. Dev Genes Evol 211:179–183

Obata Y, Kono T (2002) Maternal primary imprinting is established at a specific time for each gene throughout oocyte growth. J Biol Chem 277:5285–5289

Obata Y, Kaneko-Ishino T, Koide T, Takai Y, Ueda T, Domeki I, Shiroishi T, Ishino F, Kono T (1998) Disruption of primary imprinting during oocyte growth leads to the modified expression of imprinted genes during embryogenesis. Development 125:1553–1560

Ogawa O, McNoe LA, Eccles MR, Morison IM, Reeve AE (1993) Human insulin-like growth factor type I and type II receptors are not imprinted. Hum Mol Genet 2:2163–2165

Okano M, Bell DW, Haber DA, Li E (1999) DNA methyltransferases Dnmt3a and Dnmt3b are essential for de novo methylation and mammalian development. Cell 99:247–257

Okano M, Xie S, Li E (1998a) Cloning and characterization of a family of novel mammalian DNA (cytosine-5) methyltransferases. Nat Genet 19:219–220

Okano M, Xie S, Li E (1998b) Dnmt2 is not required for de novo and maintenance methylation of viral DNA in embryonic stem cells. Nucleic Acids Res 26:2536–2540

O'Neill MJ, Ingram RS, Vrana PB, Tilghman SM (2000) Allelic expression of IGF2 in marsupials and birds. Dev Genes Evol 210:18–20

Orphanides G, Reinberg D (2002) A unified theory of gene expression. Cell 108:439–451

Paldi A, Gyapay G, Jami J (1995) Imprinted chromosomal regions of the human genome display sex-specific meiotic recombination frequencies. Curr Biol 5:1030–1035

Pant V, Mariano P, Kanduri C, Mattsson A, Lobanenkov V, Heuchel R, Ohlsson R (2003) The nucleotides responsible for the direct physical contact between the chromatin insulator protein CTCF and the H19 imprinting control region manifest parent of origin-specific long-distance insulation and methylation-free domains. Genes Dev 17:586–590

Pardo-Manuel de Villena F, de la Casa-Esperon E, Sapienza C (2000) Natural selection and the function of genome imprinting: beyond the silenced minority. Trends Genet 16:573–579

Paro R (2000) Chromatin regulation. Formatting genetic text. Nature 406:579–580

Paulsen M, El-Maarri O, Engemann S, Strodicke M, Franck O, Davies K, Reinhardt R, Reik W, Walter J (2000) Sequence conservation and variability of imprinting in the Beckwith-Wiedemann syndrome gene cluster in human and mouse. Hum Mol Genet 9:1829–1841

Pearsall RS, Shibata H, Brozowska A, Yoshino K, Okuda K, deJong PJ, Plass C, Chapman VM, Hayashizaki Y, Held WA (1996) Absence of imprinting in U2AFBPL, a human homologue of the imprinted mouse gene U2afbp-rs. Biochem Biophys Res Commun 222:171–177

Pearsall RS, Plass C, Romano MA, Garrick MD, Shibata H, Hayashizaki Y, Held WA (1999) A direct repeat sequence at the Rasgrf1 locus and imprinted expression. Genomics 55:194–201

Pedone PV, Pikaart MJ, Cerrato F, Vernucci M, Ungaro P, Bruni CB, Riccio A (1999) Role of histone acetylation and DNA methylation in the maintenance of the imprinted expression of the H19 and Igf2 genes. FEBS Lett 458:45–50

Perk J, Makedonski K, Lande L, Cedar H, Razin A, Shemer R (2002) The imprinting mechanism of the Prader-Willi/Angelman regional control center. EMBO J 21:5807–5814

Reed MR, Riggs AD, Mann JR (2001) Deletion of a direct repeat element has no effect on Igf2 and H19 imprinting. Mamm Genome 12:873–876

Reik W, Collick A, Norris ML, Barton SC, Surani MA (1987) Genomic imprinting determines methylation of parental alleles in transgenic mice. Nature 328:248–251

Reik W, Dean W, Walter J (2001) Epigenetic reprogramming in mammalian development. Science 293:1089–1093

Rhee I, Bachman KE, Park BH, Jair KW, Yen RW, Schuebel KE, Cui H, Feinberg AP, Lengauer C, Kinzler KW et al. (2002) DNMT1 and DNMT3b cooperate to silence genes in human cancer cells. Nature 416:552–556

Robinson WP, Lalande M (1995) Sex-specific meiotic recombination in the Prader-Willi/Angelman syndrome imprinted region. Hum Mol Genet 4:801–806

Rougeulle C, Heard E (2002) Antisense RNA in imprinting: spreading silence through Air. Trends Genet 18:434–437

Rougeulle C, Glatt H, Lalande M (1997) The Angelman syndrome candidate gene, UBE3A/E6-AP, is imprinted in brain. Nat Genet 17:14–15

Rougeulle C, Cardoso C, Fontes M, Colleaux L, Lalande M (1998) An imprinted antisense RNA overlaps UBE3A and a second maternally expressed transcript. Nat Genet 19:15–16

Rougier N, Bourc'his D, Gomes DM, Niveleau A, Plachot M, Paldi A, Viegas-Pequignot E (1998) Chromosome methylation patterns during mammalian preimplantation development. Genes Dev 12:2108–2113

Runte M, Huttenhofer A, Gross S, Kiefmann M, Horsthemke B, Buiting K (2001) The IC-SNURF-SNRPN transcript serves as a host for multiple small nucleolar RNA species and as an antisense RNA for UBE3A. Hum Mol Genet 10:2687–2700

Saitoh S, Wada T (2000) Parent-of-origin specific histone acetylation and reactivation of a key imprinted gene locus in Prader-Willi syndrome. Am J Hum Genet 66:1958–1962

Saitoh S, Buiting K, Rogan PK, Buxton JL, Driscoll DJ, Arnemann J, Konig R, Malcolm S, Horsthemke B, Nicholls RD (1996) Minimal definition of the imprinting center and fixation of chromosome 15q11-q13 epigenotype by imprinting mutations. Proc Natl Acad Sci USA 93:7811–7815

Sapienza C, Peterson AC, Rossant J, Balling R (1987) Degree of methylation of transgenes is dependent on gamete of origin. Nature 328:251–254

Sasaki H, Jones PA, Chaillet JR, Ferguson-Smith AC, Barton SC, Reik W, Surani MA (1992) Parental imprinting: potentially active chromatin of the repressed maternal allele of the mouse insulin-like growth factor II (Igf2) gene. Genes Dev 6:1843–1856

Schoenherr CJ, Levorse JM, Tilghman SM (2003) CTCF maintains differential methylation at the Igf2/H19 locus. Nat Genet 33:66–69

Schweizer J, Zynger D, Francke U (1999) In vivo nuclease hypersensitivity studies reveal multiple sites of parental origin-dependent differential chromatin conformation in the 150 kb SNRPN transcription unit. Hum Mol Genet 8:555–566

Shemer R, Birger Y, Riggs AD, Razin A (1997) Structure of the imprinted mouse Snrpn gene and establishment of its parental-specific methylation pattern. Proc Natl Acad Sci USA 94:10267–10272

Shemer R, Hershko AY, Perk J, Mostoslavsky R, Tsuberi B, Cedar H, Buiting K, Razin A (2000) The imprinting box of the Prader-Willi/Angelman syndrome domain. Nat Genet 26:440–443

Shibata H, Yoshino K, Sunahara S, Gondo Y, Katsuki M, Ueda T, Kamiya M, Muramatsu M, Murakami Y, Kalcheva I et al. (1996) Inactive allele-specific methylation and chromatin structure of the imprinted gene U2af1-rs1 on mouse chromosome 11. Genomics 35:248–252

Shuster M, Dhar MS, Olins AL, Olins DE, Howell CY, Gollin SM, Chaillet JR (1998) Parental alleles of an imprinted mouse transgene replicate synchronously. Dev Genet 23:275–284

Simon I, Tenzen T, Reubinoff BE, Hillman D, McCarrey JR, Cedar H (1999) Asynchronous replication of imprinted genes is established in the gametes and maintained during development. Nature 401:929–932

Sleutels F, Zwart R, Barlow DP (2002) The non-coding Air RNA is required for silencing autosomal imprinted genes. Nature 415:810–813

Smilinich NJ, Day CD, Fitzpatrick GV, Caldwell GM, Lossie AC, Cooper PR, Smallwood AC, Joyce JA, Schofield PN, Reik W et al. (1999) A maternally methylated CpG island in KvLQT1 is associated with an antisense paternal transcript and loss of imprinting in Beckwith-Wiedemann syndrome. Proc Natl Acad Sci USA 96:8064–8069

Smrzka OW, Fae I, Stoger R, Kurzbauer R, Fischer GF, Henn T, Weith A, Barlow DP (1995) Conservation of a maternal-specific methylation signal at the human IGF2R locus. Hum Mol Genet 4:1945–1952

Solter D (1988) Differential imprinting and expression of maternal and paternal genomes. Annu Rev Genet 22:127–146

Solter D (1994) Refusing the ovarian time bomb. Trends Genet 10:346; author reply 348–349

Srivastava M, Hsieh S, Grinberg A, Williams-Simons L, Huang SP, Pfeifer K (2000) H19 and Igf2 monoallelic expression is regulated in two distinct ways by a shared cis-acting regulatory region upstream of H19. Genes Dev 14:1186–1195

Stadnick MP, Pieracci FM, Cranston MJ, Taksel E, Thorvaldsen JL, Bartolomei MS (1999) Role of a 461-bp G-rich repetitive element in H19 transgene imprinting. Dev Genes Evol 209:239–248

Stoger R, Kubicka P, Liu CG, Kafri T, Razin A, Cedar H, Barlow DP (1993) Maternal-specific methylation of the imprinted mouse Igf2r locus identifies the expressed locus as carrying the imprinting signal. Cell 73:61–71

Surani MA, Barton SC, Norris ML (1984) Development of reconstituted mouse eggs suggests imprinting of the genome during gametogenesis. Nature 308:548–550

Swain JL, Stewart TA, Leder P (1987) Parental legacy determines methylation and expression of an autosomal transgene: a molecular mechanism for parental imprinting. Cell 50:719–727

Szabo PE, Mann JR (1995) Biallelic expression of imprinted genes in the mouse germ line: implications for erasure, establishment, and mechanisms of genomic imprinting. Genes Dev 9:1857–1868

Szabo PE, Pfeifer GP, Mann JR (1998) Characterization of novel parent-specific epigenetic modifications upstream of the imprinted mouse H19 gene. Mol Cell Biol 18:6767–6776

Tada M, Tada T, Lefebvre L, Barton SC, Surani MA (1997) Embryonic germ cells induce epigenetic reprogramming of somatic nucleus in hybrid cells. EMBO J 16:6510–6520

Tanaka M, Puchyr M, Gertsenstein M, Harpal K, Jaenisch R, Rossant J, Nagy A (1999) Parental origin-specific expression of Mash2 is established at the time of implantation with its imprinting mechanism highly resistant to genome-wide demethylation. Mech Dev 87:129–142

Tang LY, Reddy MN, Rasheva V, Lee TL, Lin MJ, Hung MS, Shen CK (2003) The eukaryotic DNMT2 genes encode a new class of cytosine-5 DNA methyltransferases. J Biol Chem 278:33613–33616

Thorvaldsen JL, Mann MR, Nwoko O, Duran KL, Bartolomei MS (2002) Analysis of sequence upstream of the endogenous H19 gene reveals elements both essential and dispensable for imprinting. Mol Cell Biol 22:2450–2462

Tolhuis B, Palstra RJ, Splinter E, Grosveld F, de Laat W (2002) Looping and interaction between hypersensitive sites in the active beta-globin locus. Mol Cell 10:1453–1465

Tremblay KD, Duran KL, Bartolomei MS (1997) A 5' 2-kilobase-pair region of the imprinted mouse H19 gene exhibits exclusive paternal methylation throughout development. Mol Cell Biol 17:4322–4329

Tremblay KD, Saam JR, Ingram RS, Tilghman SM, Bartolomei MS (1995) A paternal-specific methylation imprint marks the alleles of the mouse H19 gene. Nat Genet 9:407–413

Tucker KL, Beard C, Dausmann J, Jackson-Grusby L, Laird PW, Lei H, Li E, Jaenisch R (1996) Germ-line passage is required for establishment of methylation and expression patterns of imprinted but not of nonimprinted genes. Genes Dev 10:1008–1020

Ueda T, Abe K, Miura A, Yuzuriha M, Zubair M, Noguchi M, Niwa K, Kawase Y, Kono T, Matsuda Y et al. (2000) The paternal methylation imprint of the mouse H19 locus is acquired in the gonocyte stage during foetal testis development. Genes Cells 5:649–659

Varmuza S, Mann M (1994) Genomic imprinting – defusing the ovarian time bomb. Trends Genet 10:118–123

Villar AJ, Eddy EM, Pedersen RA (1995) Developmental regulation of genomic imprinting during gametogenesis. Dev Biol 172:264–271

Vu TH, Hoffman AR (1997) Imprinting of the Angelman syndrome gene, UBE3A, is restricted to brain. Nat Genet 17:12–13

Wade PA (2001) Methyl CpG-binding proteins and transcriptional repression. Bioessays 23:1131–1137

Wade PA, Gegonne A, Jones PL, Ballestar E, Aubry F, Wolffe AP (1999) Mi-2 complex couples DNA methylation to chromatin remodelling and histone deacetylation. Nat Genet 23:62–66

Weber M, Milligan L, Delalbre A, Antoine E, Brunel C, Cathala G, Forné T (2001) Extensive tissue-specific variation of allelic methylation in the Igf2 gene during mouse fetal development: relation to expression and imprinting. Mech Dev 101:133–141

Weber M, Hagege H, Murrell A, Brunel C, Reik W, Cathala G, Forne T (2003) Genomic imprinting controls matrix attachment regions in the Igf2 gene. Mol Cell Biol 23:8953–8959

Wilkins JF, Haig D (2003) What good is genomic imprinting: the function of parent-specific gene expression. Nat Rev Genet 4:359–368

Wutz A, Smrzka OW, Schweifer N, Schellander K, Wagner EF, Barlow DP (1997) Imprinted expression of the Igf2r gene depends on an intronic CpG island. Nature 389:745–749

Wutz A, Theussl HC, Dausman J, Jaenisch R, Barlow DP, Wagner EF (2001) Non-imprinted Igf2r expression decreases growth and rescues the Tme mutation in mice. Development 128:1881–1887

Xin Z, Allis CD, Wagstaff J (2001) Parent-specific complementary patterns of histone H3 lysine 9 and H3 lysine 4 methylation at the Prader-Willi syndrome imprinting center. Am J Hum Genet 69:1389–1394

Xu GL, Bestor TH, Bourc'his D, Hsieh CL, Tommerup N, Bugge M, Hulten M, Qu X, Russo JJ, Viegas-Pequignot E (1999) Chromosome instability and immunodeficiency syndrome caused by mutations in a DNA methyltransferase gene. Nature 402:187–191

Xu Y, Goodyer CG, Deal C, Polychronakos C (1993) Functional polymorphism in the parental imprinting of the human IGF2R gene. Biochem Biophys Res Commun 197:747–754

Yang T, Adamson TE, Resnick JL, Leff S, Wevrick R, Francke U, Jenkins NA, Copeland NG, Brannan CI (1998) A mouse model for Prader-Willi syndrome imprinting-centre mutations. Nat Genet 19:25–31

Yatsuki H, Joh K, Higashimoto K, Soejima H, Arai Y, Wang Y, Hatada I, Obata Y, Morisaki H, Zhang Z et al. (2002) Domain regulation of imprinting cluster in Kip2/Lit1 subdomain on mouse chromosome 7F4/F5: large-scale DNA methylation analysis reveals that DMR-Lit1 is a putative imprinting control region. Genome Res 12:1860–1870

Yen RW, Vertino PM, Nelkin BD, Yu JJ, el-Deiry W, Cumaraswamy A, Lennon GG, Trask BJ, Celano P, Baylin SB (1992) Isolation and characterization of the cDNA encoding human DNA methyltransferase. Nucleic Acids Res 20:2287–2291

Yoder JA, Bestor TH (1998) A candidate mammalian DNA methyltransferase related to pmt1p of fission yeast. Hum Mol Genet 7:279–284

Yoon BJ, Herman H, Sikora A, Smith LT, Plass C, Soloway PD (2002) Regulation of DNA methylation of Rasgrf1. Nat Genet 30:92–96

Zhang Y, Ng HH, Erdjument-Bromage H, Tempst P, Bird A, Reinberg D (1999) Analysis of the NuRD subunits reveals a histone deacetylase core complex and a connection with DNA methylation. Genes Dev 13:1924–1935

Zwart R, Sleutels F, Wutz A, Schinkel AH, Barlow DP (2001) Bidirectional action of the Igf2r imprint control element on upstream and downstream imprinted genes. Genes Dev 15:2361–2366

Seed Development and Genomic Imprinting in Plants

Claudia Köhler, Ueli Grossniklaus

Abstract   Genomic imprinting refers to an epigenetic phenomenon where the activity of an allele depends on its parental origin. Imprinting at individual genes has only been described in mammals and seed plants. We will discuss the role imprinted genes play in seed development and compare the situation in plants with that in mammals. Interestingly, many imprinted genes appear to control cell proliferation and growth in both groups of organisms although imprinting in plants may also be involved in the cellular differentiation of the two pairs of gametes involved in double fertilization. DNA methylation plays some role in the control of parent-of-origin-specific expression in both mammals and plants. Thus, although imprinting evolved independently in mammals and plants, there are striking similarities at the phenotypic and possibly also mechanistic level.

1
Introduction

Epigenetic gene regulation refers to mitotically or meiotically heritable changes in gene expression that do not involve changes at the DNA sequence level. Therefore, epigenetic mechanisms allow a flexible regulation of gene expression in addition to hard-wired regulatory networks. Epigenetic regulation of gene expression can occur after environmental, developmental or genetic changes, e.g. after changes in gene dosage, chromosome number or ploidy levels. Genomic imprinting is a specific example of epigenetic programming that occurs prior to fertilization during gametogenesis. Imprinted genes are differentially expressed depending on which parent they are inherited from (reviewed in Baroux et al. 2002a; Ferguson-Smith et al. 2003; Gutierrez-Marcos et al. 2003). Thus, genomic imprinting is a mechanism that

C. Köhler, U. Grossniklaus
Institute of Plant Biology and Zürich–Basel Plant Science Center, University of Zürich, Zollikerstrasse 107, 8008 Zürich, Switzerland, e-mail: grossnik@botinst.unizh.ch

Progress in Molecular and Subcellular Biology
P. Jeanteur (Ed.)
Epigenetics and Chromatin
© Springer-Verlag Berlin Heidelberg 2005

leads to nonequivalent parental genomes, in clear contrast to Mendel's first law of the equivalence of F_1 hybrids. Normal embryonic development in mammals requires the contribution of both paternal and maternal genomes (Barton et al. 1984; McGrath and Solter 1984; Surani et al. 1984): disturbances in imprinting result in developmental aberrations, disease and cancer (Feinberg 1993; Reik and Maher 1997). Genomic imprinting at a specific locus was first described in maize (Kermicle 1970) and plays an important role in the development of flowering plants (angiosperms) (Lin 1982; Birchler and Hart 1987; Grossniklaus et al. 1998). The genetic tractability of plant systems facilitates the elucidation of the regulation of genomic imprinting. Specifically, plants allow the creation of interploidy hybrids, with altered ratios of parental genome contributions and, therefore, altered ratios of imprinted genes. Furthermore, in contrast to animals, genome-wide demethylation is not lethal in plants, allowing questions regarding the role of methylation in the regulation of imprinted genes to be addressed more readily. Genomic imprinting has evolved in mammals and seed plants most likely in response to similar selective pressures that maintain a fine balance between the competing interests of the maternal and paternal genomes in the regulation of embryo size and post-natal or seedling survival (Haig and Westoby 1989; Moore and Reik 1996; Tilghman 1999). In this chapter we will highlight similarities and differences between the mechanisms of genomic imprinting in mammals and plants.

2
Seed Development in Angiosperms

Unlike animals, where gametes differentiate directly from meiotic products, the plant life cycle alternates between a diploid sporophytic (spore-producing) and a haploid gametophytic (gamete-producing) generation (Fig. 1). The diploid sporophyte develops from the fusion product of two haploid gametes. Late in the development of the sporophyte, specific cells in reproductive organs undergo meiosis to form haploid spores, which subsequently form multicellular gametophytes through a series of mitotic divisions. A subset of cells in the gametophyte differentiates to form the gametes. Land plants are divided into four groups: the bryophyte group, including mosses and liverworts, a group consisting of ferns and horsetails, the gymnosperm group and the angiosperm group. Together, gymnosperms and angiosperms are known as seed-forming plants. Plants of all groups cycle between sporophytic and gametophytic generations. However, the relative sizes and nutritional relationships between the two generations vary greatly between the groups. In seed plants, the sporophytic generation is the dominant, free-living generation, whereas the gametophytes are very small and depend on the sporophyte for nutrients (Li and Ma 2002). To date, genomic imprinting has only been demonstrated in the angiosperm group. This could be due to insufficient

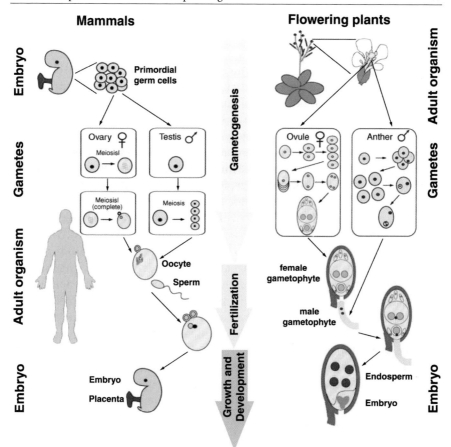

Fig. 1. Reproduction in mammals and flowering plants. Mammals: primordial germ cells are formed early during embryogenesis and migrate later into the embryonic gonads. In female embryos, the oogonia enter meiosis I and primary oocytes stop development at prophase of meiosis I. In adult females, oocytes complete meiosis I, forming the secondary oocyte and first polar body. Development of the secondary oocyte is arrested at metaphase II. In adult males, spermatocytes undergo meiosis and form spermatids that differentiate into sperm. Fertilized eggs (zygotes) complete meiosis II and extrude the second polar body. Maternal and paternal genomes remain in separate pronuclei until the first round of DNA synthesis is complete. The embryo and extra-embryonic membranes (giving rise to parts of the placenta) develop from the same fertilized egg. The placenta regulates the transfer of nutrients from the mother to the developing embryo. Flowering plants: germ cells are formed within flowers of adult plants. Formation of female gametes occurs in the ovule, whereas male gametes are formed in the anther. In the ovule the sporophytic megaspore mother cell undergoes meiosis, giving rise to four megaspores, one of which survives. This functional megaspore divides mitotically to form the female gametophyte that contains the two female gametes, the haploid egg cell and homo-diploid central cell. In the anther, meiosis of a sporophytic microspore mother cell results in four microspores. Each microspore divides mitotically, forming one vegetative and one generative cell. A second mitotic division of the generative cell yields the two sperm cells. The egg cell and the central of the female gametophyte are fertilized by the two sperm cells, giving rise to the diploid embryo and triploid endosperm, respectively. The endosperm is a nourishing tissue that surrounds the embryo and transfers nutrients from the mother plant to the developing seeds

analysis of the non-angiosperm groups or, alternatively, it is possible that imprinting co-evolved in angiosperms with the process of double fertilization.

The angiosperm ovule is the female reproductive organ central to sexual reproduction. It originates as a protrusion from the placental tissues of the carpel. Within the ovule a single cell at the distal end of the primordium differentiates into the megaspore mother cell and undergoes meiosis to form a tetrad of haploid megaspores. Only one of these survives and undergoes several rounds of mitosis to form the multicellular female gametophyte (embryo sac), which contains the gametes. This is in contrast to animals, where the meiotic products differentiate directly into gametes. In the majority of angiosperms, the female gametophyte consists of seven cells: three antipodal cells, two synergid cells, one egg cell, and one central cell that contains two polar nuclei (Drews et al. 1998; Grossniklaus and Schneitz 1998). The male gametophyte (pollen) develops from microspores, the meiotic products produced in the male organs of the flower. At maturity it typically comprises two sperm cells contained within a vegetative cell (Mascarenhas 1989). Seed development starts with the double fertilization of the egg and central cells by the two sperm cells, giving rise to the embryo and the endosperm, respectively. The two sperm nuclei are genetically identical, as are the egg and central cell nucleus in most species. However, the central cell typically contains two doses of maternal genes, generating a triploid endosperm. The developing plant embryo is surrounded by endosperm, which is thought to regulate the transfer of nutrients from the maternal sporophyte to the embryo.

3
Development and Function of the Endosperm

Amongst angiosperms, a great variety of developmental pathways produce endosperms of different size and form (Lopes and Larkins 1993; Baroux et al. 2002b). In some species, the endosperm is the major nutrient storage tissue of the developing seedling and is maintained in the mature seed. This persistent endosperm is found in some monocotyledonous species, e.g. the cereals, but is essentially absent in others, e.g. the orchids. In contrast, many dicotyledonous species have a transient endosperm, which is consumed by the developing embryo, with the mature seed containing only one or a few endosperm cell layers (Maheshwari 1950; Vijayaraghavan and Prabhakar 1984). The endosperm of *Arabidopsis thaliana* belongs to the transient type and endosperm development starts after central cell fertilization with a series of syncytial nuclear divisions. The endosperm forms three distinct domains: the micropylar domain, which surrounds the developing embryo, the central peripheral domain, and the chalazal domain (Boisnard-Lorig et al. 2001). The endosperm starts to cellularize when the embryo has reached the late heart stage of development. The cellularized endosperm cells no longer replicate, thus cellularization marks the end of endosperm proliferation (Brown et al. 1999). The embryo consumes the cellularized endosperm through an ill-defined mechanism.

The formation of an endosperm could be a reason why land angiosperms have been most successful in colonizing the planet. In gymnosperms, the female gametophyte consists of a large number of cells and constitutes the main source of nutrients for the developing embryo. In most angiosperms, this role has been taken over by the endosperm, which develops post-fertilization. What is the evolutionary origin of the endosperm? There are currently two theories (reviewed in Friedman and Williams 2004). One theory suggests that the endosperm evolved from a supernumerary embryo. This theory is based on the finding that in *Ephedra*, a taxon belonging to the *Gnetales*, double fertilization of the binucleate egg cell leads to the formation of two embryos within one ovule. Often only one of the embryos continues to develop and subsequently outcompetes and consumes the other embryo. Thus, it is possible that the endosperm in angiosperms represents a developmental transformation of an embryo into a source of nutrients (Friedman 2001). However, recent phylogenetic evidence suggests that *Gnetales* are more closely related to gymnosperm conifers than to the angiosperms with which they share double fertilization (Friedman 2001). The other theory about the origin of the endosperm argues that the endosperm is homologous to the gymnosperm female gametophyte but the endosperm acquired a sexual function. Transporting nutrients to an endosperm would limit the distribution of food reserves to gametophytes that have been fertilized (Friedman 2001). Both theories suggest that the main function of the endosperm is to provide nutrients to the developing embryo, with the formation of the endosperm coupled to fertilization. Thus, in many ways, the endosperm can be considered the functional equivalent of extra-embryonic tissues and the placenta in mammals, an analogy that will be revisited in the discussion about the evolution of genomic imprinting.

4
A Role for Genomic Imprinting in Seed Development?

Mammals require the participation of both maternal (m) and paternal (p) genomes during embryonic development. Embryos in which all genes are either maternally derived (2 m:0p) or paternally derived (0 m:2p) do not complete development, implying that functions specific to the maternal and paternal genomes are required for normal development (Barton et al. 1984; McGrath and Solter 1984; Surani et al. 1984). Similar requirements for maternal and paternal genomes do not exist in all angiosperm species. Maternal haploid embryos (1 m:0p) and paternal haploid embryos (0 m:1p) can produce rare viable seedlings in some angiosperms (Kimber and Riley 1963; Sarkar and Coe 1966; Kermicle 1969). Apomixis (Nogler 1984), the asexual formation of seeds without paternal contribution, can produce viable maternal embryos (2 m:0p) and has long been known to occur in dandelion (*Taraxacum* spp.), hawkweed (*Hieracium* spp.) and buttercup (*Ranunculus* spp.) (Murbeck 1904; Rutishauser 1954). Apomicitically derived embryos complete development only when surrounded by functional endosperm, which often is the product of sexual reproduction but can also develop autonomously in many species. These findings suggest that there is no absolute requirement for both parental genomes to ensure normal embryo and endosperm development. However, these are rare situations occurring under specific circumstances and it is not clear how much they tell us about normal seed development. In fact, it has become apparent over recent years that maternal effects on seed development are widespread and are due to genomic imprinting as well as other molecular mechanism, such as dosage effects and the activities of maternally stored gene products (Baroux et al. 2002a).

5 The Discovery of Genomic Imprinting in Maize

The concept of genomic imprinting in plants is based on genetic experiments from a few pioneering groups (Kermicle 1970; Johnson et al. 1980; Lin 1982, 1984). Kermicle (1970) first demonstrated gene regulation by genomic imprinting at a specific locus in maize. As mentioned above, the endosperm and embryo are genetic twins, a main difference, however, being the 2 m:1p genome ratio in the triploid endosperm versus a 1 m:1p ratio in the diploid embryo. Kernels of maize varieties have different color patterns that range from yellow to dark purple, with intermediate levels. The purple color is due to the accumulation of the pigment anthocyanin in the outermost cell layer of the endosperm (aleurone), among others controlled by the *r1* (*red*) locus (Brink 1956). The *R:r-std* allele (hereafter referred to as *R*) of the *r1* locus gives rise to solid purple pigmentation when inherited from the mother. The reciprocal cross, however, results in kernels with a mottled pigmentation where most cells do not express the *R* phenotype. These observations imply that anthocyanin accumulation is controlled by a maternal effect, because full pigmentation requires that *R* be transmitted through the mother. In principle, maternal effects can be caused (1) by the action gene products that were stored in the female gametes prior to fertilization but are required post-fertilization (cytoplasmic basis), (2) by uniparental gene expression due to genomic imprinting (chromosomal basis), or (3) by dosage sensitivity due to different copy numbers derived from maternal and paternal gametes in the endosperm.

Kermicle performed a series of elegant genetic experiments to investigate the basis of the maternal effect observed with *R*. To test whether it resulted from a dosage effect due to the two maternal *R* copies present in the endosperm, the paternal copy number of *R* was changed using B chromosomes, which contain a translocated additional copy of the *R* gene. However, pollen with two paternal copies of *R* crossed to a homozygous *r-g* mother plant (*r-g* being a colorless tester allele) resulted in kernels that all had a mottled phenotype. From this, it was concluded that the gametic origin of R is important for the kernel phenotype, not gene dosage. To distinguish between a cytoplasmic and a chromosomal effect, Kermicle used the K10 R strain, which produces mosaic seeds after R loss due to chromosome breakage. In mosaic kernels that received R from both parents, a mottled phenotype was observed after the maternal Rs had been lost, demonstrating unambiguously that the low expression level of paternally inherited R alleles had a chromosomal basis and, thus, was regulated by genomic imprinting.

A priori, it is not clear whether the imprinted state is the 'on' or the 'off' state of R (or a combination of the two). Kermicle (1978) isolated the *trans*-acting locus *mdr1* (*maternal derepression of r1*) and demonstrated that introducing *mdr1* through the female gametophyte suppresses the full color phenotype of maternally inherited R. In the presence of R and *mdr1*, the resultant kernels are mottled instead of fully pigmented as when R alone is maternally inherited. As *mdr1* was shown to be recessive, the proposed function of *MDR1* is to activate expression of the maternal R allele in the central cell. Therefore, the imprinted state of the R allele is the 'on' state; the non-imprinted or default state is the 'off' state. This is in contrast to many imprinted genes in mammals, where the imprinted state of a gene is often the 'off' state. As discussed below, the imprinted state of a locus being the 'on' state is becoming a common theme for imprinted loci in plants, although the number of imprinted genes studied is still small.

6
Studies on Other Potentially Imprinted Genes in Maize

For many years, genomic imprinting in plants was only studied in maize and several other candidate imprinted genes were identified. As outlined above, an unambiguous demonstration of imprinting requires that dosage effects and a cytoplasmic basis are excluded. The latter is hard to achieve even using molecular methods and has rarely been done. A demonstration of differential steady-state levels of mRNA derived from maternal and paternal alleles is not sufficient as such differences may result from long-lived, stored maternal mRNAs. Thus, active transcription after fertilization has to be demonstrated for all potentially imprinted genes that are already expressed in the female gametophyte prior to fertilization. Despite these caveats, several maize genes have been identified that are likely regulated by genomic imprinting. At the molecular level, it was shown that specific alleles of the α-*tubulin* (Lund et al. 1995b) and *zein* storage protein (Lund et al. 1995a) gene families show parent-of-origin-specific DNA methylation patterns that correlate with expression levels in the endosperm. Maternally inherited alleles were found to be demethylated and expressed at high levels during later phases of seed development. These studies provided the first link between DNA methylation and genomic imprinting in plants.

Another interesting case of a potentially imprinted gene relates to the post-transcriptional regulation of the 10-kDa zein protein (encoded by *zps10* on chromosome 9L) by a *trans*-acting mechanism mediated by the *dzr1* locus on chromosome 4S (Benner et al. 1989). *dzr1* has yet to be cloned; there are no loss-of-function *dzr1* mutants known and the work described here relies on natural allelic variants. So far, differential mRNA stability of *zps10* is the only known molecular phenotype controlled by *dzr1*. The 10-kDa zein protein accumulates to high levels in the maize inbred line BSSS53; conversely, it accumulates to low levels in the lines W64A and Mo17 (Chaudhuri and Messing 1994). Reciprocal crosses between BSSS53 and Mo17 lines reveal a maternal effect on *zps10* mRNA accumulation in the F_1 hybrids. High levels of *zps10* mRNA accumulated in the endosperm when *dzr1*+BSSS53 was maternally transmitted, whereas low levels of *zps10* mRNA were detected when maternal *dzr1* originated from the Mo17 background. Importantly, the changes in *zps10* transcript level were independent of the contribution of *dzr1*+BSSS53 from the paternal side (Chaudhuri and Messing 1994). Whereas the negative effect of *dzr1*+Mo17 on *zps10* mRNA accumulation depends on the transmission of *dzr1* through the female gametophyte and not the genetic dose, other *dzr1* alleles (e.g. BSSS53 and W64A) show a typical gene dosage response. Thus, the *dzr1*+Mo17 allele is likely regulated by genomic imprinting, although a cytoplasmic basis of the effect has not formally been excluded.

Importantly, for all these loci only specific alleles are regulated by genomic imprinting whereas others are not (e.g. the BSSS53 allele of *dzr1*). This is an important difference to imprinting in mammals, where allele-specific imprinting is very rare and usually all alleles at an imprinted locus are subject to the same epigenetic regulation (Baroux et al. 2002a). Recently, however, an imprinted locus, *no-apical-meristem (NAM) related protein1* (*nrp1*) has been reported in maize, for which alleles from four inbreds tested were expressed at much higher levels if inherited maternally (Guo et al. 2003). As *nrp1* is not expressed prior to fertilization, it is clearly regulated by genomic imprinting and, to our knowledge, it represents the first maize gene with locus- rather than allele-specific imprinting.

7
Maternal Control of Early Seed Development in Arabidopsis

Vielle-Calzada and coworkers (2000) discovered the exciting phenomenon that early seed development in *Arabidopsis* is under extensive maternal control. The offspring of *Arabidopsis* plants, containing the reporter gene *β-glucuronidase* (*GUS*), crossed with wild-type plants showed differential GUS expression depending on which parent the reporter gene was inherited from. GUS expression was detected early in embryo and/or endosperm development only if the reporter gene was inherited from the mother, whereas expression of the paternally contributed GUS was not detected early in seed development. For a few genes that were also active at later stages of development, paternally derived activity became detectable at the mid-globular stage. This phenomenon was consistently observed in different transgenic lines, carrying the *GUS* reporter gene in different genomic locations. Analyses of endogenous gene expression corroborated the data obtained with the GUS transgenes: the paternal transcripts were detected only after mid-globular stage. The differential accumulation of maternal and paternal transcripts in the fertilization products strongly suggests a parent-of-origin control of early gene expression after fertilization. There are three possible explanations for this phenomenon: (1) the detected transcripts are maternally stored and there is no active zygotic transcription until the mid-globular stage, (2) the genes are regulated by imprinting, with the maternal allele being actively transcribed after fertilization and the paternal allele being silent during early embryogenesis, or (3) a combination of the two. It is likely that all three mechanisms play a role in the maternal control of early embryogenesis. As many of the tested genes are already expressed prior to fertilization, the different possibilities cannot easily be distinguished. However, clear evidence that at least some of these genes are regulated by genomic imprinting comes from (1) the maternal-specific expression of genes that are not expressed prior to fertilization, (2) the maternal-specific expression in specific cells of the young embryo, and (3) the maternal-specific expression of genes that show a strong increase in activity after fertilization.

The genes investigated in the study of Vielle-Calzada and colleagues (2000) encode a broad range of proteins with diverse cellular functions, which would suggest that the lack of paternal genome activity is a general phenomenon affecting more than just a certain subset of genes. Of course, there are likely to be genes that are not regulated through this general mechanism, similar to genes escaping the X inactivation mechanism in mammals. Most likely these genes play essential roles during early embryo and/or endosperm development and could possibly represent genes that are specifically expressed from the paternal side. Indeed, some genes that are expressed very early when paternally inherited have been reported (Weijers et al. 2001). However, these also show differences in expression levels depending on parental origin. There seems to be considerable variability from gene to gene and from seed to seed with respect to maternal and paternal expression levels. Consistent with this, Baroux and coworkers (2002), who used a *trans*-activation system to express barnase in developing seeds, reported that the majority of embryos showed defects only at the globular stage, but a minority arrested as early as the two- or four-cell stage. This suggests that depending on the particular locus or embryo, a basal activity may be detectable at early stages but that there is a dramatic increase in activity from the paternal genome at the mid-globular stage. These observations have been confirmed in several laboratories for *Arabidopsis* (e.g. Sorensen et al. 2001; Golden et al. 2002) but also for other species. An early report using a viral promoter driving a reporter gene suggested that in maize the parental genome becomes active just a few hours after in vitro fertilization (Scholten et al. 2002). Recently, however, two maize genes have been described, which are active when inherited maternally, but, when inherited paternally, they are silent at 5 and 10 days after pollination, respectively (Danilevskaya et al. 2003; Gutierrez-Marcos et al. 2004). Using microarray analysis, an absence of paternal activity could indeed be demonstrated for a very large number of maize genes (D. Grimanelli, pers. comm.). This suggests that, as in many animals, early seed development in the angiosperms is largely under maternal control and that maternal effects are widespread. To what extent those effects are based on genomic imprinting and to what extent on cytoplasmic effects remains an important open question.

8
Intragenomic Parental Conflict and the Evolution of Genomic Imprinting

Haig and Westoby (1989, 1991) proposed a theory suggesting that genomic imprinting evolved as a consequence of an intragenomic conflict over the allocation of nutrients from mother to offspring. Most of the seed's nutrients are directly transferred from the mother plant, and larger seeds with extra food reserves produce more vigorous seedlings. However, the production of larger food reserves causes greater metabolic costs for the mother, reducing the resources available for future offspring. The pollen parent directly benefits from an increase in seed size (with more vigorous offspring) without experiencing any direct cost. Thus, the intragenomic conflict theory proposes that genes controlling growth and vigor of the seed will be subject to selective forces that bring genes promoting growth under paternal control, whereas genes that restrict growth will come under maternal control. This theory is consistent with many phenotypes that result from the disruption of imprinted genes in animals, where they are required for fetal growth and placental development (Moore and Haig 1991; Tilghman 1999; Arney et al. 2001), but none of the imprinted maize genes described above shows any effect on seed development (Messing and Grossniklaus 1999). Nevertheless, some parent-of-origin effects on the endosperm, which can be considered the functional equivalent of the mammalian placenta because it mediates the transfer of nutrients to the developing embryo, may be related to genomic imprinting (Haig and Westoby 1991; Messing and Grossniklaus 1999; Alleman and Doctor 2000).

Increasing the dosage of paternally derived genes promotes the growth of both placenta and endosperm, whereas increasing the dosage of maternally derived genes has the opposite effect (Haig and Westoby 1991; Moore and Haig 1991; Surani 1998; Arney et al. 2001). Normal endosperm development in several species depends on a 2 m:1p genome ratio (Lin 1984; Haig and Westoby 1991). In some *Arabidopsis* accessions, interploidy crosses between diploids and tetraploids result in viable triploid embryos. However, a cross of a diploid mother with tetraploid pollen (2n × 4n) results in enlarged seeds with accelerated mitosis and delayed cellularization of the 2 m:2p endosperm. In contrast, a 4n × 2n cross results in small seeds, with reduced endosperm mitosis and precocious cellularization of the 4 m:1p endosperm (Scott et al. 1998). These results are consistent with the predictions made from the parental conflict theory. However, there are many other explanations, such as dosage effects or interactions between the cytoplasm of the gametes with the nuclear genome of the resulting zygote, that are also consistent with these observations (Birchler 1993; von Wangenheim and Peterson 2004).

9
Imprinting of the MEDEA Locus in Arabidopsis

The *medea* (*mea*) mutant has a parent-of-origin effect, and was isolated in a genetic screen for female gametophytic mutants (Grossniklaus et al. 1998). The maternal inheritance of a mutant *mea* (*mea^m*) allele results in seed abortion, regardless of the paternal contribution. In a heterozygous *mea* mutant plant, half of the female gametophytes will carry a mutant *mea^m* allele and the subsequent seeds will abort post-fertilization. *MEA* has a drastic effect on cell proliferation. First, embryos derived from *mea* mutant gametophytes develop slower than the wild type but reach a much larger size than a corresponding wild-type embryo at the same developmental stage (Grossniklaus et al. 1998). In *mea* mutants, endosperm proliferation is also abnormal. Second, in *mea* mutant gametophytes endosperm can form in the absence of fertilization (Grossniklaus and Vielle-Calzada 1998; Kiyosue et al. 1999). The *MEA* locus was shown to be regulated by genomic imprinting (Vielle-Calzada et al. 1999) and, interestingly, the phenotypes observed are consistent with the expectations of the intragenomic parental conflict theory: *MEA* is a maternally expressed gene restricting growth and a mutation in *MEA* leads to the formation of delayed development and larger embryos as expected. However, the theory was developed on the premise that a female would produce offspring from different fathers. *Arabidopsis*, however, is an inbreeding plant that reproduces almost exclusively by self-fertilization. Both parents are united in one plant, which removes any parental conflict, eliminating the selective pressures thought to drive the evolution of uniparentally expressed genes. This apparent contradiction can be resolved by assuming that *Arabidopsis* has retained the imprinting system present in its out-crossing progenitors. *A. thaliana* and its nearest out-crossing relative *A. lyrata* diverged around 5 million years ago (Koch et al. 2000). Due to the generations of inbreeding, the ancestral imprinting system may have partly broken down, e.g. tolerating changes in maternal or paternal contributions resulting from interploidy crosses. Alternatively, the rate of out-crossing in the wild may be sufficient to maintain the presumed selective pressures that lead to the evolution of genomic imprinting.

The *mea* mutation is caused by the disruption of a gene encoding a *Polycomb* group (PcG) protein, similar to *Drosophila* Enhancer of zeste [E(z); Grossniklaus et al. 1998]. In *Drosophila*, the proposed function of PcG proteins is the stable repression of homeotic genes through many rounds of cell divisions, most likely by modulating the chromatin structure of their target loci (Francis and Kingston 2001). Interploidy crosses ruled out the hypothesis that the maternal phenotype was due to haplo-insufficiency of *MEA* in the endosperm. Expression analysis showed that *MEA* is expressed in the female gametophyte before fertilization and in both the embryo and endosperm after fertilization (Grossniklaus et al. 1998). Analysis of nascent *MEA* transcripts in endosperm nuclei showed that only two out of three *MEA* copies were expressed, suggesting that the paternal *MEA* copy is silent. Furthermore, allele-specific RT-PCR analysis confirmed that only the maternal copies are expressed in developing seeds (Vielle-Calzada et al. 1999). A mechanism that can explain allele-specific expression of *MEA* was recently proposed (Choi et al. 2002; Xiao et al. 2003). Choi and colleagues (2002) have shown that a putative DNA-glycosylase, DEMETER (DME), plays a crucial role in activating *MEA* expression in the female gametophyte prior to fertilization. *dme* mutants have a similar seed abortion phenotype as *mea* mutants and also show a parent-of-origin effect. *DME* was reported to be expressed in the central cell of the female gametophyte, and thus only the maternal allele of *MEA*, not the paternal allele, can be activated by DME. In plants, DNA glycosylases have been shown to be involved in DNA repair processes (Garcia-Ortiz et al. 2001). However, DME is much larger than the typical DNA-glycosylases used in DNA repair and, together with its role in regulating *MEA* expression, it seems unlikely that *DME* is solely involved in DNA repair. Instead, it is possible that DME's biochemical function is to excise 5-methylcytosine from the genome, as has been shown for related DNA glycosylases (Jost et al. 2001). However, there are conflicting reports as to whether the *MEA* promoter contains regions with cytosine methylation (Choi et al. 2002; Xiao et al. 2003) and the exact function of *DME* is as yet unclear. However, it is possible that activation of *MEA* requires a putative demethylation activity of *DME*. The activation of the *R* locus by *MDR1* in maize, which was described earlier, shows striking similarities to the activation of *MEA* by DME and one can speculate that *MDR1* encodes a protein with similar functions to DME.

Xiao and coworkers (2003) identified the *METHYLTRANSFERASE1* (*MET1*) gene as being a potential antagonist of *DME*. By screening for suppressors of the *dme* mutant phenotype, several alleles of *met1* were identified (Xiao et al. 2003). The suppressor effect of *met1* on the *dme* phenotype was reported to be dependent on the presence of a wild-type *MEA* maternal allele, suggesting that *met1* mutations act upstream of *MEA* to suppress the *dme* effect on seed viability. In the presence of mutations in *met1*, *MEA* transcription in the central cell was restored in a *dme* mutant. Therefore, the simplest hypothesis is that transcriptional activity of the maternal *MEA* allele is regulated via a methylation–demethylation-dependent mechanism that requires the activity of MET1 and DME. However, as this hypothesis is based on genetic data only, less direct models involving intermediate steps have not been ruled out.

10
Function of MEDEA During Gametophyte and Seed Development

Even though a regulatory mechanism for *MEA* expression in the central cell has been proposed (Choi et al. 2002; Xiao et al. 2003), there are still several open questions concerning *MEA* expression and function during embryogenesis. *MEA* expression was detected in the embryo after fertilization (Vielle-Calzada et al. 1999), but no paternal transcripts were detected in either early or late developing seeds using sensitive, allele-specific quantitative PCR methods (D. Page and U. Grossniklaus, unpubl.). In contrast, Kinoshita and coworkers (1999) detected paternal *MEA* transcripts in dissected late-torpedo-stage embryos by regular RT-PCR. Furthermore, expression studies using a reporter gene showed paternal *MEA* expression in the embryo in some cases (Luo et al. 2000). Wether thesse contrasting findings are due to technical differences or caused by second-site modifiers, which are prevalent among *Arabidopsis* accessions (C. Spillane and U. Grossniklaus, unpubl.), remains an open question. The regulation of imprinting in the embryo has gotten little attention until now.

Another open question is the function of *MEA* before and after fertilization. The *mea* mutant is in a class of mutants that have the ability to form seed-like structures in the absence of fertilization [*fertilization independent seed* (*fis*) mutants]. In the absence of fertilization, the central cell nucleus starts to replicate and divide, forming an endosperm without embryo development. Endosperm development is accompanied by the proliferation of maternal tissues: the seed coat and the silique wall. Members of the *fis* mutant class have been identified in independent mutant screens and reverse genetic approaches. Four genes have been described thus far: *MEDEA/FIS1* (Grossniklaus et al. 1998; Kiyosue et al. 1999; Luo et al. 1999), *FERTILIZATION INDEPENDENT ENDOSPERM* (*FIE*; Ohad et al. 1999), *FIS2* (Luo et al. 1999) and *MSI1* (Köhler et al. 2003a). Mutants in all four genes have a similar maternal-effect phenotype that leads to seed abortion, but, with the exception of *MEA*, the nature of this maternal effect has yet to be clearly determined. All four genes encode PcG proteins that most likely act together in a protein complex (Köhler et al. 2003a). The MEA protein contains a SET domain, found initially in three *Drosophila* proteins: Suppressor of variegation 3–9 [Su(var)3–9], E(z) and Trithorax (Trx) (Jones and Gelbart 1993; Tschiersch et al. 1994). The SET domain for these three proteins confers histone methyltransferase activity, methylating different lysine residues of histone H3. Specifically, homologues of the *Drosophila* Su(var)3–9 from yeast, plants and mammals have been shown to methylate lysine 9 of H3 (H3K9) (Rea et al. 2000; Jackson et al. 2002; Peters et al. 2002), whereas proteins homologous to E(z) methylate H3K9 as well as lysine 27 (H3K27) (Czermin et al. 2002; Kuzmichev et al. 2002; Müller et al. 2002). Both methylation marks are associated with transcriptionally inactive chromatin, in contrast to methylation marks applied by members of the *Trx* family at lysine 4 (H3K4) (Briggs et al. 2001; Milne et al. 2002; Roguev et al. 2001) that are associated at transcriptionally active loci (Kouzarides 2002). The presence of a SET domain within the MEA protein, as well as the similarity of the MEA–FIE complex with the E(z)–Esc complex from *Drosophila*, suggests that the MEA–FIE complex also has histone methyltransferase activity (Köhler and Grossniklaus 2002).

The ability of the *fis* mutants to undergo fertilization-independent endosperm development indicates that the MEA–FIE complex represses genes that promote endosperm development. Two days after fertilization, expression of *MEA* is reduced (D. Page and U. Grossniklaus, unpubl.), implying that the major function of the MEA–FIE complex is performed before and shortly after fertilization. A direct target gene of the MEA–FIE complex is the type I MADS-box transcription factor *PHERES1* (*PHE1*). *PHE1* expression is induced during autonomous endosperm development in *fis* mutants, supporting the hypothesis that the MEA–FIE complex is necessary for gene repression before fertilization. In wild-type seeds, *PHE1* expression is transiently induced after fertilization, which is in contrast to *fis* mutant seeds, where *PHE1* expression is stronger and remains high until the seeds abort (Köhler et al. 2003b). It remains to be seen whether the observed deregulation of *PHE1* after fertilization is due to epigenetic changes occurring previously in the gametophyte. Thus, it is possible that *PHE1* is transcriptionally repressed in the female gametophyte by epigenetic marks placed by the MEA–FIE complex. The onset of nuclear divisions after fertilization could cause a dilution of these epigenetic marks and subsequently activate *PHE1* transcription. The lack of these epigenetic marks in the *fis* mutant gametophytes would cause precocious and stronger expression of *PHE1* after fertilization. If the function of the MEA–FIE complex is to repress paternal genes that promote endosperm development, one could hypothesize that *PHE1* is paternally expressed and the MEA–FIE complex actively represses its expression.

It has been shown that the requirement of MEA function can be bypassed by a mutation in the gene *DECREASED DNA METHYLATION1* (*DDM1*) (Vielle-Calzada et al. 1999). *DDM1* encodes a SWI2/SNF2-related protein (Jeddeloh et al. 1999) with chromatin remodeling activity (Brzeski and Jerzmanowski 2003). Lack of *DDM1* function causes a 70 % reduction in genomic cytosine methylation, most likely due to a reduced efficiency of cytosine methylation after replication. *mea/mea, ddm1/ddm1* double mutants show a reduced seed abortion ratio (37 % compared to 100 % in *mea/mea* mutants; Köhler et al. 2003b), which correlates with reduced *PHE1* expression levels, suggesting that part of the effect of *ddm1* could be due to changes in the methylation profile of MEA's target gene *PHE1* (Köhler et al. 2003b). Viable *mea/mea* mutant seeds can also be obtained after pollination with hypomethylated pollen derived from plants expressing an antisense copy of *MET1* (*MET1as*) in a *mea* mutant background (Luo et al. 2000). The changed expression of paternal genes in hypomethylated pollen appears to balance the post-fertilization effect caused by the *mea* mutation. Therefore, in addition to the function of *MET1* in methylating and in silencing the *MEA* locus in the female gametophyte, *MET1* likely plays a role in the formation of the male gametes. It was shown that *MET1as* does not affect the expression of the paternal *MEA* allele (Luo et al. 2000); thus this effect is likely acting via other genes that may act downstream of *MEA*.

11
Imprinting of the FWA Locus in the Female Gametophyte

The *FWA* gene was originally identified from late-flowering mutants that show ectopic *FWA* expression. The *fwa-1* mutant does not have a nucleotide change in the *FWA* gene, but rather the ectopic expression is associated with a heritable loss of DNA methylation (Soppe et al. 2000). Recent investigations by Kinoshita et al. (2004) suggest a mechanism that explains the stability of such epigenetic marks in plants. *FWA* is expressed only in the central cell of the female gametophyte as well as in the developing endosperm, but not in either the egg cell or the embryo. This endosperm-specific expression is associated with dramatically reduced DNA methylation of 5′ direct repeats in the *FWA* promoter in the endosperm. This region is highly methylated in sporophytic tissues, pollen and embryos. After fertilization, only the maternal allele of *FWA* is expressed in the endosperm. Paternal expression can be detected when the male parent is deficient in *MET1* activity. Interestingly, mutations in the DNA methylase CHROMOMETHYLASE3 (CMT3) and the de novo DOMAINS REARRANGED METHYLTRANSFERASE1,2 did not affect the silencing of the paternal *FWA* allele. These results suggest that parent-of-origin-specific *FWA* expression is controlled specifically by MET1, the maintenance methyltransferase. Additionally, the activation of *FWA* expression in the central cell depends on DME, analogous to the mechanism of *MEA* activation by *DME* in the central cell (Choi et al. 2002; Kinoshita et al. 2004). Thus, in contrast to mammals, the maternal-specific expression of *FWA* is established by a maternal-specific activation of *FWA*, and possibly not by a paternal-specific de novo methylation. Therefore, the default state of expression for maternal alleles of *FWA*, as for *MEA*, is the silent state, which is overcome by maternal *DME* activity. Whether this is also true for the paternal *MEA* allele, or whether specific activities are required to keep it silent, is currently unknown.

12
The Role of Imprinting During Gametophyte and Seed Development

The *R* locus as well as *MEA* and *FWA* are imprinted genes where the default (non-imprinted) state of the maternal allele is the 'off' state, while the imprinted state is the 'on' state. It is possible, however, that the genetic regulation of the paternal and maternal allele differs and that both epigenetic states have to be actively established. As the chromosomes inherited by the endosperm are not transmitted to the progeny, *DME*- and *MET1*-mediated epigenetic modifications do not need to be reset for the next generation. This is fundamentally different to the situation in mammals, where epigenetic modifications of imprinted genes are reset during reproduction. For imprinting in the embryo, however, a resetting of imprints is expected to occur. There are two main differences during plant and animal gametogenesis. First, there is no predetermined germ line in plants. Cells that will differentiate into the male and female gametophytes are derived from the reproductive meristem, a pluripotent group of stem cells. The second difference is that the gametes are formed through mitotic divisions of meiotic products in plants, and are located within the male and female gametophytes. The female gametophyte contains the two female gametes, the egg cell and the central cell. The male gametophyte consists of a vegetative cell containing the two male gametes.

If the imprint is established during gametophyte development, there are four possibilities at which stage this could occur: pre-meiotic, post-meiotic, pre-mitotic or post-mitotic. If the parental imprint occurs after mitosis in only one of the daughter cells, this would lead to two non-equivalent female gametes (egg and central cell) or two non-equivalent sperm cells (Messing and Grossniklaus 1999). It is known that during pollen development, methylation differences are established between the vegetative and generative cell, which gives rise to the sperm cells (Oakeley et al. 1997). These global methylation changes could be associated with an erasure of existing imprinting marks and reprogramming of the genome, similar to the situation in animals. However, both sperms are functionally equivalent and are thought to fertilize egg and central cell randomly in most plant species (Dumas and Mogensen 1993; Russell 1993). Thus, it is unlikely that epigenetic differences lead to non-equivalent sperm cells. Therefore, epigenetic differences between cells are most likely established only in the female gametophyte prior to fertilization, e.g. by the action of a putative DNA-glycosylase like *DME*. Currently, there are no published investigations about the timing of methylation changes during female gametogenesis, which could expand our understanding of the role of imprinting during plant reproduction. However, clearly epigenetic states introduced during female gametophyte development have the potential to control differentiation processes that could control the different fate of, e.g., the egg and central cell (Messing and Grossniklaus 1999).

13
Imprinting and Apomixis

Apomixis is defined as asexual reproduction through seed; consequently, the progeny of apomictic plants are genetic clones of the mother. Introducing apomixis into a desirable genetic background would allow its propagation while keeping its genetic composition intact, giving apomixis a great agronomical potential (Spillane et al. 2004). Apomictic plants have successfully overcome the fertilization-dependent barrier of embryo development. However, many apomicts require fertilization of the central cell, forming the triploid endosperm, implying that the 2 m:1p ratio is crucial for successful endosperm development in these species (Koltunow and Grossniklaus 2003). Fertilization of an unreduced polar nucleus by a normally reduced sperm would result in an 'unbalanced' 4 m:1p genome ratio that is often lethal. In natural apomicts, two different strategies have been devised to prevent unbalanced genome ratios causing seed abortion: (1) alterations in either gametophyte development or the fertilization process, and (2) changes in the sensitivity towards unbalanced ratios of maternal to paternal genome ratios (Savidan 2000; Grossniklaus et al. 1998; Grossniklaus et al. 2001). Some species employ the first strategy: gametophyte development is altered such that the embryo sack contains four instead of eight cells, with a single, unreduced polar nucleus, and after fertilization the 2 m:1p ratio is restored (Nogler 1984). *Ranunculus auricomus* uses an alternative mechanism, where the double fertilization process is modified, with both reduced sperm nuclei fusing with the unreduced polar nuclei, restoring a 2 m:1p genome ratio (Rutishauser 1954). The second strategy is used in *Tripsacum dactyloides* and *Paspalum* spp., where a single reduced sperm fertilizes the unreduced polar nucleus, giving rise to unbalanced genome ratios with no negative impacts on endosperm development (Grimanelli et al. 1997; Quarin 1999). Furthermore, in many apomictic species of the Asteraceae, autonomous endosperm development occurs in conjunction with autonomous embryo development, resulting in a 2 m:0p genome ratio in the endosperm. How natural apomicts manage to bypass the genome dosage sensitivity in the endosperm is a fascinating, but unresolved, question.

Another experimental approach to investigating apomixis utilizes the model plant *Arabidopsis* (Grossniklaus et al. 2001). In wild-type *Arabidopsis* plants, autonomous embryo and endosperm development does not occur. In mutants of the *fis* class, seed-like structures develop in the absence of fertilization (Chaudhury et al. 1997). However, embryo development does usually not occur, although a few embryo-like structures have been reported (Chaudhury et al. 1997; Köhler et al. 2003a), and endosperm development is abnormal, as it fails to cellularize. This aberration can be overcome when, e.g., the autonomous endosperm mutant *fie* is combined with a hypomethylated genome, allowing completion of endosperm development (Vinkenoog et al. 2000). This shows that the requirement of a paternal genome for successful endosperm development can be bypassed under certain conditions even in *Arabidopsis*.

14
Possible Epigenetic Marks Distinguishing Maternal and Paternal Alleles

14.1
Chromatin Structure

In several plant species, transgene silencing is connected with modifications in DNA methylation and chromatin structure. Silencing of the *HYGROMYCIN PHOSPHOTRANSFERASE* transgene locus in *Arabidopsis* is correlated with both increased methylation and changes in chromatin structure demonstrated by an increased resistance to DNAseI and micrococcal nuclease digestion (Assaad and Signer 1992; Ye and Signer 1996). Heterochromatin is defined as chromatin that is inaccessible to DNA binding factors (and the aforementioned nucleases) and transcriptionally silent, which distinguishes it from the more accessible and transcriptionally active euchromatin (Grewal and Moazed 2003). The heterochromatic state is stably inherited through many cell divisions and, not surprisingly, plays a central role in regulating gene expression during development and cellular differentiation in many systems. Like the previously discussed DNA methylation, heterochromatinization is a potential mechanism to render parental genomes functionally distinct. Histones and their subsequent post-translational modifications play a pivotal role in the assembly of heterochromatin. DNA and histones are assembled together to form the nucleosome. The DNA is wrapped approximately two turns around a core octamer composed of two subunits from each of the following four histones: H2A, H2B, H3 and H4. Consistent with a role for chromatin structure in imprinting, different histone variants are specifically expressed during pollen development. Atypical isoforms of the core histones H2A, H2B and H3 accumulate in condensing nuclei of generative cells in tobacco (Xu et al. 1999) and the lily *Lilium longiflorum* (Ueda and Tanaka 1995). One possible role for these particular isoforms is in chromatin condensation of the sperm nuclei, similar to the role that protamine proteins play in mammals (Ueda et al. 2000).

In addition to the increased complexity generated by the use of different histone isoforms, the aminotermini of histones can contain different post-translational modifications that influence the chromatin condensation state. The two most prominent modifications are methylation and acetylation at lysine residues of amino termini of H3 and H4. Histone acetylation correlates with transcriptional activity (euchromatin), whereas decreased acetylation correlates with a transcriptionally repressed state (heterochromatin). In addition to hypoacetylation, heterochromatin is also characterized by methylation of lysine 9 of histone H3 (H3K9). As mentioned above, the SET domain protein Su(var)3–9 and its homologues methylate this residue. KRYPTONITE (KYP) is one homologue of Su(var)3–9 in *Arabidopsis* and has been shown to methylate H3K9 residues at specific target loci (Jackson et al. 2002). Interestingly, *kyp* mutants have a loss of methylation at CpNpG trinucleotides, in addition to the loss of H3K9 methylation marks. Methylation marks at CpNpG trinucleotides are also lost in the DNA methyltransferase *chromomethylase3* (*cmt3*) mutant. The DNA hypomethylation in *kyp* mutants is therefore most likely a consequence of the failure to recruit CMT3 to methylated H3K9 residues. The LIKE HETEROCHROMATIN PROTEIN1 (LHP1) homologue may mediate the recruitment of CMT3 to methylated H3K9 sites, as LHP1 has the ability to bind to methylated H3K9 residues and interacts with CMT3 in vitro (Jackson et al. 2002). However, recent results suggest that the relationship between histone methylation and DNA methylation is a complex one, involving regulatory feedback loops (Tariq and Paszkowksi 2004).

14.2
DNA Methylation During Gametogenesis

Two key events in the formation of gametes in mammals are the initial erasure of epigenetic marks in primordial germ cells and the establishment of new epigenetic marks in the germline (Surani 1998). Even though it is not completely clear whether DNA methylation patterns alone establish the imprints, genome-wide changes in methylation occur before and after fertilization in mammals.

Global changes in DNA methylation have also been observed in tobacco pollen sperm cells during maturation (Oakeley et al. 1997). Recent investigations of the *met1* mutant of *Arabidopsis* have provided compelling evidence that critical methylation changes occur during male and female gametogenesis. Saze and colleagues (2003) showed that an absence of *MET1* activity caused dramatic epigenetic diversification of the gametes. This diversity seems to be a consequence of passive postmeiotic demethylation, which led to gametes with hemi-methylated and fully demethylated DNA. These hemi-methylated sequences become fully methylated again in the zygote once *MET1* is provided. The effects of *MET1* depletion were observed in both male as well as female gametes. It is intriguing that the reduction of the gametophytic phase during plant evolution has not gone beyond two postmeiotic divisions, the minimum number of divisions required to obtain a fully demethylated genome in the absence of MET1 activity. Saze and co-workers (2003) conclude that DNA methylation provides essential information for subsequent chromatin modifications. The study of Johnson and colleagues (2002) provides an alternative explanation for these observations, as they failed to observe a direct correlation between the loss of DNA methylation and the loss of histone H3K9 methylation marks. However, the loss of H3K9 methylation marks directly correlated with the level of expression of the target gene. Johnson et al. (2002) hypothesized that the loss of DNA methylation caused the transcriptional activation of normally silenced loci, which in turn lead to a loss of histone methylation due to the transcription-coupled incorporation of the histone variant H3.3 (Ahmad and Henikoff 2002). However, the study of Saze and co-workers (2003) provides clear evidence that DNA methylation is involved during gametogenesis in plants, although the relative importance for gametogenesis of DNA methylation compared to chromatin structure is still unknown.

14.3
DNA Methylation During Seed Development

The work of Adams and associates (2000) investigated the role of DNA methylation patterns in parent-of-origin effects during seed development. Reciprocal crosses with the previously described *MET1as* lines produced endosperm with phenotypes similar to those obtained from interploidy crosses. Cytosine methylation is reduced by as much as 85 % in the *MET1as* lines (Finnegan et al. 1996). Crossing *MET1as* as a pollen donor to pollinate wild-type plants resulted in seeds that were small, containing fewer peripheral endosperm nuclei and a small chalazal endosperm. Cellularization in the affected seeds occurs earlier than in wild-type endosperm. This phenotype closely resembles a seed phenotype obtained from a 4n × 2n cross, giving rise to the hypothesis that in the hypomethylated pollen, maternal genes become activated that are normally silent. In contrast, a hypomethylated mother plant pollinated with wild-type pollen produces seeds that have a higher seed weight, due to an overproliferation of endosperm nuclei, an overgrowth of the chalazal endosperm and a delay in cellularization. In this example, genes normally expressed in the paternal gametes are postulated to be activated due to a hypomethylated maternal genome, phenocopying the seeds derived from a 2n × 4n cross (Adams et al. 2000). The phenotypes suggest that hypomethylation removes imprint marks, giving rise to either a maternalized paternal genome (hypomethylated pollen) or a paternalized maternal genome (hypomethylated mother plant). While this hypothesis conveniently explains the phenotypic similarities of seeds obtained after reciprocal crosses with either hypomethylated parents or parents with a different ploidy, it is important to remember that the global demethylation in *MET1as* lines is also accompanied by a redistribution of methylation marks that can lead to hypermethylation at some loci, as has been demonstrated for the *SUPERMAN* and *AGAMOUS* loci (Jacobsen et al. 2000). Therefore, it is impossible to draw a direct correlation between the observed effects and the hypomethylation of the genome.

15
Conclusions

To date, genomic imprinting at individual genes as discussed in this chapter has only been described in seed plants and mammals. Due to their long evolutionary separation, genomic imprinting must have evolved independently in these two groups of organisms. Nevertheless, many aspects of imprinting show striking similarities between plants and animals. On the one hand, they share a 'placental habit' and similar selective pressures may have led to the evolution of imprinting with similar phenotypic effects. These are consistent with the intragenomic parental conflict hypothesis for the evolution of imprinting but can also be explained by other hypotheses. On the other hand, molecular mechanisms involving, e.g., DNA methylation seem to have been used in both animals and plants for the regulation of imprinted genes. Future work will show how far these parallels will go.

References

Adams S, Vinkenoog R, Spielman M, Dickinson HG, Scott RJ (2000) Parent-of-origin effects on seed development in *Arabidopsis thaliana* require DNA methylation. Development 127:2493–2502

Ahmad K, Henikoff S (2002) The histone variant H3.3 marks active chromatin by replication-independent nucleosome assembly. Mol Cell 9:1191–1200

Alleman M, Doctor J (2000) Genomic imprinting in plants: observations and evolutionary implications. Plant Mol Biol 43:147–161

Arney KL, Erhardt S, Drewell RA, Surani MA (2001) Epigenetic reprogramming of the genome – from the germ line to the embryo and back again. Int J Dev Biol 45:533–540

Assaad FF, Signer ER (1992) Somatic and germinal recombination of a direct repeat in *Arabidopsis*. Genetics 132:553–566

Baroux C, Spillane C, Grossniklaus U (2002a) Genomic imprinting during seed development. Adv Genet 46:165–214

Baroux C, Spillane C, Grossniklaus U (2002b) Evolutionary origins of the endosperm in flowering plants. Genome Biol 3:1026 (reviews)

Barton SC, Surani MA, Norris ML (1984) Role of paternal and maternal genomes in mouse development. Nature 311:374–376

Benner MS, Philipps RL, Kirihara JA, Messing JW (1989) Genetic analysis of methionine-rich storage protein accumulation in maize. Theor Appl Genet 78:761–767

Birchler JA (1993) Dosage analysis of maize endosperm development. Annu Rev Genet 27:181–204

Birchler JA, Hart JR (1987) Interaction of endosperm size factors in maize. Genetics 117:309–317

Boisnard-Lorig C, Colon-Carmona A, Bauch M, Hodge S, Doerner P, Bancharel E, Dumas C, Haseloff J, Berger F (2001) Dynamic analyses of the expression of the HISTONE::YFP fusion protein in *Arabidopsis* show that syncytial endosperm is divided in mitotic domains. Plant Cell 13:495–509

Brzeski J, Jerzmanowski A (2003) Deficient in DNA methylation 1 (DDM1) defines a novel family of chromatin-remodeling factors. J Biol Chem 278:823–828

Briggs SD, Bryk M, Strahl BD, Cheung WL, Davie JK, Dent SY, Winston F, Allis CD (2001) Histone H3 lysine 4 methylation is mediated by Set1 and required for cell growth and rDNA silencing in *Saccharomyces cerevisiae*. Genes Dev 15:3286–3295

Brink RA (1956) A genetic change associated with the *R* locus in maize which is directed and potentially reversible. Genetics 41:872–890

Brown RC, Lemmon BE, Nguyen H, Olsen O-A (1999) Development of the endosperm in *Arabidopsis thaliana*. Sex Plant Reprod 12:32–42

Brown WV, Emery WHP (1958) Apomixis in the Gramineae: Panicoideae. Am J Bot 45:253–263

Chaudhuri S, Messing J (1994) Allele-specific parental imprinting of *dzr1*, a posttranscriptional regulator of zein accumulation. Proc Natl Acad Sci USA 91:4867–4871

Chaudhury AM, Ming L, Miller C, Craig S, Dennis ES, Peacock WJ (1997) Fertilization-independent seed development in *Arabidopsis thaliana*. Proc Natl Acad Sci USA 94:4223–4228

Choi Y, Gehring M, Johnson L, Hannon M, Harada JJ, Goldberg RB, Jacobsen SE, Fischer RL (2002) DEMETER, a DNA glycosylase domain protein, is required for endosperm gene imprinting and seed viability in *Arabidopsis*. Cell 110:33–42

Czermin B, Melfi R, McCabe D, Seitz V, Imhof A, Pirrotta V (2002) *Drosophila* enhancer of Zeste/ESC complexes have a histone H3 methyltransferase activity that marks chromosomal *Polycomb* sites. Cell 111:185–196

Danilevskaya ON, Hermon P, Hantke, S, Muszynski MG, Kollipara K, Ananiev EV (2003) Duplicated *fie* genes in maize: expression pattern and imprinting suggest distinct functions. Plant Cell 15:425–438

Drews GN, Lee D, Christensen CA (1998) Genetic analysis of female gametophyte development and function. Plant Cell 10:5–17

Dumas C, Mogensen HL (1993) Gametes and fertilization: maize as a model system for experimental embryogenesis in flowering plants. Plant Cell 5:1337–1348

Feinberg AP (1993) Genomic imprinting and gene activation in cancer. Nat Genet 4:110–113

Ferguson-Smith A, Lin SP, Tsai CE, Youngson N, Tevendale M (2003) Genomic imprinting – insights from studies in mice. Semin Cell Dev Biol 14:43–49

Finnegan EJ, Peacock WJ, Dennis ES (1996) Reduced DNA methylation in *Arabidopsis thaliana* results in abnormal plant development. Proc Natl Acad Sci USA 93:8449–8454

Francis NJ, Kingston RE (2001) Mechanisms of transcriptional memory. Nat Rev Mol Cell Biol 2:409–421

Friedman WE (2001) Developmental and evolutionary hypotheses for the origin of double fertilization and endosperm. CR Acad Sci III 324:559–567

Friedman WE, Williams JH (2004) Developmental evolution of the sexual process in ancient flowering plant lineages. Plant Cell 16 (Suppl 1):S119–S132

Garcia-Ortiz MV, Ariza RR, Roldan-Arjona T (2001) An OGG1 orthologue encoding a functional 8-oxoguanine DNA glycosylase/lyase in *Arabidopsis thaliana*. Plant Mol Biol 47:795–804

Grewal SI, Moazed D (2003) Heterochromatin and epigenetic control of gene expression. Science 301:798–802

Golden TA, Schauer SE, Lang JD, Pien S, Mushegian AR, Grossniklaus U, Meinke DW, Ray A (2002) *SHORT INTEGUMENTS1/SUSPENSOR1/CARPEL FACTORY*, a Dicer homolog, is a maternal effect gene required for embryo development in *Arabidopsis*. Plant Physiol 130:808–822

Grimanelli D, Hernández M, Perotti E, Savidan Y (1997) Dosage effects in the endosperm of diplosporous apomictic *Tripsacum* (Poaceae). Sex Plant Reprod 10:279–282

Grossniklaus U, Moore JM, Gagliano WB (1998) Molecular and genetic approaches to understanding and engineering apomixis: *Arabidopsis* as a powerful tool. In: Virma S, Siddq EA, Muralidharan K (eds) Advances in hybrid rice technology. IRRI, Philippines, pp 187–211

Grossniklaus U, Schneitz K (1998) The molecular and genetic basis of ovule and megagametophyte development. Semin Cell Dev Biol 9:227–238

Grossniklaus U, Vielle-Calzada J-P (1998) Parental conflict and infanticide during embryogenesis. Trends Plant Sci 3:328

Grossniklaus U, Vielle-Calzada JP, Hoeppner MA, Gagliano WB (1998) Maternal control of embryogenesis by *MEDEA*, a *Polycomb* group gene in *Arabidopsis*. Science 280:446–450

Grossniklaus U, Spillane C, Page DR, Köhler C (2001) Genomic imprinting and seed development: endosperm formation with and without sex. Curr Opin Plant Biol 4:21–27

Guo M, Rupe MA, Danilevskaya ON, Yang X, Hu Z (2003) Genome-wide mRNA profiling reveals heterochronic allelic variation and a new imprinted gene in hybrid maize endosperm. Plant J 36:30–44

Gutierrez-Marcos JF, Pennington PD, Costa LM, Dickinson HG (2003) Imprinting in the endosperm: a possible role in preventing wide hybridization. Philos Trans R Soc Lond B Biol Sci 358:1105–1111

Gutierrez-Marcos JF, Costa LM, Biderre-Petit C, Khbaya B, O'Sullivan DM, Wormald M, Perez P, Dickinson HG (2004) *Maternally expressed gene1* is a novel maize endosperm transfer cell-specific gene with a maternal parent-of-origin pattern of expression. Plant Cell 16:1288–1301

Haig D, Westoby M (1989) Parent specific gene expression and the triploid endosperm. Am Nat 134:147–155

Haig D, Westoby M (1991) Genomic imprinting in endosperm: its effect on seed development in crosses between species, and between different ploidies of the same species, and its implications for the evolution of apomixes. Philos Trans R Soc Lond 333:1–13

Jackson JP, Lindroth AM, Cao X, Jacobsen SE (2002) Control of CpNpG DNA methylation by the KRYPTONITE histone H3 methyltransferase. Nature 416:556–560

Jacobsen SE, Sakai H, Finnegan EJ, Cao X, Meyerowitz EM (2000) Ectopic hypermethylation of flower-specific genes in *Arabidopsis*. Curr Biol 10:179–186

Jeddeloh JA, Stokes TL, Richards EJ (1999) Maintenance of genomic methylation requires a SWI2/SNF2-like protein. Nat Genet 22:94–97

Johnson L, Cao X, Jacobsen S (2002) Interplay between two epigenetic marks. DNA methylation and histone H3 lysine 9 methylation. Curr Biol 12:1360–1367

Johnson SA, Nijs TPM, Peloquin SJ, Hanneman RE (1980) The significance of genetic balance to endosperm development in interspecific crosses. Theor Appl Genet 57:5–9

Jones RS, Gelbart WM (1993) The *Drosophila Polycomb*-group gene *Enhancer of zeste* contains a region with sequence similarity to *trithorax*. Mol Cell Biol 13:6357–6366

Jost JP, Oakeley EJ, Zhu B, Benjamin D, Thiry S, Siegmann M, Jost YC (2001) 5-Methylcytosine DNA glycosylase participates in the genome-wide loss of DNA methylation occurring during mouse myoblast differentiation. Nucleic Acids Res 29:4452–4461

Kermicle JL (1969) Androgenesis conditioned by a mutation in maize. Science 166:1422–1424

Kermicle J (1970) Dependence of the R-mottled aleurone phenotype in maize on the mode of sexual transmission. Genetics 66:69–85

Kermicle JL (1978) Imprinting of gene action in maize endosperm. In: Walden DB (ed) Maize breeding and genetics. Wiley, New York, pp 357–371

Kimber G, Riley R (1963) Haploid angiosperms. Bot Rev 29:480–531

Kinoshita T, Yadegari R, Harada JJ, Goldberg RB, Fischer RL (1999) Imprinting of the *MEDEA Polycomb* gene in the *Arabidopsis* endosperm. Plant Cell 11:1945–1952

Kinoshita T, Miura A, Choi Y, Kinoshita Y, Cao X, Jacobsen SE, Fischer RL, Kakutani T (2004) One-way control of *FWA* imprinting in *Arabidopsis* endosperm by DNA methylation. Science 303:521–523

Kiyosue T, Ohad N, Yadegari R, Hannon M, Dinneny J, Wells D, Katz A, Margossian L, Harada JJ, Goldberg RB, Fischer RL (1999) Control of fertilization-independent endosperm development by the *MEDEA Polycomb* gene in *Arabidopsis*. Proc Natl Acad Sci USA 96:4186–4191

Koch MA, Haubold B, Mitchell-Olds T (2000) Comparative evolutionary analysis of chalcone synthase and alcohol dehydrogenase loci in *Arabidopsis, Arabis,* and related genera (Brassicaceae). Mol Biol Evol 17:1483–1498

Köhler C, Grossniklaus U (2002) Epigenetic inheritance of expression states in plant development: the role of *Polycomb* group proteins. Curr Opin Cell Biol 14:773–779

Köhler C, Hennig L, Bouveret R, Gheyselinck J, Grossniklaus U, Gruissem W (2003a) *Arabidopsis* MSI1 is a component of the MEA/FIE *Polycomb* group complex and required for seed development. EMBO J 22:4804–4814

Köhler C, Hennig L, Spillane C, Pien S, Gruissem W, Grossniklaus U (2003b) The *Polycomb*-group protein *MEDEA* regulates seed development by controlling expression of the MADS-box gene *PHERES1*. Genes Dev 17:1540–1553

Koltunow AM, Grossniklaus U (2003) Apomixis: a developmental perspective. Annu Rev Plant Biol 54:547–574

Kouzarides T (2002) Histone methylation in transcriptional control. Curr Opin Genet Dev 12:198–209

Kuzmichev A, Nishioka K, Erdjument-Bromage H, Tempst P, Reinberg D (2002) Histone methyltransferase activity associated with a human multiprotein complex containing the *Enhancer of Zeste* protein. Genes Dev 16:2893–2905

Li W, Ma H (2002) Gametophyte development. Curr Biol 12:R718–R721

Lin B-Y (1982) Association of endosperm reduction with parental imprinting in maize. Genetics 100:475–486

Lin B-Y (1984) Ploidy barrier to endosperm development in maize. Genetics 107:103–115

Lopes MA, Larkins BA (1993) Endosperm origin, development, and function. Plant Cell 10:1383–1399

Lund G, Ciceri P, Viotti A (1995a) Maternal-specific demethylation and expression of specific alleles of zein genes in the endosperm of *Zea mays* L. Plant J 8:571–581

Lund G, Messing J, Viotti A (1995b) Endosperm-specific demethylation and activation of specific alleles of alpha-tubulin genes of *Zea mays* L. Mol Gen Genet 246:716–722

Luo M, Bilodeau P, Koltunow A, Dennis ES, Peacock WJ, Chaudhury AM (1999) Genes controlling fertilization-independent seed development in *Arabidopsis thaliana*. Proc Natl Acad Sci USA 96:296–301

Luo M, Bilodeau P, Dennis ES, Peacock WJ, Chaudhury A (2000) Expression and parent-of-origin effects for FIS2, MEA, and FIE in the endosperm and embryo of developing *Arabidopsis* seeds. Proc Natl Acad Sci USA 97:10637–10642

Maheshwari P (1950) An introduction to the embryology of the angiosperms. McGraw-Hill, New York, pp 221–267

Mascarenhas JP (1989) The male gametophyte of flowering plants. Plant Cell 1:657–664

McGrath J, Solter D (1984) Completion of mouse embryogenesis requires both the maternal and paternal genomes. Cell 37:179–183

Messing J, Grossniklaus U (1999) Genomic imprinting in plants. Results Probl Cell Differ 25:23–40

Milne TA, Briggs SD, Brock HW, Martin ME, Gibbs D, Allis CD, Hess JL (2002) MLL targets SET domain methyltransferase activity to hox gene promoters. Mol Cell 10:1107–1117

Moore T, Haig D (1991) Genomic imprinting in mammalian development: a parental tug-of-war. Trends Genet 7:45–49

Moore T, Reik W (1996) Genetic conflict in early development: parental imprinting in normal and abnormal growth. Rev Reprod 1:73–77

Mordhorst AP, Toonen MAJ, de Vries SC (1997) Plant embryogenesis. Crit Rev Plant Sci 16:535–576

Müller J, Hart CM, Francis NJ, Vargas ML, Sengupta A, Wild B, Miller EL, O'Connor MB, Kingston RE, Simon JA (2002) Histone methyltransferase activity of a *Drosophila Polycomb* group repressor complex. Cell 111:197–208

Murbeck S (1904) Parthenogenese bei den Gattungen *Taraxacum* und *Hieracium*. Bot Not 6:285–296

Neumann B, Barlow DP (1996) Multiple roles for DNA methylation in gametic imprinting. Curr Opin Genet Dev 6:159–163

Nogler GA (1984) Gametophytic apomixis. In: Johri BM (ed) Embryology of angiosperms. Springer, Berlin Heidelberg New York, pp 475–518

Oakeley EJ, Podesta A, Jost JP (1997) Developmental changes in DNA methylation of the two tobacco pollen nuclei during maturation. Proc Natl Acad Sci USA 94:11721–11725

Ohad N, Yadegari R, Margossian L, Hannon M, Michaeli D, Harada JJ, Goldberg RB, Fischer RL (1999) Mutations in FIE, a WD *Polycomb* group gene, allow endosperm development without fertilization. Plant Cell 11:407–416

Peters AH, Mermoud JE, O'Carroll D, Pagani M, Schweizer D, Brockdorff N, Jenuwein T (2002) Histone H3 lysine 9 methylation is an epigenetic imprint of facultative heterochromatin. Nat Genet 30:77–80

Quarin CL (1999) Effect of pollen source and pollen ploidy on endosperm formation and seed set in pseudogamous apomictic *Paspalum notatum*. Sex Plant Reprod 11:331–335

Rea S, Eisenhaber F, O'Carroll D, Strahl BD, Sun ZW, Schmid M, Opravil S, Mechtler K, Ponting CP, Allis CD, Jenuwein T (2000) Regulation of chromatin structure by site-specific histone H3 methyltransferases. Nature 406:593–599

Reik W, Maher ER (1997) Imprinting in clusters: lessons from Beckwith-Wiedemann syndrome. Trends Genet 13:330–334

Roguev A, Schaft D, Shevchenko A, Pijnappel WW, Wilm M, Aasland R, Stewart AF (2001) The *Saccharomyces cerevisiae* Set1 complex includes an Ash2 homologue and methylates histone 3 lysine 4. EMBO J 20:7137–7148

Russell SD (1993) The egg cell: development and role in fertilization and early embryogenesis. Plant Cell 5:1349–1359

Rutishauser A (1954) Entwicklungserregung der Eizelle bei pseudogamen Arten der Gattung *Ranunculus*. Bull Schweiz Akad Med Wissensch 10:491–512

Sarkar KR, Coe EH (1966) A genetic analysis of origin of maternal haploids in maize. Genetics 54:453

Savidan Y (2000) Apomixis, genetics and breeding. Plant Breed Rev 18:13–86

Saze H, Scheid OM, Paszkowski J (2003) Maintenance of CpG methylation is essential for epigenetic inheritance during plant gametogenesis. Nat Genet 34:65–69

Scholten S, Lörz H, Kranz E (2002) Paternal mRNA and protein synthesis coincides with male chromatin decondensation in maize zygotes. Plant J 32:221–231

Scott RJ, Spielman M, Bailey J, Dickinson HG (1998) Parent-of-origin effects on seed development in *Arabidopsis thaliana*. Development 125:3329–3341

Soppe WJ, Jacobsen SE, Alonso-Blanco C, Jackson JP, Kakutani T, Koornneef M, Peeters AJ (2000) The late flowering phenotype of *fwa* mutants is caused by gain-of-function epigenetic alleles of a homeodomain gene. Mol Cell 6:791–802

Sorensen MB, Chaudhury AM, Robert H, Bancharel E, Berger F (2001) *Polycomb* group genes control pattern formation in plant seed. Curr Biol 11:277–281

Spillane C, Curtis MD, Grossniklaus U (2004) Apomixis technology development – virgin births in farmers' fields? Nat Biotechnol 22:687–691

Surani MA (1998) Imprinting and the initiation of gene silencing in the germ line. Cell 93:309–312

Surani MA, Barton SC, Norris ML (1984) Development of reconstituted mouse eggs suggests imprinting of the genome during gametogenesis. Nature 308:548–550

Tariq M, Paszkowski J (2004) DNA and histone methylation in plants. Trends Genet 20:244–251

Tilghman SM (1999) The sins of the fathers and the mothers: genomic imprinting in mammalian development. Cell 96:186–193

Tschiersch B, Hofmann A, Krauss V, Dorn R, Korge G, Reuter G (1994) The protein encoded by the *Drosophila* position-effect variegation suppressor gene Su(var)3–9 combines domains of antagonistic regulators of homeotic gene complexes. EMBO J 13:3822–3831

Ueda K, Tanaka I (1995) The appearance of male gamete-specific histones gH2B and gH3 during pollen development in *Lilium longiflorum*. Dev Biol 169:210–217

Ueda K, Kinoshita Y, Xu ZJ, Ide N, Ono M, Akahori Y, Tanaka I, Inoue M (2000) Unusual core histones specifically expressed in male gametic cells of *Lilium longiflorum*. Chromosoma 108:491–500

Vielle-Calzada JP, Thomas J, Spillane C, Coluccio A, Hoeppner MA, Grossniklaus U (1999) Maintenance of genomic imprinting at the *Arabidopsis medea* locus requires zygotic DDM1 activity. Genes Dev 13:2971–2982

Vielle-Calzada JP, Baskar R, Grossniklaus U (2000) Delayed activation of the paternal genome during seed development. Nature 404:91–94

Vijayaraghavan MR, Prabhakar K (1984) The endosperm. In: Johri BM (ed) Embryology of angiosperms. Springer, Berlin Heidelberg New York, pp 319–376

Vinkenoog R, Spielman M, Adams S, Dickinson HG, Scott RJ (2000) Genomic imprinting in plants. Methods Mol Biol 181:327–370

Von Wangenheim KH, Peterson HP (2004) Aberrant endosperm development in interploidy crosses reveals a timer of differentiation. Dev Biol 270:277–289

Weijers D, Geldner N, Offringa R, Jürgens G (2001) Seed development: early paternal gene activity in *Arabidopsis*. Nature 414:709–710

Xiao W, Gehring M, Choi Y, Margossian L, Pu H, Harada JJ, Goldberg RB, Pennell RI, Fischer RL (2003) Imprinting of the *MEA Polycomb* gene is controlled by antagonism between MET1 methyltransferase and DME glycosylase. Dev Cell 5:891–901

Xu H, Swoboda I, Bhalla PL, Singh MB (1999) Male gametic cell-specific expression of H2A and H3 histone genes. Plant Mol Biol 39:601–607

Ye F, Signer ER (1996) RIGS (repeat-induced gene silencing) in *Arabidopsis* is transcriptional and alters chromatin configuration. Proc Natl Acad Sci USA 93:10881–10886

Subject Index

A

ACF   15
– ACF1   17
additional sex combs   34
alpha-tubulin   243
antisilencing function 1   7
apomixis   241, 253
ASF1   7, 10
ASH1   47
ASH2   47
ASX   34
ATP-dependent chromatin remodeling fac-
     tors   4
ATPdependent chromatin remodelling   37
aza-deoxycytidine   17
– 5-aza-2'-deoxycytidine   178

B

beta-globin gene   184–188, 190, 194
Brahma   47
Brahma Complex   48
BRDT   79, 80, 82, 83
Brg1   4
Brm   4, 47
bromodomain   50

C

CAF-1   7, 9, 12
Capturing Chromosome Conformation   44
CDYL   78, 82, 83
Cellular Memory Module (CMM)   50
CENP-A   11
CHRAC   15
chromatin   2, 32, 123, 136
– architecture   216
– assembly   7
– assembly factor 1   7
– immunoprecipitation   49
– remodeling   158
– structure   188, 192, 194
chromodomain proteins   41

Clr4   19
confocal microscopy   72
CpG islands   170
CREB-binding protein   47
cystein-rich   76, 84

D

dCBP   47
de novo methylation   151
DECREASED DNA-METHYLATION1 (DDM1)
     250
DEMETER (DME)   248
demethylation   152
developmental expression/regulation
     184–188, 190, 194
differentially methylated regions (DMRs)
     161
differentiation   252
dMi-2   34
DNA methylation   99, 212, 251, 255
– and disease   162
DNA methyltransferases   170, 171
DNA replication   5, 215
dosage compensation   123
double fertilization   238
dRing/sex combs extra   34
Drosophila   32
dzr1   244

E

E(PC)   34
E(Z)   34
early embryo development   153
embryogenesis   35
embryonic stem cells   107
endosperm   240, 253
– evolution   241
engrailed gene   38
enhancer of *Polycomb*   34
enhancer of *zeste*   34
epi   1

epididymis   76, 84
epigenetic   124
– control of gene expression   157
– inheeritance   21
ESC   34
extra sex combs   34
Extra SexCombs/Enhancer of Zeste
    [ESC/E(Z)] complex   33

F
FACT (Facilitates Chromatin Transcription)
    49
fasciata   10
female gametophyte   240
functional imprints   223
FWA   251

G
GAF   47
GAGA factor   47
GAGAG sequence   38
gametogenesis   239
gene silencing   157, 159
genome formatting   223
genome-wide prediction of PREs   38
genomic imprinting   237
germ cells   153

H
H1t   69, 70
H2A variants   68
H3 variants   68
H3t   71
HBO1   14
heat-shock protein cognate 4   34
Heterochromatin Protein 1 (HP1)   2, 41
heterochromatin   2, 93
– replication   11
higher order chromatin structure   19
higher-order folding   198–200
Hir1   12
Hir2   12
histone
– acetylases   4
– acetylation   13
– acetyltransferase (HATs)   3, 48
– chaperones   7
– code hypothesis   39, 71, 77, 79, 81, 82, 92,
    176
– deacetylase (HDAC)   41, 78–80, 82, 83
– deacetylation   13

– fold   70, 82
– H3.1   12
– H3.3   12
– methylation   41
– methyltransferase   4, 49, 94, 250
– modification   194, 195, 197, 214
– modifying enzymes   3
– ariants   11, 254
HMG box-containing factors   41
homeotic genes   31
HP1   2, 10, 20, 74, 78
HR6   80
HSC70-4   34
hypersensitivity to nuclease digestion
    43
hypomethylated pollen   256

I
ICF syndrome   173
IGF2   84
imprint establishment   252
imprinted X-inactivation   109
imprinting cycle   216
imprinting evolution   208
intergenic transcription   192, 193, 195–198,
    300
interploidy crosses   247
ISWI   4, 14

K
Kaiso   176
KRYPTONITE (KYP)   254

L
LCR   185–188, 190, 193–195, 198–201
LINE/L1 element   77
locus control region   185–188, 190, 193–195,
    198–201
long-range regulation   198–200
looping mechanism   44

M
maintenance methylation   152
maintenance of chromatin structure through
    cell division   54
maintenance of determined states through
    mitosis   55
mammalian PREs/TREs   56
maternal control   245
maternal derepression of r1 (mdr1)   243

MeCP2   175
MEDEA (MEA)   247
methyltransferase   74, 78
METHYLTRANSFERASE1 (MET1)   248, 255
Mi2   4
MSL   124, 133, 141

N
NAP 1   7
no apical meristem related protein1 (nrp1)
      244
non-coding RNA   42, 131, 132
nuclear compartmentalization   45
nuclear organization   199–201
nucleosome   45
– assembly protein 1   7
NURF   15

O
Origin Recognition Complex   14

P
p300   14
Pairing Sensitive Silencing   37
parental conflict   246
paternal expression   245
PC   34
PcG reponse elements (PREs)   31
PcG-mediated silencing   36
PCL   34
PCNA   9, 18, 20
PH   34
PHERES1 (PHE1)   250
PHO   34
PHO5   81
PHOL   34
placental habit   257
plants   237
pleiohomeotic   34
Polycomb   34
– group (PcG)   31
– group (PcG) proteins   248
– Repressive Complex 1 (PRC1)   33
– -like   34
polyhomeotic   34
polytene chromosomes   35
Position Effect Variegation   36
posterior sex combs   34
primary imprints   218

proliferating cell nuclear antigen   9
PSC   34

R
recombination   66, 68, 69, 73
red1 (r1)   242
Replication Timing   6, 100
RNA   20
– interference (RNAi)   42
Rpd3   34
RSF   15

S
S phase   6
SCM   34
seed   238
sex comb on midleg   34
single-strand break (SSB)   75
SNF2H   17
Snf5 related 1   47
SNR1   47
spermatogonia   66, 69, 70, 72–74
spreading of PcG factors   43
SU(VAR)3-9   4, 19
SU(Z)2   34
SU(Z)12   34
suppressor of zeste 12   34
suppressor of zeste 2   34
Suv39h2   74, 78
SV40   5
Swi6   19

T
telomeric sequence   77
TH2A   70, 82
TH2B   70, 82
TH3   70
Thritorax-like gene   38
timing of DNA replication   54
transcription   138
– elongation   45
– initiation   45
Trichostatin A (TSA)   78
Trithorax   47
– Acetylation Complex (TAC1)   48
– group (trxG)   31
TRX   47
trxG response elements (TREs)   31
Tsix   114

U
uH2A   73, 79, 80

V
variant histones   97

W
WSTF   18

X
X chromosome   123, 133
– inactivation   1, 56
X-inactivation center   91
Xist   101
XY body   68, 73

Z
Z   47
zein   243
Zeste   47

Printing: Mercedes-Druck, Berlin
Binding: Stein+Lehmann, Berlin